Business Information Systems

PEARSON EDUCATION

We work with leading authors to develop the
strongest educational materials in Business,
bringing cutting edge thinking and best learning practice
to a global market.

Under a range of well-known imprints, including
Financial Times Prentice Hall, we craft high quality
print and electronic publications which help
readers to understand and apply their content,
whether studying or at work.

To find out about the complete range of our
publishing please visit us on the World Wide Web at:
www.pearsoned-ema.com

Business Information Systems

SIXTH EDITION

HD Clifton
DC Ince
AG Sutcliffe

FINANCIAL TIMES
PRENTICE HALL

An imprint of **PEARSON EDUCATION**

Harlow, England · London · New York · Reading, Massachusetts · San Francisco · Toronto · Don Mills, Ontario · Sydney
Tokyo · Singapore · Hong Kong · Seoul · Taipei · Cape Town · Madrid · Mexico City · Amsterdam · Munich · Paris · Milan

PEARSON EDUCATION LIMITED

Edinburgh Gate
Harlow
Essex CM20 2JE
England

and Associated Companies throughout the World.

Visit us on the World Wide Web at:
www.pearsoned-ema.com

First published as *Business Data Systems* by Prentice Hall 1978, 1983, 1986 and 1990
Fifth edition 1994
Sixth edition 2000

ISBN 0 130 82960 9

British Library Cataloguing-in-Publication Data
A catalogue record for this book can be obtained from the British Library

Library of Congress Cataloging-in-Publication Data
Clifton, H.D. (Harold Dennis), 1927–
    Business information systems / H.D. Clifton, A.G. Sutcliffe. —
6th ed.
        p.   cm.
    Includes bibliographical references and index.
    ISBN 0–13–082960–9
    1. Business—Data processing.   2. Systems analysis.
I. Sutcliffe, Alistair, 1951–   . II. Title.
HF5548.2.C5427   2000
658′ .05—dc21   99–42791
                                                 CIP

10 9 8 7 6 5 4 3 2 1
05 04 03 02 01 00

Typeset by 35 in 9.5/13pt Stone Serif
Printed and bound by Rotolito Lombarda, Italy

Contents

List of figures

List of tables

Acknowledgements

We are grateful to the professional institutions below for permission to reproduce past examination questions.

    The Association of Chartered Certified Accountants
    The British Computer Society
    The Chartered Institute of Management Accountants
    The Institute of Chartered Accountants in England and Wales
    The Institute of Chartered Secretaries and Administrators

It should be pointed out that the solutions to the exercises are attributable entirely to the authors and do not necessarily reflect those of the respective professional institutions.

HDC
DCI
AGS

1 Business and management information

AIMS

After reading this chapter you should:

- understand the nature of business organizations;
- be familiar with six main work areas found in business organizations;
- understand the nature of business information;
- understand the concept of management information;
- be familiar with the role of decision support systems and executive information systems;
- be comfortable with the concept of an expert system and understand the subject of artificial intelligence;
- understand the relationship between business systems theory and the modern information system;
- be familiar with the relationships that exist between information systems and users.

1.1 BUSINESS ORGANIZATIONS

All concerns that are in some ways constrained by money and resources can be regarded as business organizations. These include manufacturing and commercial companies, central and local government departments, administrative organizations, financial institutions, and service agencies. All of these organizations, hereafter referred to as 'organizations' or 'companies', are involved with information systems (IS), i.e. data processing (DP) and systems analysis. They vary considerably in size, from huge international corporations and government departments down to small 'back street' companies and private individuals. The factors common to all of these should be a clear understanding of the purpose and aims of the organization, and a systems approach to the solution of problems.

Solving business problems is rarely a simple matter of designing an algorithm or mathematical equation. There are no immediate solutions to most business problems; businesses involve a complex mix of people, policy and technology, and exist within the constraints of economics and society. A solution is a method of alleviating the burden of a task or of avoiding a calamity that might ensue from difficulties encountered. It is more likely that effective results derive from a well-organized system carefully planned and assiduously operated. Business problems are usually associated with the need for information to control the business's activities.

What are the aims of business organizations? Why do businesses exist and for what purposes? There are, no doubt, many metaphysical answers to these questions but in the pragmatic sense they are related to the control over money, people and resources. A company financed by privately owned share capital aims to maintain or increase its profitability and thereby maximize the long-term value of its shares. The vagaries of the stock market are beyond the scope of this text but profitability is a prime aim if a company is to continue in existence. Other long-term aims of companies are expansion, diversification, and monopolization of products and markets. These aims are facilitated through the use of IS to improve the company's competitive advantage. Business information systems are now a critical part of the profitability and even survival of many companies.

Organizations that are publicly owned, i.e. government controlled, are usually regulated by the need to keep within their operating budgets. Their objectives may be decided as a result of political, social or economic considerations but, in the end, they are operated within monetary constraints. This calls for the provision of adequate information about their activities and environment.

Readers unfamiliar with business organizations are advised to read References 1.1–1.3.

Business work areas

A business consists of components, such as departments, units, and teams, which carry out specific jobs of work to contribute to the organization's goal. The need for information and hence a means of processing raw data rapidly and accurately applies to a wide range of work areas in business. Figure 1.1 outlines the way in which work areas connect together and contribute to the financial and management accounting systems of a 'typical' manufacturing company.

Perhaps the most obvious information need is a company's financial position as derived from its business transactions. Closely associated with financial accounting is management accounting, i.e. the control over a company's manufacturing costs in relation to its productive output.

Accounting information stems from book-keeping procedures, each and every financial transaction contributing to some extent to the financial position of a company. There are numerous items of expenditure and income that have to be accounted for in arriving at a company's annual balance sheet and profit/loss statement. With IS, therefore, we are aiming for a means of gathering all the financial transactions, and processing them accurately and economically in order

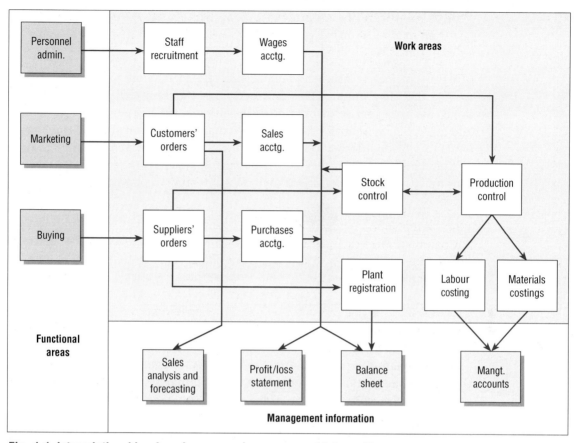

Fig. 1.1 Interrelationship of work areas and management information

to provide not only the statutory information required of an organization but also further information to improve its efficiency and profitability. References 1.1, 1.6, 1.17 and 1.21 provide useful knowledge for readers unacquainted with accounting principles.

The main work areas inherent to most companies and organizations are:

- wages accounting;
- sales control and accounting;
- purchases control and accounting;
- management (cost) accounting;
- stock control; and
- production control.

These areas are outlined below in order to provide readers unfamiliar with business with some basic understanding of the procedures and problems.

Wages accounting

A paramount feature of all business enterprises is the necessity to employ and remunerate a workforce. The workforce usually comprises people of a wide range of skills – manual, technical and managerial. The factor common to all the workforce is that they must be paid and hence the organization must be geared to doing this in a precise manner. This necessitates the existence of a personnel administration system allied to a properly organized payroll system. The former exists to provide information regarding the employees' personal attributes, capabilities and terms of employment. The payroll system is the means of knowing how much to pay employees on each occasion, and accounting for any monies otherwise expended.

Payroll systems can be regarded as consisting of the following parts:

■ Computation of each employee's gross earnings, wage or salary.
■ Computation of deductions due to tax, pension contributions, etc., and net pay.
■ Accounting for amounts deducted from earnings.

The complexity of the first part varies and is dependent upon the employee's terms of payment. In the case of staff, their monthly salary is usually a twelfth of their annual salary plus commission or bonus to which they may be entitled, thus little computation is entailed. On the other hand, certain types of manual workers have complicated wage structures and so several calculations are necessary. For instance, earnings may depend upon hours worked in various time periods, with a different hourly rate for each period (dayworkers). Pieceworkers are so called because all or some of their earnings depend upon the amount of work done, i.e. output achieved, either individually or collectively.

The computation of net payments is more standardized than that of gross wages, being mainly the calculation of deductions such as income tax (PAYE in the UK), national insurance contributions, superannuation (pension) payments, and so on.

The accounting aspects of wages fall into two main categories:

1 The distribution of monies to employees as either cash, cheques or credit transfers.
2 The entry of the correct amounts into the nominal ledger to cover subsequent payments to tax authorities, pension funds, and so on.

The principal documents involved in wages accounting are as follows.

Pay advice

A small document (payslip) for each employee showing the precise details of the pay for the period. Typically, the pay advice shows the gross earnings, and in some cases how this is composed, each amount deducted and the net amount.

Also shown is indicative information such as the employer's number and name, department number, tax code, national insurance number, week number and/or the date.

Payroll

This is a list containing for each employee substantially the same information as on the pay advice but amplified to include the amounts-to-date, e.g. the gross wage paid up to and including this period since the start of the tax year. The payroll is usually listed in a sequence such as employee number within department number to facilitate reference to it subsequently.

Bank direct credit list

Since most salaried employees have their net pay transferred directly into their bank accounts, it is necessary to send a list to the company's bank detailing the amounts to be transferred to the employees' accounts.

Cheques and Giro credits

In cases where an employee is not paid in cash nor by credit transfer, a cheque or Giro credit has to be drawn.

Deductions analysis

This shows the total amount of each deduction made from wages for the period and to date. It may also be analysed into departmental totals.

Cost analyses

These are less standardized than the other payroll documents. As they are for the purposes of cost accounting, their contents depend entirely upon the nature of the company's activities. There is likely to be a number of cost analyses but typically in a manufacturing company a cost analysis contains the period and year-to-date gross amounts paid to each labour group for each current job in the factory.

End-of-year documents

These are statutory documents required by the Inland Revenue and the employees in order that they may check that the correct amount of income tax has been deducted.

Audit and control documents

Payrolls need to be carefully controlled and checked as large amounts of money are involved. The various documents described above need to be supplemented by sets of totals so that if a discrepancy arises, it is not too difficult to trace. This is particularly important where cash is involved.

Wages records

It is necessary to maintain a record for each employee containing full and accurate details of pay. This record must be kept up to date in terms of amendments as well as the updating of the totals-to-date. Amendments include changes to an employee's gross wage, tax code, department, bank, and so on. The wages records have also to be amended to incorporate details of new employees and remove details pertaining to leavers. The wages records are connected with the personnel records in that the personnel department informs the wages department of changes in personnel.

Sales control and accounting

Customer order control entails procedures for ensuring that orders from customers/clients are received, recorded and acknowledged in an efficient and organized manner. This applies no matter whether the orders are received through the post, by telephone, via sales representatives, or by means of communication systems such as electronic mail or teletext.

At a later stage, order control is necessary to ensure that orders are actually fulfilled, i.e. customers receive the correct goods on time and at the right destination.

Particular features of order control include:

- Dealing with enquiries from prospective purchasers re prices, availabilities and delivery dates.
- Checking, in conjunction with the accounts department, the credit ratings of customers placing orders on credit.
- Comparison of orders received against quotations given previously (if any) in terms of price and availability.
- Checking that goods ordered are valid, i.e. exist, and querying dubious orders with customers.
- Handling customer complaints and queries regarding price and delivery; these may be passed on to the sales representatives in certain cases.
- Checking the fulfilment of orders via the dispatch department.

Sales accounting is a function of the accounts department involving the monetary side of customers' orders. In this respect customers fall into two main types:

1 Companies or persons who have accounts with the company.
2 Customers who pay CWO, COD or against pro forma invoices. Note that we are not concerned here with over-the-counter sales in shops and supermarkets.

Some firms' customers are all of the first type, e.g. government departments, local authorities, accredited companies and persons. Other firms deal only with customers of the second type, e.g. mail-order customers. A number of firms have both types of customers.

The main documents connected with sales accounting are as follows.

Customer orders

These must hold all the information needed to identify the customer, the goods ordered, the delivery address and, perhaps, the price and date of delivery. Identification is likely to entail catalogue numbers, commodity codes, account numbers, etc.

Sales invoices

A sales invoice is sent to the customer after delivery of the goods and acts as a request for payment. Copies are often used as dispatch notes and packing notes. An invoice needs to contain sufficient information to clearly identify the goods, their prices and values, discounts, packing and carriage charges, and VAT (sales tax).

Sales statements

Statements are sent to customers at regular intervals, e.g. monthly, in order to show their indebtedness or credit position. Each statement lists the sales and payments transacted since the previous statement, and shows the current financial position, i.e. the amount owing or in credit. The contents of statements are derived from the sales ledger records.

Sales analysis reports

A very wide variety of sales analyses are created within the totality of company marketing. Essentially they show sales quantities and values summarized into various groupings. The purpose of sales analysis is to find what has been sold, where, when and to whom. This information can then be used to forecast future sales demands and to plan marketing activities. Typical groupings are commodities, commodity groups, sales areas, regions, zones, sales periods and customer categories. These are combined together in numerous ways to suit the needs of the sales manager, e.g. commodity groups within sales areas for each of the past twelve months.

Purchases control and accounting

Purchasing involves the procedures for ensuring that all the materials, components, tools, equipment and other items needed by the company are available at the right time, right place, and right price. The precise nature of the purchasing function depends upon the type of items purchased, but generally the following procedures apply:

- Obtaining quotations of price, delivery date and quality from prospective suppliers.
- Placing orders with suppliers, monitoring delivery and chasing overdue orders.

7

- Checking goods received (by goods receiving department) for correct type, quality and quantity.
- Passing details of accepted goods to the accounts department for subsequent payment of suppliers.
- Checking suppliers' invoices and statements for accuracy before making payment.
- Accumulating information about suppliers' prices, delivery achievements and quality.

The main documents associated with purchases accounting are:

Purchase orders

These must contain sufficient information for the supplier to be able to supply the correct item(s) to the right location or person at the appropriate date. Absolute identification of the required item depends to some extent on the supplier. If the supplier has a clear catalogue showing all the goods clearly labelled and coded, then there should be no problem. On the other hand if the requirement is non-standard or made-to-specification, great care is necessary to ensure that the correct item is ordered and received.

Goods received notes (GRNs)

A GRN is a record of receipt of a certain quantity of an item. It is made out at the time of receipt and then validated after the goods have been inspected.

Purchase invoices

These are received from suppliers after the goods have been received. They must contain sufficient detail to relate to the goods supplied and to the purchase order. Known discrepancies between the order and the invoice should be noted on the invoice, e.g. items omitted or short.

Remittance advices

These are documents sent to the suppliers advising that payment is being made to them. The information thereon is primarily the amount, date and method of payment, and any reference numbers that apply to the remittance.

Cheques

A cheque to accompany the remittance advice has to be drawn unless payment is by bank credit transfer.

Credit transfers

A list is sent to the company's bank instructing them to transfer monies to the accounts of the suppliers listed.

Purchases analyses

In some cases it is beneficial to analyse the company's purchases in various ways. The three main reasons for this are as follows:

1 To measure the efficiency of suppliers.

2 To cost the purchases of the departments.

3 To cost the purchases for work done, i.e. job costing.

Purchase records

A record must be kept for each supplier to show the details of his or her account, i.e. the purchase ledger. The main information in each record is the supplier's account number, name, the current balance and the transactions (invoices and remittances) outstanding for the period in question.

Management (cost) accounting

Management accounting, often referred to as costing, is a complex subject and so only a superficial explanation is given here. The general idea is that costing ensures that the organization's activities are accomplished at acceptable costs and that all cost discrepancies are accounted for. Thus cost expenditures are compared with standards and the differences between actual and standard costs are analysed.

Standard costing entails the setting of predecided estimated costs, i.e. standards, and the regular comparison of these standards with the actual costs incurred. The difference between standard and actual costs is termed a 'variance' and analysis of variances provide management with valuable information for control and planning.

Variance analysis is complicated owing to the different ways in which the variances may be broken down. A profit variance, for instance, may be split into a sales variance and a cost variance.

The sales variance is then broken down further into volume variance, i.e. the difference between the standard sales quantity and the actual quantity sold, and the price variance, i.e. the difference between the standard price and the actual price. Similarly, the cost variance is broken down into materials and labour, which in turn are broken down into finer detail such as materials prices and usages; wage rates, and labour efficiency. Further information on standard costing is available in Reference 1.7.

Costing applies particularly to production, in which small rises in costs can reduce or completely eliminate profitability. The two largest production costs are labour and materials, and the former is so high that in developed countries automation is more economic for many types of production.

The analysis of cost data provides information that helps management to decide methods of production and materials to use. One of the complexities of costing is the apportionment of overheads, for instance, the distribution of overhead costs such as rent, heating and administration between production jobs so that their true cost and hence their selling price can be determined.

Other aspects of costing include marginal costing, break-even analysis, and budgeting:

■ Marginal costing is the determination of the extra cost involved in producing an item above a certain level of production. Thus if the fixed overheads have been absorbed, a marginal cost tends to be less than the standard cost.

■ Break-even analysis relates turnover to costs so that it is possible to decide at what level of turnover the operation becomes profitable and, above this level, what profit is made.

■ Budgeting or budgetary control is a means of controlling expenditure and other monies by making forecasts of future expenditure, income, sales, etc. This enables provisional accounts and costs to be devised so that an advance estimate of the financial situation is available. Departments are then expected to conform to their budgets or otherwise to justify deviations.

The aforementioned costing procedures demand considerable processing of data so that suitable reports are available to management, and hence the need for a computerized information system.

Stock control

Stock (inventory) control is closely connected with stores control. The former is the monitoring and decisions regarding the items held in stock, the latter is the methods of operating the warehouse or stores.

Stock is a buffer between supply and demand. The supply may be either from the company's production or from external sources, or both. Demand may come from customers or from internal sources. In any event it is important that when a demand arises, there is sufficient stock to meet it with minimum delay. However, when an item is so expensive or unmanageable that it is not economic or practical to stock it, it must be purchased or manufactured as and when required. Furthermore all stock costs money. Companies therefore try to minimize the money tied up in stock and still meet demand. Keeping ahead of this dilemma has given rise to 'just in time' inventory control, in which computer systems monitor supply and demand constantly and trigger rapid re-supply when necessary. These systems can only exist with close co-operation between the company and its suppliers, backed up by information systems.

Inventory theory (modelling) takes us into the realms of operations research, which involves mathematical and statistical principles. In a nutshell, inventory theory covers the methods of minimizing the overall cost of stockholding. It takes into consideration the costs of storage, reordering, capital tied up in stock, stock depreciation and deterioration, and the loss of profit caused by out-of-stock situations.

More mundanely, stock control involves the maintenance of records relating to stock levels, issues, outstanding orders, reorder levels, and so on.

The factors entering into stores control are the size and weight of stock items, and also their value, flammability, stability (chemical and physical), and identification. These factors determine the need for handling equipment, security, packaging, labelling and storage space (both indoors and outside).

From the information aspect an important requirement is stock valuation, i.e. the book value of all stock-in-hand at a certain time. This valuation often follows a physical stock check carried out manually but may also stem from computer-held records. It involves reconciliation between the records and the physical count. In any event, the figures should be accurate, allowing for stock losses, deterioration and changes in value, as stock valuation contributes to the firm's annual balance sheet.

From the IS aspect much depends upon the number of stock items, their turnover and dispersal. If there is a large number of items being issued from and received into several dispersed stores, there is obviously a significant problem in capturing and subsequently processing the data.

Production control

This subject covers a wide field since there are many different manufactured products and methods of production. Essentially, production planning covers what to make and how to make; production control ensures that the planning is achieved by monitoring and control of the machines, processes and people which carry out production.

From the IS aspect, the engineering industry poses the need for most information. This is particularly so for industries that purchase materials, make and/or buy components, and assemble products through various levels of subassembly. Examples are the car, machine tool, domestic appliance, aircraft, and entertainment equipment industries.

The information required for production control purposes includes:

■ material requirements for each time period;
■ quantities of components and subassemblies to be made by each period;
■ the amounts of equipment, machines, tools, jigs, etc., needed for each stage;
■ the amount of each labour category needed during each period;
■ the loading of each production unit in each period;
■ the progress of each job and reasons for delays.

If the products and production processes are complex, as is usually the case, the above information may be difficult to acquire.

In order to determine the quantities of materials, components and subassemblies needed each period, a huge amount of computation may be necessary. There

11

could be a large range of products each comprising several levels of assembly. Each level entails determining the precise quantities of all items at a lower level, resulting in perhaps tens of thousands of components and materials being needed at the lowest level.

Similarly, the components may be made through a series of manufacturing operations involving different machines and types of labour. In view of the high cost of labour and of most machines, it is advantageous to keep these facilities fully occupied. This again involves very considerable computation so that machine and labour resources can be assigned to tasks in the correct production sequence and to tie in with the availability of materials and components, etc. Production and manufacturing are subject to increasing automation as use of robots and computer control machines spreads. Increasingly manufacturing is becoming computer-integrated manufacturing (CIM) in which the process of machine control (CNC – computerized numerical control), monitoring, and error handling is integrated with other information systems so that raw material consumption and production output is automatically notified to other systems. This requires a computer network which runs throughout the factory and its administration.

Although this is an abbreviated explanation of a big subject, it should give the reader some notion of production control problems.

It is evident from the above descriptions of the work areas that they generate a great amount of raw data, and that their control requires a considerable amount of information. The comparison of data and information is discussed in the next section, for the present we are concerned with the interrelationship of work areas through their common use of data and information. Every work area has links with some or all of the others, e.g. stock control receives data from the purchasing department concerning materials and passes information to production control about components required.

The interrelationships between work areas give rise to the concept of 'integration' of activities and information. This concept is especially pertinent to the sharing of data, e.g. stock data is of use in purchasing, sales and production control. Information systems designers are well aware of this common use of data, so consequently they attempt to create systems that allow every application to have access to the data in an up-to-date condition.

The above requirements mean that in practice it is necessary to use a computer-based system which has the capability of storing all the records. Such an arrangement is termed a 'database'; this concept is explained in more detail in Section 5.3. These aims have proved difficult to realize for many organizations, and, in recent years, there has been an acceptance of the fact that centralized computer systems are not always the answer. This has led to the growth of more autonomous systems but with some degree of interconnection. These are known as distributed processing systems (Section 3.3).

Similarly, many departments of a company have their own microcomputers, connected to main computers by means of a network system (Section 3.3).

1.2 BUSINESS INFORMATION

Before discussing business information it is worthwhile considering what is meant by 'data' and by 'information' in this context.

Data are the raw figures appertaining to the routine activities of the organization. These figures in themselves do not enable decisions of any consequence to be taken, and in order for data to be more useful they need to be 'processed' to provide information.

Processing is a broad term but essentially can be taken as meaning the conversion of data into information; that is to say, turning not very useful figures into usable facts and figures. This generally means a reduction in amount; a lot of data is 'distilled' into a smaller amount of information. For example, the data might stem from the huge number of transactions of a bank, and the resultant information is the bank's balance sheet. The computerized systems which carry out this data processing are entitled information systems (IS).

■ Levels of information

Information can be regarded as being of several levels, the number of which depends upon the framework within which the information is used. It is also true to say that one person's information is another person's data. In other words, the difference between information and data is only meaningful in relation to the level under consideration.

For instance, the population statistics submitted by local government authorities to national government is information at the local government level and data at national government level. By combining and processing the statistics from local authorities, national government is able to produce national-level information.

Five levels of information are sensible for most business situations, as follows:

1 *International information*, e.g. projected world food resources over the next few years based upon land usage and the availability of seed and fertilizer, also information regarding world natural resources, weather patterns, currency exchange rates, population statistics and energy demands.

2 *National information*, e.g. a country's balance-of-payments statistics derived from import/export figures and invisible earnings; also industrial output, employment statistics and population trends.

3 *Corporate information*, e.g a company's balance sheet derived from summaries of the various accounts, stock evaluation and plant register; also sales analyses, productive output and market trends.

4 *Departmental information*, e.g. a department's expenditure analysis derived from its individual expenses, a branch's turnover based upon its sales transactions.

5 *Individual information*, e.g. a worker's take-home pay derived from his or her rate of pay, hours worked, tax liability and social security deductions; also a sales representative's sales total for a certain period.

■ Categories of information

There are three main categories of business information, and these are concerned with the usage of the information. These categories also relate to the levels of information described above. Higher levels of information tend to be associated with strategic information and lower levels with tactical and operational information.

Strategic information

This relates to long-term planning policies and is therefore of most interest to top management. Government-wise, strategic information includes population studies, national resource availabilities, foreign investments and balance-of-payment statistics. Company-wise, it includes market availability and penetration figures, projected raw material costs, product developments, manpower changes and new technologies. The meaning of 'long-term' depends on the level of the strategic information. At international level 'long-term' could mean twenty years or more; at corporate level it is likely to be closer to five years.

The nature of strategic information should facilitate long-term planning and will probably include projections and forecasts. It is improbable that an organization's strategic information is entirely computer-produced. Strategic information tends to be held in models, i.e. collections of information organized to produce certain predictions about the world when fed input information from a lower level. Inputs may often come from people as guesses or opinions, and from external sources, also from lower level tactical information systems. Strategic information systems therefore represent the most refined view of data and their use is in planning and decision support.

Tactical information

This is used in shorter-term planning, i.e. months rather than years, and is of more interest at departmental level. Included in tactical information are sales analyses and forecasts, cashflow projections, production resource requirements and the annual financial statements.

These types of information are generally based on data arising from current activities, such as history reports for trend analysis, aggregation reports to compare, for instance, one department with another, and a variety of classification reports to refine operational data. It is also probable that some tactical information stems directly from external sources with little or no need for processing. As an example, a change to a competitive company's product range could have a definite effect on future sales that would not show up in a computer-prepared sales forecast. In tactical planning it is therefore wise to combine information from several sources before making decisions.

Operational information

This applies to the short-term, perhaps hourly, running of a department. It includes current stocks-in-hand, outstanding and overdue purchase orders, work-in-progress levels, and customers' to-follow orders. Operational information is generally derived quickly from current activity data. There are three main needs at this level: monitoring information to ensure some process is behaving to plan; exception reporting to alert people when something has gone wrong; and status reporting, essentially monitoring at a specific time. Exception reporting is often connected with the need for emergency action and so demands rapid preparation. Operational information is usually of direct interest to fewer people than is tactical information but is more specific to those persons. This implies that operational information must be closely geared to the needs of its recipients.

SAQ 1.1 Each of the items below represents information. What sort of information is it?

1 A forecast of the demand for airline tickets produced by a travel company.

2 The number of complaints during the previous week received by a company which manufactures electronic goods.

3 The increase in sales over the past three months achieved by a particular store within a department store group.

4 The number of 999 calls received by the emergency services today.

Solution

1 is strategic information, 2 is operational information, 3 is tactical information and 4 is operational information.

In Chapter 2 we look at the way information systems have evolved from basic systems which process operational data to those which produce strategic information.

Quality of information

When considering the desirability of providing information, the following points should be taken into account.

Brevity

Too much detail can result in the overlooking of vital facts. Each recipient of information should be provided only with his or her needs. This suggests that a large amount of information should, wherever possible, be split into smaller packets tailored to suit individual recipients. The aim of brevity also indicates a need for specificity – the information should go straight to the heart of the matter, especially if immediate actions are called for.

Accuracy

The degree of accuracy of information relates to its usages. There is no point in striving for great accuracy if this is of no consequence. For example, the marketing director is not interested in the value of sales accurate to within a penny, the nearest hundred pounds probably suffices; the financial accountant, however, needs accuracy to the exact penny. Generally, the degree of accuracy of information is known to its recipient; if doubt might exist, the degree of accuracy should be stated on the report.

Timeliness and up-to-dateness

Information should be as timely and up to date as is necessary for its effective use. Speed in creating information increases its cost and so it is unwise to prepare it unnecessarily quickly. On the other hand, information that arrives too late or is out of date is useless, and so the time, effort and cost are wasted.

Timeliness and up-to-dateness are allied. Timeliness implies that the information is delivered punctually after its preparation. Up to date means that it is accurate as at a certain stated date/time.

In some situations timeliness takes priority over up-to-dateness, e.g. the sales manager wants the previous day's sales figures first thing next morning, even though they do not include the last hour's sales. On the other hand up-to-dateness might be paramount. For instance a financial report must be fully up to date as at the exact end of the financial year but need not appear until several days afterwards.

■ Targeting for action

Information calling for action must be directed to the person(s) who initiate the appropriate action. It is important that it goes directly to that person and not via a chain of unconcerned managers. Such a communication chain devalues the information by making it less timely and it erodes the time to perform the necessary actions.

The information should comprise sufficient facts and figures to enable effective and immediate action to be taken, e.g. an out-of-stock report must contain the information for expediting a reorder. Where all the required information cannot be included in the report, reference to its whereabouts should be incorporated.

Rarity

The value of a piece of information depends upon its unusualness. 'Man bites dog' has news value but not vice versa. Thus information reports should emphasize the unusual, as described in the following section under 'Exception reports'.

1.3 MANAGEMENT INFORMATION

The term 'management information system' has become synonymous with computer-based DP systems. This is misleading, however, since it is quite possible to operate an efficient DP system that does not provide true management information. An example of this situation is a payroll system that provides all the necessary figures for paying the employees, and the tax, pension and insurance deductions, but no real information for managerial purposes.

Management information arises, among other places, from the routine data processed regularly in the organization. This is illustrated in Figure 1.1, and although this is a simplification of a company's operations, the contributions of the work areas to management information can be seen. Management information, in turn, contributes to policy-making by managers. The other significant factor is environmental information; this is often of a diverse and subjective nature. A manager combines external and internal information before making consequential decisions.

An example of this is the sales director's decisions in regard to future marketing policy. These cannot be based entirely upon internal information derived from customers' orders because, although sales of a product range may be increasing, it is evident from intelligence of competitive developments that this is unlikely to continue. And so a subjective judgement is made by the sales director as to what market share the company's product range can continue to enjoy. This judgement, together with the actual sales figures, enables a reasonable forecast of future sales.

Personal computers have brought many managers closer to the means of obtaining information by on-demand queries rather than having to wait for printed reports. Managers are becoming increasingly specific in their demands for computer-produced information, and will be less inclined to accept 'information' based on the IS staff's judgement of what is required. Younger managers, brought up with a more technological background, combined with the introduction of more friendly software, have induced a user-driven relationship in which information must be, and is, produced quickly on demand to meet the challenges of managing business in an ever changing world.

There is nowadays widespread usage of graphically presented information. Computer software is readily available for displaying coloured charts, graphs, histograms, etc., including those with a three-dimensional effect. Graphical presentation enables managers to detect information that is of particular interest or that calls for further investigation. Discriminating and sensible employment of colours and shapes (icons) makes a strong impact, allowing the user to assess quickly the information represented. Direct access to information is referred to as end-user computing, and provision of facilities to help this has given rise to concepts of the 'information centre' (see Section 1.6).

■ Management reports

The information received by managers and their usage of it vary from one organization to another. Nevertheless management reports fall broadly into the following categories.

Analyses

An analysis is a grouping of information to provide a summary of results within the various groupings. A typical example is the analysis of sales under headings such as commodity group, sales area, customer classification, and period; in other words, what, where, who and when.

Because of the vast amount of information available from analyses it is advisable to be selective. Selection may be in terms of headings, or of the items included under those headings. For instance, the four headings above may be combined into many different analyses, far more than could be utilized in practice. The other means of selection is to include only certain items within each heading, e.g. only selected sales areas.

It is beneficial if the selection of analyses and/or items can be varied on request so that no unnecessary information is prepared and presented. This might entail variations from week to week and from one recipient to another, but these requirements need not prove insurmountable. Although the aim of providing variable and selectable information might seem overambitious, this is not so if a suitable approach is adopted from the start. In other words, the system is designed to provide all possible analyses and the particular requirements selected by entering parameters.

Generally, managers at a higher level require more highly summarized information than those under them. The marketing director, for instance, may want sales figures for the whole country analysed into commodity groups, whereas a branch manager needs the sales figures for his or her area only analysed by individual commodities. In a situation like this, there is obviously scope for creating the two analyses concurrently and then printing separate information for each recipient.

Analyses can be thought of as being multi-dimensional. The simplest, a one-dimensional analysis, is a straightforward list of summarized items, e.g. the sales in each area. A two-dimensional analysis could be sales of each commodity group within each area. Next, by adding the periods we get a three-dimensional analysis, and so on. There is, of course, no theoretical limit to this but practical considerations usually restrict analyses to not more than three or four dimensions.

SAQ 1.2 Describe some of the analyses which a department store group might make on their sales figures.

Solution

Some typical analyses are:

- An analysis of those sold items which are popular and which are not popular.

- An analysis of the sales achieved by individual stores.

- An analysis of the sales achieved by each department in each department store, for example the Sports departments.

Predictions and forecasts

A prediction is information based upon previous figures projected into the future. A forecast is based upon conjectural information, i.e. subjective judgements as to the effect of various factors. So-called sales forecasting is usually based upon past sales figures and, strictly speaking, ought to be called sales prediction.

It is important with predictions and forecasts that managers understand what is involved in their preparation, otherwise the danger is that they come to rely too much on mathematical techniques and not enough on judgements based on other information. In business there are few situations that lend themselves completely to mathematical forecasting, as unpredictability often clouds the meaning of forecasted information. Nevertheless there are situations that benefit from forecasting and, when this is the case, software proves valuable.

The presentation of forecast information should incorporate an indication of its decreasing accuracy as it looks further into the future. This is important, otherwise the manager may think that next year's forecast is no less valid than tomorrow's.

Optimizations

It is often the case that many alternatives are possible within a situation, only one or a few of which are the best. An optimization technique is applied to the raw data in order to select mathematically the optimum arrangement.

The question then arises as to what is meant by the 'best' or 'optimum' and the answer is, 'it depends what you want'. That is to say, the manager, making use of an optimization report, decides what he or she wants bearing in mind the conflicting constraints. If the least-cost arrangement is required this must be evaluated against the longer time and lower quality that might well be incurred. Nothing comes free and so it is usually a case of selecting a small number of alternatives, perhaps taking other factors into account as well as those employed in the techniques. Typical optimization techniques are linear programming, queuing theory and simulation; these are part of operational research (OR) methodology, itself part of management science. These techniques are applied to problems such as inventory modelling, assignment, scheduling deliveries and allocation.

Linear programming: Methods for determining the optimum amalgamation of several variables of known characteristics to obtain results subject to predecided constraints. For example, the blending of ingredients of known dietary characteristics and costs in order to produce a foodstuff meeting certain nutritional standards at minimum cost.

Queuing theory: The analysis of times spent waiting for service taking into account the number of service points and the amount of traffic (persons or things needing service). An example is the analysis of queuing in a large supermarket in order to decide the number of tills needed. Another example is the analysis of queues of data messages awaiting transmission through a computer network.

Simulation (modelling): The modelling of a process or situation mathematically so as to determine the outcome of various combinations of influential factors, e.g. vehicle flow patterns are simulated by the use of random numbers in order to facilitate road network planning.

Decision theory: The application of statistical probability theory in deciding the best policy to adopt to achieve an objective, e.g. a number of methods have various probabilities of success and lead to other tasks. Which is the best set of methods to adopt in order to maximize success?

Game theory: The application of statistical logic to competitive situations such as bidding for contracts.

Replacement theory: Deciding the best point of time to replace equipment subject to wear and tear in order to minimize the probability of sudden failure and also to minimize replacement costs.

Typical problems which are amenable to OR techniques are:

■ Inventory modelling, to determine the stockholding levels and reorder quantities that give minimum overall stockholding costs. Taken into account are the separate costs such as capital invested, reordering and space occupied, and also the demand pattern of the item (see 'Stock control', Section 1.1).

- Assignment and allocation of work, machines, services, etc., for planning the optimum utilization of resources, taking into consideration the characteristics of the resources and the requirements of the project (e.g. allocating skilled workers to tasks).

- Scheduling, i.e. allocating resources within time and possibly spatial ordering (e.g. delivery of orders by a van driver, answering of service requests from customers).

The reader is reminded that References 1.15 and 1.19 contain more facts on OR.

Management reports can also be categorized as below. These groups are complementary to those described above.

Regular reports

These are prepared at regular times, e.g. every week, and are usually associated with a cycle of activities, e.g. a payroll. Some people question the genuine need for regular reporting as more information becomes available on demand via end-user computing. This is discussed in Chapter 8; for the moment we will accept that a genuine need exists.

The recipient of a regular report becomes completely familiar with its contents and layout. Nevertheless a regular report must be clearly headed and annotated to fortify the recipient's understanding of the information.

It is important that an arrangement exists for the feedback of ideas, complaints and other comments about regular reports, to prevent a situation arising where all recipients believe that everyone else is entirely satisfied and so the report is never modernized. Feedback also covers cessation of the need for a report, so that a manager does not continue to receive superfluous information. Regular reports are satisfactory provided there is a continuing need, but the recipient must not be swamped by too frequent information. It is more effective to supply information only when it is actually needed and containing only what is required at that time.

Exception reports

The principle of exception reporting is based upon the idea that 'no news is good news'. In other words, if no decisions or actions are necessary, then the situation goes unreported. This philosophy also applies to 'management by exception', i.e. managers should direct their attention to exceptional items and situations as the means of maintaining control.

The problem is in deciding what is exceptional. This necessitates careful thought, and the information requirements must be fully determined before exception reporting is implemented. Exceptions are when a measure exceeds a threshold, but this may be within a time period or only when this event occurs a given number of times in a period. A fundamental aspect of exception reporting is the capability of altering the exception parameters. With inflation, for instance, it is advantageous to use relative parameters instead of absolute values. For example, the exceptional cost would be that which is over twice the average, thereby counteracting the effect of inflation.

Exception reports, perhaps more than other types, should contain sufficient information for the recipient to deal with the problem. It is likely that an exception report induces a request for further information. An extension to exception reporting is diagnosis, when the report gives information not only about the exceptions, but also supplies supporting information to help users understand why the exception has occurred.

SAQ 1.3   A company keeps the items it sells in a warehouse. Orders are made by customers and result in the items being withdrawn from that warehouse. Can you think of an exception report that needs to be generated for this part of the system?

Solution

The major exception is when an item goes out of stock, or more realistically when it goes below a quantity which requires reordering to take place.

1.4 DECISION SUPPORT SYSTEMS

An extension of management reports and dealing with exceptions is decision support systems (DSSs). The goal of a DSS is to support the decision-making process although not necessarily to provide sufficient information to make the decision a *fait accompli*. In fact, by the nature of many business decision situations, it is unlikely that the DSS could do that in any case. DSSs enable managers to retrieve information *ad hoc* and as straightforwardly as possible in order to facilitate current decision-making.

DSSs are most effective in risk, i.e. probability, situations where the manager is faced with a number of alternative actions. Ideally the DSS, if given estimates of relevant costs, times, workloads, etc., is capable of assessing all or some of the outcomes of the alternatives. If, for instance, the situation was such that OR techniques could be applied, this would be done automatically by the DSS and the optimum result presented to the manager. In a straightforward case, which would be unusual, the need for management decision might be eliminated as the OR technique had made the decision for him or her.

There are usually too many factors, however, for the DSS to be able to come up with a single answer. Consequently it is necessary for the DSS to have access to a wide range of software and a database management system (Section 5.3) so that facilities needed to support a particular type of decision can be utilized.

Examples of such facilities include the following:

- OR techniques such as those explained in the previous paragraph.
- Network analysis (PERT) for decisions calling for project planning and cost estimations.
- Spreadsheets for cost and other numerical analysis, to enable 'what if' and sensitivity analysis.
- Statistical techniques such as trend analysis, correlation analysis and sampling techniques.
- Database searching and analysis in order to extract relevant information and analyse or summarize it. For this purpose a structured query language (SQL) might be built into the DSS.

A DSS is interactive to a much greater extent than most management information systems. This is a vital characteristic owing to the wide nature of user's requirements. The usual arrangement is for the DSS to record the user's requirements and subsequently to analyse the problem with an algorithm or model-based technique and then possibly measure the degree of success in prediction. The DSS could then be adapted manually to improve the efficacy.

DSSs come in a variety of different types, and function at a variety of levels in the management of a business.

Operational

These systems help junior managers and clerical staff make decisions about control of the day-to-day running of the business, such as inventory control of stock levels, display layout in supermarkets.

Tactical

These DSSs help decision-making with a longer time horizon, i.e. decisions which effect the running and direction of the business for weeks or months ahead. Tactical decision-making, however, still refers to changing a business within the limitations of some plan and this level is generally the responsibility of middle management. Typical tactical DSSs might be changing a pricing policy, selecting suppliers, or planning a manpower allocation schedule.

Strategic

This level refers to decision-making which alters the course of the business. Strategic DSSs rarely model the business as it is, furthermore, because their main use is in planning the future business direction: they contain models of the business as it was, the competition, and the business as it may be in the future. These systems are used by middle and senior managers to model the business and its environment and to take decisions on policy and future direction.

All three levels can be supported by systems in a variety of ways, from simple information provision, leaving the manager to make the decision, to more sophisticated systems that analyse data and actively help decision-making. Simple systems just attempt to retrieve data appropriate for a particular decision. More sophisticated DSSs employ some algorithms to analyse data. Typical of these systems are optimization DSSs using linear programming and other OR techniques with historical data to determine, for instance, the optimal inventory level for products. The next step is to add a model to the system which describes part of the world being analysed. Model-based DSSs contain a description of part of the business such as inventory, product marketing, manufacturing process, fixed assets, etc. Various parameters or variables can be input into the model to test assumptions. The system then outputs metrics and predicted values for the domain being modelled, e.g. manufacturing time and unit cost in a manufacturing process model, cost price, market price, cost of advertising and market share for a marketing DSS. Finally, DSSs can merge with expert system technology in intelligent DSS. In these systems a set of rules are combined with the more usual algorithmic/model-based system. The rules act as an intelligent pre-processor for input variables. The intelligence is the experts' decision-making knowledge of the domain which is analysed as rules of the type 'IF situation THEN modify variable or use different calculation'. I–DSSs have a wide variety of applications in market analysis price sensitivity and process planning.

The presentation of the output from a DSS is important since it must convey the user's requirements in an easily assimilable way. This means that a variety of methods has to be available, especially in visual form. A wide range of graphics software is available enabling the presentation of information in the form of graphs, histograms, etc. (see Sections 4.4 and 9.5). Graphical presentation gives the user the chance to see results quickly and to get the general idea. A growing area is simulation DSSs in which a graphical picture of the business is presented as a diagram showing, for example, parts of the manufacturing plant with material and information flows between them. Other simulations can show inventory levels in warehouses, pie charts of sales displayed on a geographic map, etc. The main function of simulation interfaces is to allow interactive modelling and decision-making. The manager can change various values using sliders and immediately see the effect of change as it is propagated through the model. In more advanced versions the model can be edited by changing the graphical layout. This can be backed up by printouts giving the same information in tabular and/or graphical form.

Examples of business scenarios calling for the employment of DSSs include the following:

■ The allocation of sales offices and representatives to areas taking into account the market potentials of the areas, the representatives' past performances (track records), predicted changes in markets and products, costs of premises and staff redeployment costs.

- Production planning taking into account factors such as existent factory/plant loads, new orders, availabilities of raw materials, and machine capabilities and reliabilities.

- Stock (inventory) planning of expensive goods giving consideration to present stocks, likely future demands, availability of space in the stores, capital tied up, price history of the goods and restocking lead times.

SAQ 1.4 The three decision support systems below are all used by businesses. Are they operational, strategic or tactical?

1 A system used to predict the levels of stock in a warehouse at the end of the day.

2 A system used to track the pattern of sales in a department store, for example which departments are increasing their sales and which are not.

3 A system which uses past business information to predict the optimal geographical spread of sales staff and the products they sell in a company that sells computers.

Solution

1 is operational, 2 is tactical and 3 is strategic.

Executive information systems

These are a variation of the DSSs theme which try to produce integrated computer systems to suit senior managers. Unfortunately senior managers have until recently been very resistant to direct use of computer systems for a variety of reasons. One problem is that senior managers are very busy, so it is difficult to gain much time for analysing their requirements. Other barriers have been resistance to using a keyboard, poor usability of many systems, managers' lack of time and patience to learn new systems, and finally the reluctance of some managers to use technology, which they associate with their secretaries. In spite of this, use of computer systems by senior management is growing and executive information systems may become more important.

Executive information systems (EISs) provide decision support, with information retrieval, powerful display capabilities for business graphics, and communications. The idea is to give a manager all the necessary computer facilities at his or her fingertips. Operation and the user interface is the key to successful EISs. Many features of EISs have grown out of integrated office automation systems and include the following functions:

- Information retrieval facilities linked to corporate databases, but this has to be more than a query language such as SQL (structured query language).

User-friendly means of data retrieval by menus and form filling are necessary, as well as considerable pre-processing of information.

■ Display facilities for business graphics.

■ Presentation facilities to make overhead projector slides from word processed and graphical material. Microsoft's Powerpoint is an example of this kind of product.

■ Multi-media facilities – ability to display media as voice, sound, film, video.

■ Decision support systems integrated with display and presentation facilities.

■ Communication support via electronic mail, fax, and autodial phone numbers.

■ Organization support, electronic diary and meeting management, agenda and daily *aide-mémoire* functions.

EISs are still largely at the development stage because of the problem of providing all these facilities with simple yet powerful user interfaces.

■ Expert systems (ESs)

Expert systems are computer programs that mimic human expertise in a particular subject area. Such systems are capable of limited reasoning and can appear to make intelligent decisions. In reality, expert systems are just programs containing complex logic and rules. ESs are progeny of the research into artificial intelligence that had little or no effect on the conduct of business until the late 1970s when the first generation of expert systems appeared.

An ES purports to be knowledge-based rather than data-based: in other words, it is far less reliant on human instruction in the form of system design and computer programming, and is capable of making its own judgements and forming its own solutions within a given field of knowledge. When presented with a problem, an ES does not follow a predetermined set of instructions, as does a conventionally programmed computer, but utilizes sets of rules which provide the solution when used with a process of problem-solving called inference, i.e. it is told what to solve but not how to solve. ESs are the embodiment within a computer of knowledge-based processes deriving from the skills and experience of human experts. Such a system should also be capable of justifying its line of reasoning in a way that is understandable to the user.

There are three main components of expert systems:

1 *The knowledge base.* This contains a set of facts (or assertions) which the system is given. During reasoning the system may add further facts from its own inferences.

2 *Inference mechanism.* A process that acts upon the knowledge base to solve the problem. Various forms of inference are possible but two of the most common are forward and backward chaining.

3 *User interface and explanation facility*. The user interface gathers user input in the form of data and facts which are input into the inference process. Results are also communicated back to the user. The explanation facility explains the results of the system's reasoning to the user.

SAQ 1.5 What do you think might be in a knowledge base for a system used by a bank?

Solution

Typical rules would be those which determine the size of overdraft a customer should be given, rules which determine the size of a loan, and rules which determine whether a customer should be treated as a special customer and allocated a personal account manager.

In order to understand these components let us take an example of an expert system which is used in a production application and which advises staff on when to reorder items for which there may be a danger of being out of stock. The knowledge base would contain rules such as:

> *If the item has sold more than 10 per cent of its average stock level over the past week and the item has appeared on the order forms of at least two of our major customers then order enough of the item to last four months.*

Here the rule states what condition should happen and the result of that condition. The knowledge base consists of many of these rules (often hundreds or even thousands) and the job of the expert system is to interpret them. You may think that this is straightforward. Unfortunately, it is not: these rules are developed by asking staff who carry out functions such as stock reordering and can be contradictory and incomplete. This means that the inference mechanism used, whereby an expert system makes a decision, is usually the major component of the processing that occurs. Instead of executing the rules as if they were a program the inference mechanism works its way through them finding an optimal set which supports some advice that the expert system is to provide.

Often the information in an expert system is incomplete; in order to complete this information the user interface will prompt the user for extra information which may be required to resolve two sets of contradictory rules. This process is carried out via the user interface.

The final component of an expert system is the explanation facility. Expert systems often come up with results which surprise the user. In order for the user to be satisfied that the expert system has worked correctly and is not in error, perhaps because some of the rules it uses are wrong, there is a need for some explanation facility. This entails the user interface providing an explanation of its decisions. Typically for the example above used in, say, a builder's merchant, the user interface might state that:

I recommended that we reorder the brick item 55xxcv because although we have much stock (10,200 bricks) there has been a major activity in brick sales over the past week from customers who, once they order bricks, carry on ordering for some time.

Expert systems are mainly used to support the human decision-making process. Some examples of their use in business and commerce are shown below.

- Financial dealers in the city use expert systems in order to determine the right mix of stocks and shares for a client who wants to acquire a portfolio of these shares for a specific purpose, for example the customer may want a long-term investment.

- An airline would use an expert system to help in the scheduling of personnel on flights in order to satisfy complex constraints such as the fact that pilots do not fly for more than the legally specified time and that personnel are available at one airport in order to staff a plane which is landing at that airport and proceeding on another leg of its journey.

- An expert system might be used by a wholesaler to determine the source of the cheapest items that they stock. This price may be based on a wide variety of factors such as cost of transportation, bulk deals and experience of poor products delivered from particular suppliers.

Developing expert systems

In order to develop ESs, special computer languages, such as Prolog and Lisp, are used. These languages are 'declarative' rather than 'procedural' such as COBOL and Pascal. The difference lies in how the program executes. In a declarative language a program may be considered as a set of rules.

Applications and limitations of ES

Expert systems have been used in a wide variety of business applications. Problems such as diagnosis, scheduling and planning are suitable applications. Some examples of where expert system technology can be used are: selecting an appropriate pension plan for a client; predicting where to drill for oil from geological survey reports; advising on selection of pesticides; planning supermarket displays; and planning delivery schedules. ESs are successful in small, narrow domains where knowledge is well structured, but they cannot address problems which involve general or common-sense knowledge. Another important aspect of some ESs is their ability to learn in a limited fashion from the feedback provided by users after they have acted upon its advice. That is to say, advice that results in a satisfactory outcome for the user is strengthened, whereas unsatisfactory results cause the advice to be weakened or deleted. However ESs cannot learn in the way people can. The system is utterly dependent on the knowledge it was provided with in the first place. Machine learning systems can add new rules and categories of facts but these systems are in the realm of artificial intelligence. ESs are limited to inferring new facts and reaching conclusions on the basis of knowledge they were designed with.

ESs can be applicable to particular aspects of business such as financial planning, marketing strategies and portfolio management. At the moment the capabilities of ESs are way ahead of the business user's abilities to exploit them. As has been only too apparent over the past thirty years or so, the business user has only slowly and hesitatingly come to grips with computer systems. The main problem is the difficulty for the user to translate his or her expertise. Users cannot become knowledge engineers unless powerful knowledge engineering tools are developed. As current tools require expert developers this means the knowledge acquisition bottleneck is likely to remain for some time. In some companies the situation is so fluid that by the time the ES has been set up, and sufficient feedback entered into it, the problems will have altered so radically that the knowledge in the ES would be irrelevant. However, when stable well-known bodies of expertise can be found ESs have been successful. The ESs not only allow the expert's knowledge to be used many times but can also provide the basis for training new experts in intelligence tutoring systems.

1.5 BUSINESS SYSTEMS THEORY

Theoretical contributions come from two sources. First is general systems theory which has been concerned with understanding the nature of all systems in the world – biological, physical or social. The second influence is theories of business systems and their functions drawn from management science.

The essential lessons from systems theory are:

- Systems are an organized set of components, and have some purpose and a boundary. All aspects of the world can be viewed as systems, from cells in plants and animals to physical systems of atoms and molecules and social systems such as businesses, government departments, societies and countries.
- Systems in turn are composed of sub-systems, e.g. animals are composed of organs which are composed of cells, or a business is composed of subsidiaries which have departments and so on.
- Systems exhibit behaviour which can either be predictable (deterministic) or less so (probabilistic).
- At any level in a hierarchy of systems, one system will contain components which show connectivity; in other words, the components are organized in some way to communicate with each other.

Although systems may be viewed in a hierarchical manner, i.e. each level being decomposed progressively into smaller parts, this is not the complete picture.

At any one level the whole may not be the sum of its parts. This is the concept of 'holism' and 'emergent properties' which assert that we cannot understand everything about a system by decomposing it into its parts. Something is lost in the decomposition that belongs to the system itself. In a business this may be the policy and the image. A policy of customer care and image as a responsive organization are not facts which can be understood from 'top down' analysis by decomposition. There are 'emergent properties' belonging to the system as a whole.

Systems theory has given rise to a particular analytic approach to business problems often called soft systems methodology (SSM). It holds that business problems can only be understood by looking at the nature of the problem and the system as a whole. Indeed, the starting point for SSM is to assume that business problems are unstructured and the first job of the analyst is to figure out the nature of the problem. To that end, SSM provides a framework for organizing thought about business problems enshrined in a nice mnemonic, CATWOE. The framework encourages questions along the following lines:

- **Customers**: Who are the clients of the system? Who does it serve?
- **Actors**: Who are the main agents, usually people, who are responsible for the system and its components? What is their relationship to the activities in the system?
- **Transformation**: What does the system do? What is the activity it carries out to change something? What does it transform from its inputs to its outputs?
- **Weltanschauung** (from the German 'world view'): What is the image of the system to the outside world? What attitude does the system encourage, wish to encourage?
- **Ownership**: Who owns the system and its components? What is the nature of the ownership (e.g. shareholders, trustees)?
- **Environment**: How does a system relate to its environment? What are the influences of the environment on the system and vice versa? How does this impact on its behaviour?

SAQ 1.6 A system has been developed to track the sales of a major chain of department stores. Who are the customers, actors in the system and what is the main transformation?

Solution

The customers are people who come into the store and buy goods, the actors are the management of the company responsible for monitoring the performance of the stores and the managers of individual stores. The main transformation is that of changing bought-in goods to cash.

Besides guiding the approach to understanding business systems, SSM is a method in its own right to problem-solving. Another influence from the systems background has been socio-technical systems analysis. The main point from this is that systems are a composite of human activity (social systems) and technical (automated or computerized systems) which achieve a job of work. Only by understanding the relationship between people and computerized systems can the whole system be successfully designed. Introducing computer systems changes the relationship between people. Computers often enforce procedures, make people follow certain working practices, and stop informal communication, fixes, and non-standard ways of doing things. Frequently this does not improve matters. Organization often succeeds because people are flexible and fix problems in an *ad hoc* manner by 'bending the rules'. The downside of this is that if rules are bent too far, inefficiency, deception and fraud may be the result. Introduction of information systems therefore has to run the gauntlet of too much control and the consequent risk of hostile user reaction or too little control, with the dangers that that entails.

Socio-technical approaches advocate early involvement of users in the design team and more radical views that users should be the designers rather than computer professionals. Typical criteria for consideration would be: seeing users as 'stakeholders' in a system, trying to analyse their needs, motivations and likely commitment to a system; investigation of the power relationships which exist and may be changed by the computer system; attention to privacy concerns of workers, impact of changes to working practices. Socio-technical methods encourage the development and assessment of alternative solutions to business problems, each with different mixes of automation and human activity. The alternative solutions are then assessed against a list of criteria to help select the solution that will best fit the users' needs and social environment within the constraints of available resources.

Business theory, as may be expected, covers nearly every aspect of business life and organization, from personnel management to cost accounting, and common components such as marketing, manufacturing and procurement. However, until recently the link between business theory and information technology has been indirect. Computer systems were implemented to fulfil the well-known business objective of cost reduction and improving control. This has changed with the increasing appreciation that computer systems are in themselves a weapon in the business marketplace; more details about this can be found in Chapter 2. For instance, a supplier offering a customer a computerized service for communicating and processing orders over a network tends to lock the customer into the supplier's network and hence products, to say nothing of generating revenue from the service itself. This strategy has become a vital competitive weapon in airline booking systems, supply of car parts, marketing of holidays and many other business sectors.

Business theories of competitive advantage have resulted from and helped shape these trends. Influential among these theories is Michael Porter's theory of com-

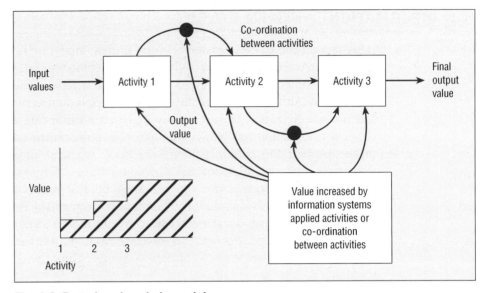

Fig. 1.2 Porter's value chain model

petitive advantage and in particular his value chain model of business activity (see Figure 1.2). This views business systems as a series of activities which add value to a product, whether a manufactured good or a service. Each activity has input at a certain cost, consumes resources which can be valued, and then produces output at a higher cost. Hence each activity can be quantified in terms of the values it adds to the company's product. Of course, not all activities add value; some may be services which are run at a net cost. However, the model allows assessment of the contribution of added value of a company as a whole as well as from each activity.

Porter also suggests that computer systems can be used to improve added value, not only by the obvious business motivators of reducing costs but also by improving communication and co-ordination between activities so that they run more efficiently. Business models can be evaluated to see if computerized support is being employed to reduce costs, add value by improving product quality, making a service better, improving co-ordination and thereby reducing raw material stocks, making a product more rapidly, and so on. In this way business models can be used to point out where new computer systems should be developed and to help assess the contribution of existing ones. This work in strategic planning of information systems will become increasingly important as companies use information technology more effectively to compete with each other and internally to improve profitability and product value.

1.6 INFORMATION SYSTEMS AND USERS

This section brings together a number of factors impinging upon users as a consequence of their becoming involved with IS. By user, or end-user as they sometimes are known, is meant anyone who, in the course of his or her work, receives information from or supplies data to an IS. All users should be able to exert some measure of control or influence over the IS, and certain users must be allowed considerable influence and control. This is tantamount to saying that all managers and most other clerical and administrative staff are, in fact, users, however indirectly.

Data processing is merely a service to its users. The preparation of information achieves little on its own; it is its subsequent use that brings about benefits. These facts should be borne in mind by the user and more so by the IS professional. How can users and, in particular, managers, relate more effectively to IS and, in so doing, improve the efficiency of the organization and the quality of their jobs?

■ User requirements

Users and system developers should give continuous thought to how IS can help them to meet their responsibilities and hence contribute to achieving corporate objectives. By being positive in this respect they have nothing to lose and something to gain personally. The questions in each person's mind must be: 'Can this work be done more effectively and/or economically by using a different method?' 'Does the work as done at present entail repetitive drudgery or duplication of effort that could be alleviated by using a machine?' 'Is the existing computer system friendly enough, i.e. easy to operate and communicate with?'

These questions inevitably propel people's minds towards thinking about new methods. There has been a tendency in the past for IS methodology to be dictated by the limitations of technology and by the IS department's interpretation of what was required. This led to a machine-orientated approach in which the user was largely the outsider.

Those days are now gone as indicated by the ready acceptance of personal computers and the increase in end-user computing. Users are more determined to participate at all stages and thus obtain information that really meets their needs. Participation is encouraged by the growing use of prototyping, i.e. users being allowed to see and discuss what they will be required to do and what they will receive in the way of information (Section 7.9).

■ User/IS staff co-operation

Assignment briefs are discussed in some detail in Section 8.1, but for the moment we can say that an assignment brief is a request for certain things to be done. The danger with formal requests is that they engender a formal atmosphere and this is the opposite of what is really needed. The user and the IS specialist should 'sit at the same side of the table'. By co-operating and if necessary admitting to lack of knowledge on certain points, the parties move towards a strong position to create an efficient and comprehensive system.

Co-operation includes a willingness to explain the technicalities of one's trade. It is sometimes the case that a systems analyst and an accountant, for instance, have little understanding of the other person's job and are unwilling to impart any information about their own. The attitude, sometimes subconscious, is, 'Why allow someone to pick my brain and gain my hard-earned knowledge and experience?'

Another attitude is to keep quiet and so avoid exposing one's ignorance. 'Keep quiet and let people think you a fool rather than speak and prove you are.' The danger with this attitude is that each party, unaware of the other's lack of understanding, proceeds on the basis that the problems are understood, only to discover later that this was not so.

▪ Information centres

In the classic paradigm, the IS professional is the servant of the user and provides analysis, design and implementation of a computer system. With the growth of end-user computing and more powerful development tools which allow users to develop systems themselves, the role of the IS department has changed towards advisers and educators. Information centres support personal computing and information access for end-users by provision of fourth-generation languages, rapid prototyping tools, report generators, etc. The information centre staff provide a service of education and advice, helping users with their problems, running seminars on new products, training on methods and development techniques, and collaborating in development projects run by users. Another important function is to facilitate user access to corporate databases by providing support for communications, network and data retrieval.

▪ User knowledge and training

It goes without saying that a professional person must have in-depth knowledge of his or her profession. It is perhaps less obvious that a professional needs some knowledge of others' expertise. Returning to the systems analyst and the accountant, the former needs to understand the broad principles and the terminology of accountancy. Even though the accountant may be willing to explain carefully the accountancy aspects of the problem, he or she cannot be expected to give the systems analyst an accountancy course. Similarly, the accountant needs to understand the general concepts of IS and its more common terminology.

In regard to IS terminology, it is difficult even for the professional to keep up to date. The plethora of buzzwords, i.e. jargon, terms, acronyms and names, is overwhelming if one attempts to remember them all. Fortunately this is not really necessary as much of the jargon passes quite quickly into disuse.

There are numerous training courses. The difficulty with certain courses is their orientation towards one make of computing equipment (hardware) or one method of using it (software). This does not matter too much as long as the user does not allow him- or herself to be indoctrinated. The main topics of a user training course should be:

- The general capabilities and limitations of computers and data communication systems.

- The broad approach to handling business applications by IS methods.

- The capabilities and *modus operandi* of some of the readily available applications software (Section 6.4).

Additionally, if the user is a go-it-alone microcomputer user:

- The capabilities of and practice with specific application packages, preferably those that the user will be using and on the same model of microcomputer.

- The methods of capturing, organizing and editing source data.

Personnel

Users of computer systems must accept that some changes will occur to their work patterns and they must be willing to face up to these. New methods involve the need for retraining, as described above, and user personnel may find themselves learning something entirely different from their present work. In an organization in which there is some choice, managers must judge who are the most suitable persons to retrain. It frequently turns out that middle-aged staff, who have probably done a satisfactory job with the present system for many years, are unfortunately the least eligible for retraining. They often cannot cope with the radical changes necessitated by a new system, and with the reorientation of thinking that goes with it.

These are problems for managers to handle and, by knowing their staff's and their own capabilities, the best decisions can be made. Such decisions may entail early retirement, transfers, retraining and occasionally recruitment of staff. The decisions are most poignant in small companies in which there are few staff members from whom to choose. Socio-technical design draws attention to the human issues of IS deployment and gives a mechanism for trade-off analysis in terms of the advantages of computerization in terms of cost reduction and job enrichment and the disadvantages in impoverishing work practices, adverse regimentation in control, and increased employee stress and dissatisfaction.

■ Steering committees

Steering committees tend to be associated with larger organizations and so their membership calls for careful selection from amongst the numerous staff. Like any other committee a steering committee is not an end in itself – its attitude requires flexibility and its meetings should be based on a definite agenda.

In a smaller firm, the steering committee is likely to be less formal and could consist of just a few managers or representatives of the departments most affected by the introduction of a new system. Participatory design recommends that steering committees are changed towards a user-centred focus. Ideally, user managers should be in charge and user representatives should set the agenda. Users with more computer knowledge may be elected as local experts to act as a communication bridge between IS professionals and the user community, although local experts do run the risk of being seen as 'one of them' rather than 'one of us'. User-centred design also advocates users being seconded to the design team as domain experts and being involved in the technical design as far as possible.

Purposes of the steering committee

The purposes of a steering committee are:

■ To decide and maintain an overall policy with regard to IS.
■ To represent the views of individual departments but also to bear in mind the interests of the organization as a whole.
■ To involve users in the design process and its management.
■ To promulgate systems development by setting up study groups and initiating systems work, to maintain thereafter a watch over the work of the project teams and systems department, and to receive and evaluate their reports.
■ To maintain a watch over the IS department as regards its cost, efficiency and, most importantly, its service to the end-users.
■ To act as the official negotiating body with suppliers of expensive hardware and software.
■ To report to the board of directors or top management with recommendations for systems developments.
■ To appoint senior systems and IS personnel.

Constitution of the steering committee

A typical constitution consists of the following:

■ a senior manager (preferably from a user department) as chairman;
■ managers of the user departments;

- the senior member of the systems team;
- the senior member of the IS department;
- a representative from any external organization involved, e.g. consultants and suppliers; and
- co-opted members as and when necessary.

SAQ 1.7 If you were to constitute a steering committee which is to oversee the implementation of a system which helps the sales staff of a department store to check on stock availability, which members of staff from the store would you place on the committee?

Solution

The manager of a store should be on the committee and at least one head of a store department. Also a senior member of staff from the stores head office concerned with sales should be on the committee.

EXERCISES

Exercise 1.1   Interrelationships of work areas and management information

Consider the functional areas, work areas and management information appertaining to a self-financing college, and redraw Figure 1.1 so as to show their interrelationships. If you are in doubt, use your imagination based on the contents of Figure 1.1. Remember that the college is self-supporting and therefore operates on a profit-making basis.

Exercise 1.2   Information categories

Suggest three lots each of strategic, tactical and operational information which might be of use to:

(a) an electricity company;

(b) a supermarket; and

(c) a government education department.

Exercise 1.3   Soft systems methodology and CATWOE categories

Video International hires video tapes of films to hotels who then transmit videos to guests via internal cable TV networks. The company is owned by Gnome Films, Inc., and has two operating divisions, one for Europe and the other for the Middle East. Films are hired from distributors who charge a rental fee based on the popularity of the film and the duration hired. Video International (VI) has contracts with hotels to supply a set number of films as specified by the hotel. Films are hired in blocks of one or more weeks and it is usual for hotels to offer guests a choice of 4/5 films. Hotels impose constraints on the type of film they wish to accept. Some hotels have a policy on non-violent films, some films may offend religious values in the Middle East, while others accept films with specific running lengths. In addition, all hotels do not wish to be allocated the same film twice. Hotels may also change their film preferences from time to time. VI tries to satisfy its customers. However, it also wishes to minimize stock, so there are conflicts when hotel managers all want the same film and it is not economic to make copies for everyone.

The problem is to satisfy the demand for films for the available titles within constraints imposed by individual hotels. Currently a booking log is maintained which shows a table of hotels by weeks and the films allocated each week. The hiring history of each hotel has to be examined to determine which films they have not received. Films are allocated to hotels and the appropriate number of copies are made for the demand. Video copies are delivered to hotels. Sometimes video tapes break and the copy has to be replaced. Records of the hotel video booking log have to be updated, showing which film-copies have been allocated to each hotel for each week. Occasionally hotel managers may object to a film which they have been allocated, in which case a replacement has to be found at short notice.

The stock controller keeps records of hotels and their preferences, films (which are updated each month as new films arrive and old ones are returned to distributors), and allocation of films to hotels. The operations manager would like to improve client satisfaction by better matching of film stock to their needs. Another wish is on-line query facility to find out which films are currently allocated to a particular hotel and a report on damaged video copies by hotel. Revenue is calculated from the booking logs; however, billing is not within the remit of the investigation.

Analyse the above system description according to the soft systems methodology approach to produce a root definition and brief entries for the CATWOE categories.

Exercise 1.4   Sales order entry

Chart and describe the main activities typically forming the procedures in a sales order entry system.

Exercise 1.5   Management information

Outline the nature of information likely to be used at each of the following levels of a business enterprise:

(a) top management;

(b) middle management;

(c) supervisory management.

Exercise 1.6   Information and data

'One person's information is another person's data.' Discuss this statement in relation to levels of information and management.

Exercise 1.7   Information centres

The concept of an Information Centre (IC) has become popular over the last few years.

(a) What is an IC?

(b) Why have ICs become important to organizations?

(BCS, 1D, April 1992)

Solution 1.1   Figure 1.3 shows the basic interrelationships. Do not be perturbed if your answer is somewhat different from this; the important thing is that you have understood the principle behind the exercise, i.e. the fact that work areas are interrelated and connect with management information.

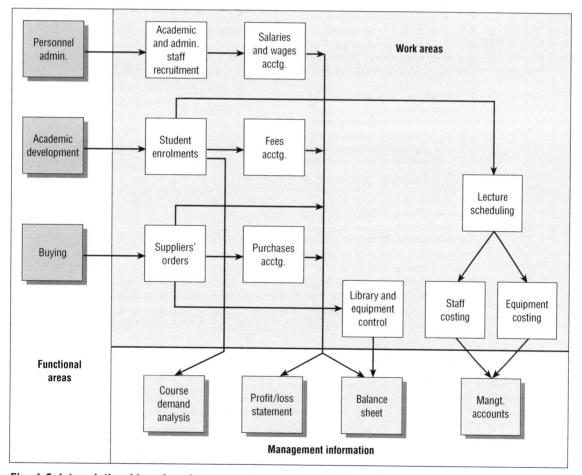

Fig. 1.3 Interrelationships of work areas and management information in a college

	Electricity co.	Supermarket	Education dept
Strategic information	1. Long-term fuel costs and availability 2. Technical developments in generating plant 3. Long-term demand forecasts based on population and industry	1. Consumer attitudes to products based on standards of living 2. Long-term availability of foodstuffs 3. Effect of town planning developments	1. Actual and predicted birth rates 2. Teacher training resources 3. School building needs based on demand and legislation
Tactical information	1. Seasonal fluctuations in demand for electricity 2. Changes in fuel costs 3. Maintenance and replacement of equipment	1. Seasonal variations in demand 2. Seasonal variations in supplies 3. Shortage of items due to industrial action	1. Current age-group populations of children 2. Requirements of books and equipment 3. Teacher availability and requirements
Operational information	1. Equipment breakdown 2. Labour disputes 3. Power generation demand and balancing	1. Out-of-stock items 2. Immediate price increases 3. Staff absenteeism	1. Staff illness 2. Pupil transportation problems 3. Pupil absenteeism

Fig. 1.4 Examples of information categories

Solution 1.2 Figure 1.4 shows examples of information categories, illustrating solution 1.2.

Solution 1.3 *Root definition*: Video International (VI) supplies its client hotels with a video rental service according to their preferences. Films are allocated to hotels to ensure variety and minimal duplication. VI strives to ensure that the film allocation meets with the clients' needs and constraints. Video copies of films are made to satisfy demand while trying to minimize excessive stock. Videos are loaned to hotels, returned and damaged copies repaired. Balancing inventory with clients' demands requires careful management.

Customers: Hotels who have rental contracts with VI and indirectly the hotel guests who view the videos.

Actors: The stock controller clerk who allocates films, the stock manager and hotel managers.

Transformations: Copying films, allocating video-copies to hotels, return of videos.

Weltanschauung: To provide a responsive service for the hotel manager's needs.

Ownership: VI is a subsidiary owned by Gnome Films; the relationship with the film distributors is not clear.

Environment: Films must not be loaned which offend the law and culture of the clients' countries.

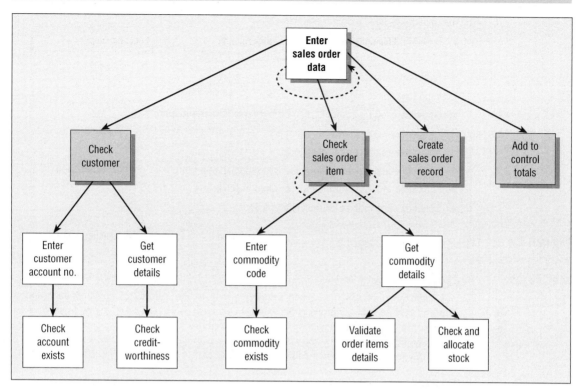

Fig. 1.5 Simplified structure chart of sales order entry procedure for Exercise 1.4

Solution 1.4   Refer to 'Sales control and accounting' in Section 1.1.

The main points are:

■ Method of order data capture – mail, phone, sales reps., electronic mail, etc.

■ Method of input – keyboard, OMR, data transmission, bar codes, WIMP.

■ Check customer for validity and credit-worthiness.

■ Check goods ordered for validity and reasonableness in terms of quantity ordered, price (if included in order), date for delivery, existence of options (colours, sizes, etc.).

■ Check stock in hand, allocate stock to order, record sales order.

■ Query any discrepancies with customer.

■ Arrange for to-follow items, i.e. outstanding orders.

Figure 1.5 depicts these procedures in the form of a structure chart.

Solution 1.5   (a) Information for top management

■ Must facilitate decisions in regard to strategic long-term planning.

■ Needs to relate to the objectives of the organization as a whole.

■ In broad terms in the first instance, supported by further, more detailed information if required.

■ Overall position quickly assimilable.

■ Important points highlighted, and backed up by extra detail.

■ Graphic presentation can be effective.

(b) Information for middle management

- ▪ Related to shorter-term tactical planning.
- ▪ Relevant only to the manager's own sphere of responsibility.
- ▪ Sufficient detail for comparison with other results.
- ▪ Important points highlighted.
- ▪ Graphics useful but should also be available in numeric form.

(c) Information for supervisory management

- ▪ Applicable to short-term decisions.
- ▪ Immediate-action information clearly distinguishable from more general information.
- ▪ Formatted in a standard layout for rapid assimilation.
- ▪ Contains suggestions for necessary actions.

Solution 1.6   Refer to Sections 1.2 and 1.3.

Solution 1.7   (a) Refer to Sections 1.6 and 9.1.

(b) The importance of ICs has arisen primarily owing to the inability of conventional systems analysts to cope with the growing demand for computer services. These include fresh work for computerization and a continuous need to make changes to existing applications. ICs are seen as a way out of this dilemma by enabling users to plan their own computer work, and then to apply it by means of high-level software such as 4GLs.

References and further reading

1.1   Atrill, P. *Accounting and Finance for Non-specialists* (Prentice Hall, 1997).
1.2   Cashmore, C. and Lyall, R. *Business Information* (Prentice Hall, 1991).
1.3   Dahl, A. and Lesnick, L. *Internet Commerce* (New Riders, 1995).
1.4   Durkin, J. *Expert Systems* (Prentice Hall, 1994).
1.5   Elliot, G. and Starkings, S. *Business Information Technology Systems, Theory and Practice* (Addison Wesley, 1997).
1.6   Emmanuel, C. *Accounting for Management Control* (International Thomson, 1997).
1.7   Finlay, J. *An Introduction to Artificial Intelligence* (University College London, 1996).
1.8   Fleming, I. *Accounting for Business Management* (McGraw-Hill, 1997).
1.9   Harry, M. *Information Systems in Business* (Financial Times, 1997).
1.10 Henning, K. *The Digital Enterprise* (Arrow, 1999).
1.11 Jackson, M. and Twaddle, G. *Business Process Implementation* (Addison Wesley, 1997).
1.12 Kandel, A. *Fuzzy Expert Systems* (John Wiley, 1996).
1.13 Laudon, K.C. and Laudon, J.P. *Essentials of Management Information Systems* (Prentice Hall, 1998).
1.14 Laudon, K.C. and Laudon, J.P. *Information Systems and the Internet* (Dryden Press, 1997).
1.15 Lewandowski, A., Serafini, P. and Speranza, M.G. *Methodology, Implementation and Applications of Decision Support Systems* (Springer-Verlag, 1992).
1.16 Luger, G. *Artificial Intelligence* (Addison Wesley, 1997).
1.17 Pizzey, A. *Accounting and Finance* (Cassell, 1994).
1.18 Russell, S.J. *Artificial Intelligence: a Modern Approach* (Prentice Hall, 1995).
1.19 Sauter, V.L. *Decision Support Systems* (John Wiley, 1997).
1.20 Simon, E. *Distributed Information Systems* (McGraw-Hill, 1996).

1.21 Thomas, A. *An Introduction to Financial Accounting* (McGraw-Hill, 1996).

1.22 Turban, E. and Aronson, J. *Decision Support Systems and Intelligent Systems* (Prentice Hall, 1997).

1.23 Turban, E. *Decision Support and Expert Systems* (Prentice Hall, 1995).

1.24 Wilkinson, J.W. *Accounting Information Systems* (John Wiley, 1997).

2 Computers and business – their interrelationship

AIMS

After reading this chapter you should be able to:

- outline the differences between data processing, information and strategic systems;
- show the increasingly intimate connection between the technical aspects of system development and the business objectives and strategy of a company;
- detail one case study which demonstrates the leverage that a computer system provides to a company's business.

2.1 INTRODUCTION

Chapter 1 described the general relationship between a business organization and computer systems. The aim of this chapter is to show how computer systems have become intimately involved with the objectives of a business to the point where many businesses would be incapable of operating or at least keeping ahead of the opposition without computer systems.

Information systems are embedded in some enterprises: the enterprise may be one in which profit and the growth of profit is of the highest priority to management, or an enterprise such as a hospital where the main aim is providing an increasingly efficient service. No matter what the aim of the enterprise, since the 1970s there has been an increase in the use of information systems as a strategic business tool.

Before looking at the relationship between information systems and business objectives it is worth looking at the way computer-based systems have developed since the 1960s.

2.2 THE CHANGING FACE OF INFORMATION SYSTEMS

The growth of information systems for the most part can be partitioned into three distinct phases:

- the development of data processing systems;

- the development of management information systems;

- the development of strategic information systems;

- the development of transformational information systems.

In the context of this book a data processing system carries out some basic function which somehow improves the internal efficiency of a company. The classic example of a data processing system is one which processes the salary and wage records of staff and administers the payments associated with these records on either a weekly or monthly basis. In the 1960s this was one of the first areas that a company computerized, often dedicating a single computer to the task. The processing of a payroll is a routine operational matter. The thing which differentiates it from both management information systems and strategic information systems is that it has no *direct* effect on a business. There may be some indirect effect: savings made from computerizing the payroll might have been used for investment purposes; however, the main purpose of such a system is that of making some internal business process more efficient.

The next stage in the development of computer-based systems occurred in the 1970s and 1980s. By then, many companies had implemented a number of data processing systems and had collected together a large amount of data which could potentially be useful to them in their business. For example, a typical system developed in the early 1970s for companies which were in the business of selling items was one which automated the ordering and invoicing of these items. Such systems were initially developed for purely operational reasons: in order to reduce the manpower required to administer sales. However, in the 1970s it was realized that such data provided useful management information which could help a business operate better and produced larger profits. For example, data on sales over a number of years could be used to predict how much stock was needed in a company warehouse in order to minimize the amount of capital tied up in idle resources. Such data could also provide information about which customers were ordering more items and which were ordering fewer, and enable sales staff to concentrate their attention on those customers whom it was felt needed some special attention such as providing larger discounts on bulk buys.

The 1970s saw an increased development of such management information systems. In the early days of information systems development this often consisted of writing software which integrated already existing operational systems into one large information system, for example a purchasing system might be integrated with a system for administering a warehouse. However, by the mid-1980s larger and larger bespoke systems were developed which affected the whole of a company's operations.

The third phase of systems development is occurring now and started approximately in the 1990s. It is unclear when the exact starting point occurred, and it is debatable why the transition occurred and what were the important drivers. In general, though, the increased availability of microcomputers, some small enough to be held in the palm of the hand, and the increase in vogue of management theories which placed service above cost reduction led to the current phase of strategic information systems development. The management theories which drive the current wave of strategic information systems development effectively say that in order for a business to stay competitive priority should be placed on those activities which produce some added value to a customer and that the primary driver should not be purely a reduction in operating costs.

The main differentiating factor between a strategic information system and a data processing system or an information system is the fact that the strategic information system changes the external business practices of a company and that this change is much more visible than the effect of the other two types of system.

As an example of three of the different types of system consider the history of the computerization of supermarkets. In the early days of supermarket growth, systems were developed which were purely data processing systems: typically these administered the payroll, paid suppliers of goods, administered the warehouses in which items were stored and processed the cash received from the customer.

A typical management information function which was then implemented was that of tracking fast-moving goods and producing trend analyses of customer buying preferences. From this data the supermarket could carry out actions such as reducing the orders they made for less popular items or, in order to resuscitate sales for a particular item, promote special offers.

The 1990s has seen an archetypal strategic information system function being deployed: that of the processing of loyalty cards. Such cards, which resemble credit cards, are given free to customers and are swiped through a magnetic reader when a customer completes the purchase of a collection of items. A customer who uses such a card is then periodically sent a voucher which can be used in exchange for goods at a supermarket. Loyalty cards have been made possible only by the increasing availability of cheap hardware devices such as magnetic strip readers and the increased embedding of microcomputers in point-of-sale equipment.

When a supermarket has a loyalty card system in place there is a wealth of information that management can use:

■ The item buying habits of individual customers can be pinpointed. Previous supermarket information systems were only able to produce aggregate figures such as the fact that the demand for a certain product was falling. Loyalty cards enable the pattern of spending by a single customer to be discerned and advantage taken of it. For example, a supermarket may discover that a certain customer does not spend much on fresh fruit and then enclose a special money-off voucher for certain fruit items which can be redeemed over some short time period.

- The timing of buying patterns can be discerned. A stock control system for a supermarket which processes the replenishment of shelves, an example of part of a management information system, can provide only very broad-brush information about the pattern of sales of items; for example, it can provide details of sales within a particular week. However, a loyalty card system can discern the pattern of sales within a day. For example, information from a loyalty card system might enable individual supermarkets to discover that the sales of convenience foods are much higher after five o'clock because working parents tend to use a supermarket in the evening. A manager might then place a special portable display in the entrance to the supermarket providing details of this week's special offers on such foods. During the day when the pattern of sales would concentrate on basic items such as bread this display might be replaced by another display which headlined special offers on these items.

- The relationship between items in a shopping basket can be discovered. For example, a loyalty card system would be able to ascertain that one category of item is often bought when another category of item is bought. For example, on Friday night it might be discovered that packs of beer would be purchased at the same time as jumbo packs of crisps or savoury snacks. This information might be used to move a portable display of such snacks close to the beer shelves of the supermarket.

This sort of system changes the way that a company does its business: it affects both the workers and the customers of the business. In the case of the supermarket, individual store managers have more information which is much richer and customers see an improved service from the supermarket which, at the same time, benefits it financially. This is not the case with data processing and management information systems where the effect is often indirect or peripheral to a business: customers might see a greater variety of items for sale because of the efficiencies achieved by a stock holding system, but effectively they see little change in the way that a supermarket operates.

The final sort of business information system is the transformational information system. This is a system which is only enabled by technology and allows a company to carry out totally new business functions without much effort. An example of this type of system is an on-line auction system. Such a system is used to solicit bids for items over the Internet and can be easily used by a company which, although it might have sold products before, had not involved itself in any auction-based selling. A transformational system differs from a strategic system in that the latter is involved with existing functions, while the former is associated with new business functions which could be implemented very quickly, normally using software packages. Transformational systems are so called because they can have the effect of transforming the whole way a company does its business.

Most systems in existence are either data processing, management information or strategic systems so the remainder of this book concentrates on them.

SAQ 2.1   Four systems are described below, categorize them as a data processing system, management information system or strategic information system.

1 A system which issues cheques to the suppliers of raw materials of a shoe manufacturing company.

2 A system which administers a telephone banking system adopted by a high-street bank.

3 A system which keeps track of the debits and credits of a bank.

4 A system which enables a bank to keep track of the monthly spending habits of its customers.

Solution

1 is a data processing system, 2 is a strategic information system (assuming, of course, that the bank is a traditional one which does its business via conventional branches), 3 is a data processing system and 4 is a management information system.

2.3 TYPES OF STRATEGIC SYSTEM

There are a number of points to be made about this division into three categories of management information system, data processing system and strategic information system:

■ That while strategic information systems may have an enormous effect on the profitability of an enterprise, management information systems and data processing systems may also have comparable effects. For example, companies who automated their stock control and delivery services in the 1980s and who were able to deliver commodities to customers much quicker than their competitors achieved large increases in profitability.

■ That a company's systems are often an amalgam of all three categories: strategic information systems sit on top of management information systems which, in turn, sit on top of data processing systems. For example, a bank which adopts a new way of working via on-line or telephone banking still employs basic data processing systems to administer functions such as issuing standing orders, and management information systems to carry out tasks such as discerning market trends in the way that customers use the bank. Indeed, it is a truism that the best strategic information systems need to rely on high-quality data processing and management information systems and that the quality of a strategic system is directly related to the underlying systems on which it draws. Because of this, companies who decide to invest in a strategic information system often have to carry out considerable redevelopment of their data processing and management information systems in order to increase the quality of the information that they provide to their strategic information system.

- That there is nothing technologically different from the vast majority of strategic information systems to data processing and management information systems; occasionally one finds some advance piece of technology such as a data mining tool being used, but in general the underlying technology is the same.

- That the process of developing a strategic information system differs in only one respect from that of developing information systems or data processing systems in that the front end of a project designed to develop a strategic information system addresses a number of non-technical issues such as what new computer-based services should an enterprise supply which will provide a competitive edge.

Ward and Griffiths (Ward and Griffiths 96) have identified four uses of strategic information systems:

- Those systems in which information is shared with other agencies such as customers and suppliers.

- Those that integrate existing data in what might be a set of disparate systems to enhance profitability.

- Those that produce new information-based services and products.

- Executive information systems that provide high-level management with information which can be used for strategic planning.

Information sharing systems

Two examples of the first use are in banking and wholesaling. The last five years have seen a tremendous increase in the availability of home banking services where a customer who, in the past, would have gone to a bank branch now uses the telephone for transactions such as finding the balance of an account and initiating standing orders. This is an application which still has a long way to go in functional terms: the growth of the Internet will mean that in the near future all the transactions that a customer could carry out *could* be computerized with the home PC offering the entry point.

The second example is in wholesaling where a company, say a food wholesaler, provides access to its internal databases so that customers can, by using their own computers, browse through stock availability and special offer information with their own computer and order any items they want. Similarly, the wholesaler might allow its suppliers to browse through its databases in order to discern whether any items are low on stock and may even have an agreement with the supplier that when an item falls below a certain stock level a predefined amount of that item is delivered.

■ Integrated systems

Such systems often integrate a number of existing systems in order to radically change some business function. For example, a company which sells agricultural products to farmers might have two separate systems: one which carries out the ordering and another which administers stock with the ordering of items being processed by phone. Integrating both these systems and allowing sales staff to visit farmers to discuss their needs, predict delivery time and alert them to special offers changes one aspect of the business processes of the company.

■ Information-based product systems

These are systems which produce information that can either be sold or can enhance an existing product. Two examples of this use of a strategic information system can be found in the building trade and on the Internet.

One of the features of the Internet is that it consists of a large amount of unstructured information: billions and billions of words of text. Even in the early days of the Internet users had major difficulties in locating specific information. The response to this from IT companies was the development of search engines: computer programs which, when you interacted with them, asked for a series of words that characterized the type of information search you wanted to carry out. For example, if you wanted to search for entries on the Internet on the music of Edward Elgar you would type in the words *Elgar* and *Music* and it would then return with the documents devoted to this topic. Initially, search engines were developed by universities; however, more and more of them are being developed by financially acute companies who offer services free to users yet charge for any company that wants to advertise on the part of the Internet in which the search engines are embedded. Search engines are just the tip of a particular iceberg; increasingly information companies are offering new services to customers which involve the trawling of information from the Internet.

The second example of a system which provides new information-based services involves building supplies companies who offer a form of quantity surveying service to their customers. For example, when a builder wants to develop houses he or she provides the building supplies company with the detailed plans. A computer program is then used to produce a listing of all the building items, such as joists, slates, window frames, that are needed and then costs the list. The builder is provided with free information that would have cost hundreds of pounds if he or she had used a quantity surveyor. This information is free provided the building supplies company is used to supply the raw materials.

Executive information systems

These systems often provide market demand information to a company and allow the high-level management of that company to plan a number of years ahead. Often these systems require the use of both internal and external databases. Past information systems have provided data which was too raw and this form of strategic information system aims to provide a lot more context. Probably the best example of this type of system is the loyalty card system described earlier. Here, the buying trends of customers can be used to make decisions such as where to open a new store. For example, when you apply for a store card you provide a postal code on the application form which is recorded in the computer. Each use of the loyalty card enables the supermarket to track the number of journeys made by a customer to particular stores and to decide whether a new store could be developed near a large group of customers who have to travel a fair distance to an existing store. Moreover, the effect of building this store and attracting customers away from existing stores can also be predicted.

SAQ 2.2 An optician places a computer and a digital camera in each of his or her stores. When a customer comes in to buy a pair of spectacles the camera takes a picture of him or her and the computer enhances the picture by showing what the customer would look wearing the various spectacles that the optician stocks without the optician needing to go to the major trouble of finding them. What sort of system would you regard this as?

Solution

It is a form of information-based product system, where the product is that of information regarding how the customer looks in certain spectacles.

2.4 BUSINESS PLANNING

Before looking at the fit between information systems and business planning it is worth looking at how planning has been carried out since the 1980s. The way that companies have planned over this period can be partitioned into four phases: short-term finance-based planning, predictive planning, competition-based planning and innovation-based planning.

Short-term finance-based planning

This is typical of the type of planning carried out by companies in the 1960s and early 1970s. It involves looking at past financial performance on a department-by-department basis and, based on past financial figures such as the annual amount

51

of sales and the annual change in the cost of raw materials, trying to set budgets and sales targets for one or perhaps two years in the future. The whole focus in this form of planning is the involvement in a discourse which is purely financial in its vocabulary.

■ Predictive planning

This is typical of the type of planning carried out in the 1970s and early 1980s. It is similar to the previous category. However, the timescale is often longer, usually five years, and starts to involve some minor strategic questions. A company that carries out this form of planning usually consults financial figures from a number of years past and attempts to predict trends and what external factors will be important in a planning horizon which could extend to five years in the future.

■ Competition-based planning

This form of planning has been carried out by large companies since the mid-1980s. The previous two forms of planning were heavily focused on internal factors: for example, how well sub-groups such as departments fared financially as compared with other departments. This form of planning marks the first attempts by a company to look outside itself. It involves appraising what competition exists in the industry sector within which the company works: what competitive strategies are succeeding and what strategies are failing. Based on this form of appraisal a company will then start planning strategically. Typical decisions that this involves include:

- Closing down or selling off parts of a business which it is felt will not realize the level of profit necessary to sustain the business.
- Marketing the products that are produced by the company in different sectors. For example, the United Kingdom education sector is moving over to the eventual point where students will have to be self-funding. This has provided a number of opportunities for investment companies selling fifteen-year maturing financial products to the parents of such students.
- Investing in underdeveloped parts of a company, which it is felt could provide a cutting competitive edge.

■ Innovation-based planning

This form of planning is driven purely by the process of developing new products and services which will, at least in the medium term before other companies catch up, provide a significant competitive advantage. The previous three forms of planning have gradually focused on the world outside a business; this form of planning focuses totally on this. High-level management in such a company will be continually asking questions such as:

■ What demographic trends can we see which will require us to add to our portfolio of leisure and sports products? For example, there is going to be a bulge of retired healthy workers with good pensions over the next fifteen years. What sorts of activities will they be undertaking?

■ In what way will the lifestyles of our customers alter over the next ten years: for example, what are the implications of an increase in 50 per cent of their owning PCs?

■ The government arrangements for student fees will mean that many students who leave university will have a high level of debt: are there any innovative financial products which we can develop for such a growing sector?

■ Remarks on the four planning phases

There are a number of things worth saying about these forms of planning which will be important as this chapter progresses.

First, a company which involves itself in the fourth form of planning will often have major demands on its IT systems over a short period of time. For example, banks and other financial institutions are capable of dreaming up a financial product which can be marketed in a time horizon of a few weeks. This places a huge short-term demand on IT departments which will often need to deploy systems to support such products in shorter times than would be normal.

Second, that if you take a time slice through industry and business you will find a continuum of companies all at different stages in terms of their business planning. Even today you will find companies working at the lowest level of annual financial planning. In general it is those industries where product growth is highest – industries like the software sector – in which innovation-based planning is at its most febrile.

Third, that the most successful companies combine the best elements of each of the four phases of planning that we have discussed. Indeed, it is almost impossible for a company which carries out innovation-based planning not to exist without a well-oiled form of annual financial planning. If you look at the most profitable companies in the world you will find a four-tier planning pyramid consisting of planning processes drawn from the four phases described here with each level drawing upon information provided by the lower levels.

2.5 DEVELOPING A BUSINESS/IS STRATEGY

The aim of this section is to look at some of the processes involved in developing an information systems strategy which matches the sort of strategic planning described in the previous sections.

There are a number of tasks which need to be carried out before developing such a strategy:

■ The business strategy of the company needs to be analysed: which markets are regarded as important, which markets are declining, what changes in the customer base are expected and how is the company going to react to them.

■ The current way that the company does business needs to be addressed and evaluated with respect to the deliverables from the first step.

■ There is a need to determine the critical success factors for the business, what are they and how can they be quantified.

■ There is a need to identify the processes which add value to the products or services which the company provides. For this a technique known as value chain analysis can be used.

■ There is a need to create an architecture which shows how the information systems resources and IT resources could fit together to satisfy the demands being placed on the company by its strategic plans.

A number of these steps are described in more detail below.

■ Investigating the business strategy

This is the first step and involves a number of activities. Depending on the company, the business strategy may be embedded in a variety of forms: in a well-organized company its business strategy will be documented in corporate documents; in other companies this strategy may be held as a mix of formal documents, informal documents and ideas in senior staff members' heads. Typical statements which might emerge from an analysis of a business strategy are:

■ We will aim to take advantage of the growth of home computing by developing packages which enable customers to calculate their own tax obligations.

■ In the next five years there will be a growth in Internet shopping. We intend to take advantage of this by developing a series of Internet sites which are dedicated to each of our core products.

■ There will be an increase of 8 per cent of retirees with large quantities of disposable income over the next five years. We intend to expand our leisure village concept in order to cater for this potentially profitable market segment.

Ward and Griffiths have identified a number of components to a business strategy:

- *A mission.* This is some broad statement of how the company sees itself. For example, 'We want to be the company everybody thinks of when they think of exotic holidays'.

- *A vision.* This is a statement which will enable staff and customers to envision how the company would work in the future. This would be rather more detailed than the mission.

- *Goals.* These are broad statements of how the company is to improve. For example, a statement such as 'We will increase our sales to medium-sized companies in the oil servicing industry'.

- *Objectives.* These are refinements of the goals which are measurable. For example, a supermarket company may have as its goal an increase in its revenue from the sales of up-market foods. The goals might envision the supermarket categorizing the items that it sells into a number of categories which represent the degree of up-marketness they possess and then make statements such as: 'We intend to increase the level of sales of level 4 goods by at least 12 per cent and the level of sales of level 5 goods by at least 15 per cent'.

- *Strategies.* These are the ways and means whereby an objective will be met. For example, a supermarket group may decide that in order to cater for customers who are to buy up-market goods more it would have to build more supermarkets in a different geographical area than it has done in the past.

- *Critical success factors.* These are things which have to be right in order for the business objectives of the company to be achieved. For example, a manufacturing firm, competing with companies who are reducing their defect rates, will regard the reduction of defect rates as a key critical success factor.

- *Business drivers.* These are forces for change that a company needs to respond to. These include drivers such as an increased market share or a reduction of costs.

Documenting business processes

Once a business strategy has been discerned, the next step is to document the individual processes that make up a business. This fairly mundane process involves the documentation of the tasks and documents required for a business process, the decisions that need to be made and the people making them, and the outputs from these processes. A typical low-level process would be the approving of a loan in a banking system and a typical high-level process would be the determination of medium-term business plans for the bank.

■ Documenting the external business environment

This looks at the way that the world outside the business is working and how it may work in the future. This involves looking at:

- The effect that new technology is having and might have on the sector in which the business is working. For example, a bank may look at the spread of home computers and its effect on conventional bank-branch business.
- The effect of government policy changes. For example, the recent decision by the British government to place more of the burden of paying for degree studies on students and their parents might be examined by finance companies.
- The way that competitors are developing their business. This may be regarded as a threat or as something to be ignored.

■ Examining and documenting the current information systems provision

Here the company evaluates what current applications have been computerized, which systems are in place, what the skills of IT personnel are and which databases are currently available. It should also include systems which are under development. From this study a number of important conclusions might emerge. For example:

- That there is some duplication of data between a number of applications.
- That there are major differences in the time that data is processed between applications.
- That some business processes are well supported by IT, while others are less well supported.
- That a number of core processes are little supported by IT.

■ Determining the gap between business strategy and IT provision

This is probably the most difficult part of the whole process. It involves comparing what IT resources the company has to the way in which the business will conduct itself over its strategic horizon. Almost certainly this will result in a number of plans for change. Typically the gap analysis will have identified:

- Business processes which are carried out manually that can be computerized. These will range from those processes which are routine to those which could greatly affect the business in the future.

- Business processes which are computerized but which are slow in duration. For example, a process may be supported by a form of batch processing and the company requires that process to be supported by some interactive software.

- That information requirements of management are either not satisfied or only incompletely satisfied by the information systems that are currently being operated.

- Strategic requirements which cannot be satisfied by the current set of applications.

It is from this gap that a company's IT strategy should develop. It will require identifying new applications, new software, new hardware and new skills which the company should foster.

2.6 AN EXAMPLE

To conclude this examination, a small case study will be presented. It involves a large, successful travel company. Such a company sells a large number of hotel and villa based holidays. The selling of these holidays is carried out by tele-sales staff at the travel company and also by travel agents.

The company has two systems which it maintains; both are data processing systems with only a small part of one of them providing some management information. The first system carries out the invoicing of customers or travel agents. It produces a form which provides the basic details of the holiday together with financial information such as the cost of the holiday and, if the holiday was booked via a travel agent, the commission that the travel agent has earned.

The second system uses some of the data provided by the first system in keeping a database of occupancy data which details when the villas and hotel rooms used by the company are booked and when they are free. This system is used by the tele-sales staff at the travel company. Such staff will receive calls from both customers and travel agents.

The system works well; however, the main problem is the fact that the travel agents and customers find it very laborious dealing with tele-sales staff; a process which can take some time, particularly during the peak booking periods of December to February: travel agents complain about the queues that build up during this period in their shops with some customers becoming so frustrated that they leave without booking a holiday. This not only affects the bookings of the travel company but also the amount of commission that the travel agent earns.

Initially, the travel company decided to introduce a simple system which enabled travel agents to book holidays from their own computers.

SAQ 2.3 What sort of system would this be: data processing, management information or strategic?

Solution

It is a strategic system as it affects the way the company is working.

However, before committing themselves to the new system the company commissioned a study from a consulting company which specialized in providing advice for the travel industry. The company was given a brief to look at trends both inside and outside the travel industry which might affect the IT provision of the company.

The consultants made a number of points in their report:

■ The proposed new system was an excellent choice that enabled the company to share its data with its travel agents.

■ There were a number of trends which would affect the company's customer base over the coming decade which could be taken advantage of in developing new systems.

■ The customers of the travel company were usually in a social class which had a high incidence of computer owning. Moreover such customers were showing an increasing use of the Internet.

■ The next five years would see an increasing penetration of digital television. This again would be followed by an increase in the use of the Internet, as it was envisaged that digital television sets would also include some form of Internet browsing.

■ Customers seemed very happy with the services provided by the company, but one criticism was made by both those customers who used travel agents and those who booked directly with the travel company: they always felt under pressure that they should book a holiday then and there because of the time it was taking to discover vacancies, flip over the pages of the brochure that the company issued and calculate the eventual price.

■ The company was not sufficiently different from its competitors to have much prospect of profit growth in the medium to long term.

■ Before the company starts further computerization it should revamp its business strategy. This was contained in a document which was now over a decade old and, even without further computerization, looked somewhat dated.

The result of this was that the company renewed its business strategy documents. This included a new mission statement, 'We will be the company that customers will turn to first when booking an overseas holiday'; a vision which included the company sharing more information with its travel agents and customers and also providing more information; some goals, which included an increase in sales to certain social classes and an increase in the number of holidays offered and object-ives such as a 10 per cent annual increase in turnover over the next five years.

Once the business strategy was in place the company looked at its IT strategy. Its current IT provision was provided by a large mini-computer with a series of rather ageing terminals serving tele-sales staff, all programming had been done using contractors. Most of the software was developed using the ageing pro-gramming language COBOL and the vast majority of processing consisted of the generation of financial reports or interactive processing which checked holiday availability.

The company had identified a number of business requirements which emerged from the revision of its business strategy:

- A requirement to share more information about holidays with its customers, both travel agents and tele-sales customers.
- A requirement to allow the booking of holidays from home or from a travel agent's office to take place.
- A requirement to provide better communication between the company and its travel representatives.
- A requirement to enhance the role of its travel representatives.

The final two requirements came from the fact that the company, like many travel companies, experienced a high turnover of staff who acted as travel representatives in the resorts in which the company offered holidays. It was clear that the longer a representative stayed with the company and gained more information and experi-ence of a particular holiday destination, the more enthusiastic customers were about the company; indeed, a number of customers cited the excellent service given by experienced representatives as the main criteria in not only booking a holiday with the travel company but also the reason for picking a particular resort. It was in the company's interest to provide an enhanced role for representatives as a strategy of paying more than competitors had not seemed to have worked.

Notice that all the requirements detailed above are expressed in broad terms without any explicit reference to computer technology.

The travel company was sufficiently impressed with the report that they then decided to translate these business requirements into outline system requirements. They took the radical step of making the decision to use Internet technology.

One technical vision that the company had was of travel agents and customers using an Internet browser to access a very large document which served the same purpose as the paper brochure that the company currently used. Also the vision encompassed both travel agents and customers employing the same browser to book holidays.

A second technical vision was that of increased communication between the company, travel agents, customers and travel representatives. For example, they envisaged embedding the e-mail addresses of staff such as the marketing director and individual resort representatives within Internet-based documents.

From the business and technical vision came the general requirements for their system. The system consisted of a number of sub-systems:

■ A communication and group-working system which provided better communication between staff working for the company. This used a form of software known as group-working software. This communication system was not only implemented at the head office of the company but also in individual resorts. Where the company employed resort representatives directly such staff were given portable computers and a modem. Where the company used an agency for looking after their customers it insisted that the agency was equipped with a computer. The travel company provided the software for communication to such companies. This system enabled resort representatives to query individual databases which contained the details of customers who were to arrive at the resort.

■ An Internet-based booking system. This consisted of a layer of software which was added to the basic booking system that the company already had implemented, an example of a strategic system requiring a solid data processing system underneath it. In the first phase of implementation this allowed travel agents to book holidays without recourse to tele-sales staff. This system was a computer-based reflection of the brochure that the company issued annually, containing photographs, travel information and costs.

■ An Internet-based brochure. The brochure was placed on a computer accessible to staff and customers alike. The system also contained a simple calculator which allowed the customer to determine the cost of a particular holiday and also a query facility which enabled the customer to discover when there were holiday vacancies. The travel company envisaged that eventually this system would merge into the system provided for travel agents and enable individuals to book holidays over the Internet.

2.7 IMPLICATIONS

The closer integration of business planning and IT planning has a number of ramifications for IT developers:

■ It has become more difficult to provide a cost justification for information systems. In the past it was moderately easy to cost the savings made from a data processing system. For strategic systems it is almost impossible as it is often based on uncertain information such as the response of customers to a new product or service.

■ Senior management involved in IT development have to have a greater awareness of business strategy; for example, most IT directors have a good knowledge of the day-to-day workings of a company based on their development of basic data processing and information systems, but are often excluded from discussions about business strategy.

■ That IT budgets are set over a relatively long term rather than annually and that these budgets are set according to strategic business requirements. This is in contrast to companies where an annual amount of funds is budgeted and where its spend is determined by individual departmental needs on a month-by-month basis.

■ An increasingly high proportion of funds devoted to the education and training of all management on the capabilities of modern IT systems.

■ A separation of budgeting for IT into two budget sub-heads: the maintenance of current systems and the development of future systems.

2.8 SUMMARY

This section has deviated from the technical orientation of the previous chapter. It describes a growing trend within large companies: that of business strategy and IT strategy becoming more closely integrated. That, and the development processes described in early sections, while vitally important to a company, are incomplete without some major business-oriented analysis taking place prior to the establishment of technical requirements.

References and further reading

2.1   Bloomfield, B. *Information Technology and Organisations* (Oxford University Press, 1997).

2.2   Cash, J.I. *Corporate Information Systems Management* (Richard Irwin Inc., 1992).

2.3   Galliers, R. *Strategic Information Systems* (Butterworth Heinemann, 1994).

2.4   Kahaner, L. *Competitive Intelligence: From Black Ops to Boardrooms – How Businesses Gather, Analyze and Use Information to Succeed in the Global Marketplace* (Charles Scribners, 1997).

2.5   Luftman, J. *Competing in the Information Age* (Oxford University Press, 1996).

2.6   Ritchie, R. *Information Systems in Business* (International Thomson, 1998).

2.7   Roberts, W. *Strategic Management and Information Systems* (Pitman, 1997).

2.8   Shields, M. *Management Information Systems* (Pitman, 1998).

2.9   Ward, J. and Griffiths, P. *Strategic Planning for Information Systems* (John Wiley, 1996).

2.10 Willcocks, L. *Strategic Sourcing of Information Systems* (John Wiley, 1998).

2.11 Wilson, D.A. *Managing Information* (Butterworth Heinemann, 1993).

3 Computer technology

AIMS

After reading this chapter you should:

- understand the basic architecture of the digital computer;
- be familiar with the main categories of computer;
- be familiar with the main categories of software;
- be able to understand the main ideas behind data transmission;
- be familiar with the main network topologies.

3.1 COMPUTER ARCHITECTURE

Computer architecture and processing

Digital computers represent numbers, and data in general, as electrical pulses and as magnetic spots. This means that number representation is exact, as are also the instructions that are executed in processing the data. 'Processing' is a wide term meaning any series of calculations, aggregations, tests and rearrangements applied to data in order to extract useful information from it.

The instructions (program) are stored in magnetic form in the computer's main store (memory), which consists of semiconductor chips. As shown in Figure 3.1, the instructions are transferred one at a time from the main store to the control unit. Here they are decoded and the control unit then sends control signals to all the other units. The peripherals have their own control units that 'interface' with the main control unit situated in the central processing unit (CPU). The CPU comprises the main control unit, the arithmetic unit and the main store.

Mainframes

Large computers are known as mainframes. These computers can handle a large number of programs at once and do so in two ways. First, many mainframes have several processors, usually up to 8, hence programs can be allocated to processors. Secondly, a single processor can be time-sliced, that is, several programs

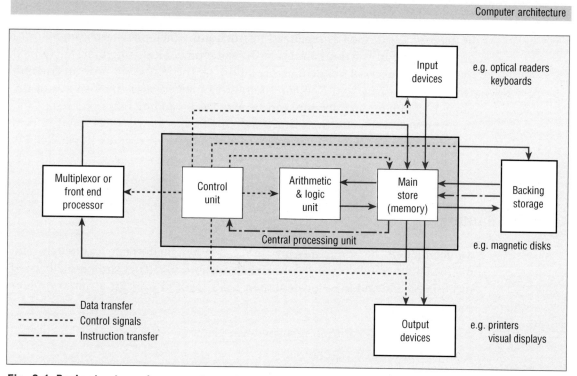

Fig. 3.1 Basic structure of a computer system

are run on the same processor but each program gets only a small fraction of its time. This is 'time sharing' in which several programs appear to be running at the same time. This is an illusion, however, because the processor is so fast that we do not notice the time slices delaying our program. In this way mainframes can run many programs on one processor and serve many users at once. As long as the load does not become too great all the programs run fast enough. However, when the load becomes excessive, response time suffers as the operating system tries in vain to satisfy all the requests for a limited processing resource.

Supercomputers

These are specialized, faster and more expensive mainframes used primarily in scientific and engineering applications (e.g. weather forecasting, oil well simulations), although these machines are now being used for more standard commercial applications when a large volume of computing resource is required. Supercomputers give a vast amount of raw processing power for calculations and logical operations. Power is often rated in Giga Flops (a billion floating point operations per second), which is similar to MIPS (millions of instructions per second) used for mainframe rating, but for arithmetic operations. The power is achieved as with mainframes by having several processors, but these use extra tricks to increase speed, such as pipelined instructions, which allow each processor to process several instructions almost at the same time. Most computers are limited by the bottleneck of sequential operation of the processor (one instruction after another). This 'von Neuman' bottleneck, named after the inventor of the modern computer architecture, can be avoided by adding extra processors and by tricks such as pipelining instructions. However, a more radical solution is to have hundreds or thousands of separate processors and compute in parallel, i.e. concurrently.

■ Minicomputers

Minicomputers lie some way in power between mainframe computers and microcomputers (discussed below and in Section 3.2). They are normally used in business and industry for dedicated tasks. Typical uses for a minicomputer include:

- monitoring and controlling a chemical plant;
- stock control in a warehouse;
- storing and providing traffic conditions for users of in-car navigation systems.

Minicomputers are now so powerful that many users, often using microcomputers, can connect up to them. Minicomputers were ubiquitous in the 1970s and 1980s. However, much of their processing has been taken over by powerful upper-end microcomputers.

■ Microcomputers

These are also known as personal computers or PCs. They can range in power from those which are used in the home to those which act as servers providing some service in a business such as dispensing files to users who require them.

■ Backing storage

The principal backing storage of a computer is nowadays mostly magnetic disks. CD ROMs and fast access tapes are also used, especially for archival data and for backup purposes. Backing storage is less expensive for holding large amounts of data than are main store media but is much slower in operation and can only work via the main store.

In the 1970s and up to the mid-1980s the main way to access a computer (usually a mainframe) was via a terminal. These were hardware devices which had little processing power; all they tended to do was send messages to another computer. During the 1980s these terminals gained more and more processing power and could carry out quite complex tasks such as validating data for correctness. The rise of distributed processing (discussed on page 77) has reduced the need for terminals. They have been replaced by microcomputers which communicate with other microcomputers and which carry out many of the functions required in a system without recourse to other computers.

Terminals and peripheral devices

Many computers have terminals linked to them, in which case the CPU is usually assisted by a front-end processor. These are virtually CPUs in their own right and control the terminals after the style in which the CPU controls the peripherals. More is written about data communications in Section 3.3.

Peripherals are diverse both in their nature and their capabilities. A peripheral differs from a terminal in that it is sited fairly close to the CPU, and generally fulfils only one function, i.e. input, output or backing storage. Chapters 4 and 5 provide more facts and figures regarding input/output and storage.

Terminals are a computer's interface with environmental activities, and fall into two main categories. The first is the batch terminal intended for remote job entry (RJE), and consequently needing high input/output speeds to cope with the large volume of data. A batch terminal could include any fast autonomous input and/or output device(s) such as an optical reader, a magnetic tape drive or a line printer.

The second category is the interactive terminal intended for manual input and low-volume output, such as the file enquiries, seat booking and on-line programming. An interactive terminal normally consists of a keyboard together with a visual display unit (VDU) and a serial printer. It is also possible to have a cluster of VDU terminals sharing one printer. Terminals can be categorized as general-purpose or dedicated. General-purpose terminals usually comprise a keyboard for input and a serial printer and VDU for output; this enables them to be employed for a wide variety of applications. Dedicated terminals often incorporate hardware devices specifically designed for a particular application.

The more flexible, i.e. intelligent, terminals have a stored program and their own built-in processor, so are effectively a small computer in their own right. They are programmed to reduce the mainframe's processing load, and they may also be employed to control the data of a group of non-intelligent terminals. As personal computers have spread, the role of intelligent terminals has diminished.

3.2 PERSONAL COMPUTERS, WORKSTATIONS AND MINICOMPUTERS

Computers on people's desks, in briefcases and even in pockets are becoming part of everyday life as miniaturization reduces size while increasing power. Definitions of mainframe, minicomputer and microcomputer have blurred as technology advances, so this section deals with a variety of computers which are smaller than mainframes and designed for use by one person or at most a small number of users.

■ Personal computers (PCs)

Personal computers, also known as microcomputers, started life because the reduced cost and greater miniaturization of electronics created microprocessors or 'chips' which enabled skilled amateurs to build their own personal computers from components. Familiar names such as Apple, Dell, and Compaq started this way. Personal computers, as the reader is no doubt aware, are desktop computers available in many walks of life, from primary school to advanced research laboratories. They have a wide range of sizes, capabilities and costs.

Decreased cost and the miniaturization of electronics has led to the insertion of microprocessors into a wide range of equipment. It has also allowed the development of a large range of microcomputers, and these have by now entered many facets of business, technology, education and domestic life. The available range of microcomputers and the number of manufacturers and suppliers are greater than for larger computers. This has brought about differences in the marketing methods, and microcomputers are often sold by mail order or over the counter.

However, the microprocessors used in the microcomputers are drawn from a much more limited range, which means that the operating system software is often written for the microprocessor rather than the computer. This need not, of course, be of great concern to a business user since other factors such as software availability, backing storage and output devices are usually of more interest. As things stand at present, a typical business microcomputer is defined essentially by the following attributes:

- It is a small desk-top computer, occupying little more space than a typewriter, and is movable.
- The cost is relatively low – currently (1999) ranging from £500 to £2000.
- It is a transaction processing device, i.e. one transaction is keyed-in at a time via a keyboard similar to that of a typewriter.
- The main components are a keyboard, a microprocessor, a monitor, a hard disk, a floppy disk drive, a CD Rom drive, and a printer.

- A good range of packaged software is available combined with simplicity of programming in the form of high-level languages.
- It can be linked quite simply into a network system.

The above characteristics make PCs amenable to most small businesses. This is particularly true as regards size and cost, i.e. they are manageable and economic, and so sales are now tens of millions worldwide.

PCs have spawned their own characteristics and terminology, which refers to their components, most of which are also found in larger computers.

Operating systems

An operating system looks after the compiling and interpreting of the high-level languages used with microcomputers and also the control of programs and files stored on the floppy disks. The most common operating systems for microcomputers are the various editions of Windows–NT, Windows (Windows 95, Windows 98), and UNIX. As already mentioned, certain microcomputers can work with several different operating systems, either optionally or as a matter of course. This flexibility gives them greater appeal because prospective users who have existent programs and files are likely to want to transfer them onto new hardware without amendment. Most operating systems now allow the user to do several things at once (multi-tasking), for instance run a print job in the background while continuing to work on a spreadsheet.

An operating system is a program which administers the hardware of a computer and makes its resources available to the user. An operating system has a number of functions:

- *It controls basic input and output to and from the computer.* When you type in words at a keyboard it is the operating system which reads these words and makes them available to the program that you are running.

- *It detects hardware failures.* Many operating systems are able to detect when a piece of hardware malfunctions, for example when memory is corrupt, and can report it to the user.

- *It shares out resources to the users.* For example, a minicomputer may have a number of users interacting with it; the operating system will allocate memory for each of the programs being run by the users.

- *It maintains and manages storage space.* For example, it will administer the file space used and required by users, and allow each user to create files which are symbolically named.

- *It keeps a system secure.* Many large operating systems contain features such as passwords which prevent unauthorized access to the resources of the system.

There are a number of operating systems which are available. Most of these are windows-based in that the user is presented with a desktop consisting of a number of windows in which programs and files are stored. The most popular micro-computer operating system is Windows 98 (and its previous version Windows 95). The most popular multi-user operating systems are Windows NT and UNIX. The latter exists in a number of different forms marketed by a variety of hardware and software companies. One recent advance has been the rise of LINUX. This is a version of UNIX which can be used for microcomputers and which is gaining a large number of devotees. One big advantage of LINUX is that it is in the public domain: the source code is available for reading and the software itself costs nothing.

Software

A large amount of microcomputer business software is now on the market. For the most part it splits into five categories. These are:

- Spreadsheets, e.g. Excel.
- Database management, e.g. Access and SQL Server.
- Word processing, e.g. WordPerfect and Microsoft-Word.
- Desktop publishing, e.g. QuarkXPress and Pagemaker.
- Integrated software, e.g. Corel office and Office 97.

The above is sometimes called horizontal software as it runs across a wide range of business users and applications. Vertical software is aimed at specific types of companies or users and again there is a massive amount available. The various vertical software covers, for instance, antique dealers, hotels, estate agents, solicitors and farmers. Vertical software includes stock control, sales and purchase accounting, payroll, production planning and job costing software.

Large microcomputers

The more powerful microcomputers are capable of handling a number of users at the same time and also several concurrent tasks. Multi-user microcomputers may have several monitors at which the users may be doing either the same or different work from each other. This facility ties in with networking (Section 3.3) such as the use of Ethernet as a local area network system (LAN). Multi-user PCs tend to run either the UNIX operating system or Windows NT. Most micro-computers, and especially the larger ones, have plenty of spare ports and slots, thus allowing for considerable expansion in terms of processing capabilities and attached peripherals. This enables a user to start with a fairly small microcomputer and to expand it as necessary into a multi-tasking, multi-user, network-linked machine.

Portable PCs or 'lap-tops'

Portable PCs, also called 'lap-tops' or notebook PCs, are small, lightweight and battery-powered. This allows them to be carried around and used in cars, trains, airliners, etc. The display is usually a panel providing twenty to thirty lines of eighty characters. If a modem and communications interface are included, a portable microcomputer becomes a means of data capture and/or message acceptance in conjunction with a distant minicomputer or mainframe.

Portable PCs are connectable to various types of printers and have hard disks, CD Rom drives and 3.5 inch diskette drives in a similar way to ordinary PCs.

Workstations

These are also personal computers and were originally developed for the scientific and engineering markets. As in PCs the heart of a workstation is a high-performance microprocessor, quite often a RISC chip. The size, appearance and components of workstations are similar to those of PCs. The boundary between workstations and PCs is hard to define, so the difference is one of degree and the type of application which workstations address. The following features are usually enhanced in workstations compared with more general-purpose PCs:

- faster microprocessor, especially RISC chips;
- more use of high-powered co-processor chips for maths and graphics screen handling;
- large monitors, typically 20 in. with high-resolution 1024×1024 pixels;
- good graphics facilities, wide-ranging colour palette;
- operating system typically UNIX or Windows NT;
- large memory capacity, a minimum of 256 Mbytes; and
- network facilities built in.

The workstation market grew with the development of graphics, CAD/CAM, desktop publishing and scientific computing. As networking assumed more importance, these machines became specialized into network server machines with large file stores and client workstations with less filestore but plenty of processing power dedicated to one person's use. Users now have the processing power equivalent to a mainframe sitting on their desks!

Minicomputers

During the early 1970s the cost of integrated circuits fell dramatically and their speeds of operation increased, more compact and less expensive machines were developed to meet the IS needs of most middle-size companies. These 'minicomputers' have the same basic structure as the larger (mainframe) computers and do not introduce any fundamentally new concepts in terms of their logic, storage media and mode of control. As with all computers, the technology changes with time and in many cases the distinction between a minicomputer and a mainframe on one hand and minicomputer and workstation on the other is somewhat blurred. However, typically minicomputers have the following characteristics.

Size and environment

Minicomputers are more compact than mainframes: the CPU occupies less space as it is not so powerful and has a smaller amount of memory. The ruggedness and lower heat emission of a minicomputer enables it to be installed in a less protected environment. There is no need for air-conditioning, and it can be installed close to its end-user, such as in a factory or a vehicle. This favours decentralization of the IS functions in an organization by facilitating distributed DP.

Cost

The lower development costs combined with high sales and mass production result in minicomputers being less expensive than the larger machines. This is less true of the peripherals, however, unless they are of lower speed or capacity. The prices of basic minicomputers are around a tenth of the cost of mainframes.

Modular construction

The compactness of a minicomputer's circuitry results in easy expansion of its CPU. A minicomputer is upgraded by merely inserting different microprocessors and circuit boards instead of replacing the complete CPU as with a mainframe computer.

Networks

The late 1970s saw a swing away from centralized computers towards the notion of distributed DP. That is to say, processing power being more accessible to users owing to the hardware being sited in their actual workplaces. The low price of computers, and particularly PCs and workstations, together with their increasing power and sophistication has resulted in these devices not only being more widely used but also being linked together to form local area networks (LANs). The main advantage of networking is that a user is no longer working in isolation but is plugged into a powerful and perhaps extensive grid of computing facilities.

Thus data and programs can be shared between users, and intercommunication becomes possible through the transmission of messages over the network (Section 3.3).

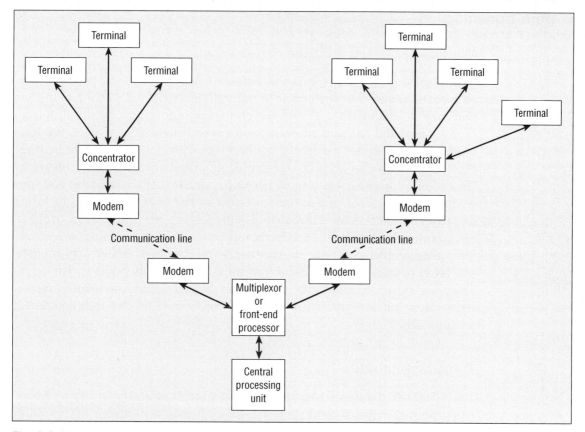

Fig. 3.2 Basic structure of a data communications system

3.3 DATA COMMUNICATIONS

Communications is an extensive subject in its own right, encompassing not only data transmission but also sound and video transmission via telephone lines, radio links and satellite links. With the increasing use of digital transmission for sound and video, computers have moved into the communications and entertainment areas. In a book of this size and content it is not practical to include more than superficial explanations of the various aspects of communications, nevertheless this section provides the reader with a general understanding of data communications as far as it impacts on business IS. Figure 3.2 shows the main components of a data communications system. In the context of business, data communications means the transmission and control of business data as it is moved from one point to another. There follow descriptions of the main concepts, equipment and techniques employed for these purposes.

▪ Data transmission

Modems

As explained previously, computers represent data by electrical pulses. Unfortunately telephone lines are unable to transmit pulses faithfully owing to the fact that they were designed solely for speech transmission; this results in distortion of the pulses and consequent errors. Since speech transmission takes the form of variations in electric current at audio frequencies along the telephone lines, the computer's pulses must be transmitted in a similar way. This is achieved by the use of a 'modem', i.e. a modulator/demodulator. At the sending end, this device converts pulses into variations either in the level (amplitude) or in the frequency of the electric current transmitted along the wire. At the received end another modem does exactly the opposite. Thus the sending and receiving equipment perceive only pulses, whereas the telephone lines carry, in effect, two levels or frequencies of sound (one for each of the bits 0 and 1). These processes are known as amplitude modulation and frequency modulation respectively. Some modems are small enough to be housed inside other equipment (card modems).

Transmission rate

The rate at which bits are transmitted along a line is measured in bauds. A baud is in effect 1 bit per second, thus a 2,400-baud line transmits a maximum of 2,400 bits in a second. As examples of transmission rates, British Telecom's Datel systems offer lines, either private or public switched telephone network (PSNT), with rates up to 50,000 baud. At the higher rates transmission is in synchronous mode and generally over private lines.

A means of attaining high speed is parallel transmission. This is where each of the data bits is sent along its own circuit, thus necessitating as many wires as there are bits in parallel. This method is used mainly for short connections such as between a computer and its high-speed peripherals, e.g. disk drives.

Multiplexing

Multiplexing is a means of combining together data from several sources so that it can be transmitted along one communication line.

Thus if several terminals situated in fairly close proximity need to communicate with a distant computer, they are connected to a multiplexor which in turn is connected to a modem. At the computer end this arrangement is in reverse, which means that the combined data from the terminals is separated by another multiplexor into individual lots as if they had come along their own lines. In other words the multiplexing is 'transparent' to the user, as are the modems.

There are two main methods of multiplexing – time division and frequency division. Time division multiplexing means that either the bits or the characters being transmitted to or from several terminals are interleaved during transmission. For instance, a bit is taken from each terminal in turn and transmitted in synchronous mode by the multiplexor. At the receiving end the other multiplexor separates the bits and passes them to the computer as complete pieces of data.

Frequency division multiplexing entails several carrier waves, each of which carries one lot of data being transmitted. Because each carrier wave has a different frequency, it is possible to separate them at the receiving end and so distinguish between the various lots of data.

A statistical multiplexor (statmux) is an intelligent device that is capable of judging each terminal's instantaneous requirements for line capacity and allocating line time accordingly. In this respect it has close similarities to a concentrator. It also corrects errors caused by line faults and noise, and provides data formatting and network management.

Multiplexors are often integral with modems and concentrators.

Protocols

A protocol is an 'agreement' whereby devices can communicate in a fully understood manner. Owing to the diverse nature of hardware devices, they usually cannot communicate with each other except by means of a protocol unless they happen to come from one manufacturer. The factors entering into data transmission and covered by the protocol include the mode of transmission, i.e. synchronous or asynchronous, speed of transmission, full or half-duplex, format of the data, and error detection and correction procedures.

A commonly used protocol is binary synchronous communications (BSC) also called 'Bisync'. With this protocol, transmission is synchronous, half-duplex and block-formatted using ASCII or EBCDIC coding. Bisync was originally intended for use in a polled environment in which the computer confirms that each block of data from a terminal has been received correctly.

Another protocol is high-level data link control (HDLC) of which there are several variants such as synchronous data line control (SDLC) and X-25. Important features of HDLC protocols are the address field and sequence numbers incorporated into the data. The address allows all devices to be treated equally in that the data travels along with other data until its address is reached. The sequence number enables packets to be put into the correct order if they arrive out of sequence as a result of being transmitted over different routes (see 'packet switching'). HDLC is therefore suitable for statmux techniques in which many different levels of data share a single line. One of the most important protocols is TCP/IP. This is the protocol that the Internet uses and is described in the next section of this chapter.

There are many other protocols devised by manufacturers and communication organizations throughout the world. In an attempt to attain some degree of compatibility the International Standards Organisation (ISO) has put forward standards for a framework for all protocols. This is known as the open system interconnection (OSI) model and contains seven layers each with a specific function such as the physical interface, message transport, error control, control messages, message sections, presentation control and application transparency. OSI can be regarded as a reference ground for protocol and hardware designers so that they are able to introduce new devices and methods with a reasonable level of conformity to existing and other new equipment.

In situations where two devices wish to communicate but employ different protocols it is possible to overcome this problem by using a protocol conversion computer or function.

Document transmission protocols

The ISO–OSI model laid the groundwork for electronic data communications. Electronic mail and X-25 services on ISDN (international switched data network) have become widespread because of international agreement on data transmission protocols. X-25, the packet switching network, provides two services, datastream, in which packets arrive and are assembled in their transmission order, although no time guarantee of arrival is given, and virtual call, in which the stream of packets between two computers is essentially continuous hence they appear to be connected in a continuous two-way conversation. In spite of its advances, X-25 was still primarily concerned with delivering the message rather than bothering about its content. This limitation means that word processing documents and graphics could not be exchanged between computers.

More recent protocols, DCA–DIA (document content architecture and document interchange architecture) and X-400 have dealt with this problem. These fill in the upper layers of the OSI model – the presentation and application layers – by controlling how the content of a document should be encoded and decoded for transmission between computers. This allows word-processed documents to be exchanged between machines with all the usual formatting commands of margins, tabs, fonts, bold text, etc. Furthermore digitized pictures and graphics can be exchanged.

Data transmission media

Telephone lines

As mentioned earlier in this section, data can be transmitted along a line (also termed a link or channel) at various speeds. The simplest and oldest type of communications channel is a pair of copper wires, i.e. a telephone line. This method is still used extensively and is satisfactory provided the low rates of transmission are acceptable.

Coaxial cables

When higher rates of transmission are necessary, ordinary wires cannot transmit without distorting the electrical pulses representing the data. Coaxial cable, similar to that used for connecting the aerials of domestic televisions, has a much better performance at higher transmission rates. It is employed as the transmission medium both for local and for intercontinental data communications.

Fibre optics

The principle of fibre optics is that data and other information is transmitted in the form of light through very fine glass fibres. The light is actually red or infrared and is contained within the glass fibre owing to the principle of total internal reflection. In other words, the light is transmitted through the fibre after the style of water in a hosepipe, bends in the fibre making no difference. The glass fibres are contained within a strong waterproof cable.

The light is passed into and taken from the fibres by means of transducers, i.e. devices that convert electrical pulses to light and vice versa. Transducing is obviously necessary in order to interface electrically operated equipment such as peripherals and word processors to the optically operated fibres.

The advantages of fibre optics are:

- High bandwidth, i.e. a powerful data transmission capability.
- Low cross-talk, i.e. no interference between adjacent fibres.
- Low attenuation, i.e. little loss of signal strength.
- Freedom from interference by external electrical and electromagnetic equipment.
- High reliability.
- Safe because no heat, sparks or electrical voltages are created.
- Economic because glass is less expensive than the copper used in ordinary wires.

The high bandwidth allows data to be transmitted at several hundred megabits per second. The low attenuation permits transmission over distances of up to 5 kilometres without the need for repeaters, i.e. reamplification of the light intensity. The most usual mode of transmission is time division multiplexing and, as each of the large number of high bandwidth fibres in a cable can carry 32 channels, the transmission capability of just one cable is enormous. Fibre optics communication can be used within computers as the bus for high-speed communication as well as for longer distance communication between computers. A standard fibre-optic distributed data interchange (FDDI) has been proposed to control how data is exchanged over optical media.

Microwave networks

A microwave link is an ultra high frequency (UHF) radio transmission between two line-of-sight points. These points house radio transmitters and receivers (transceivers) known as repeaters. They are sited at strategic positions so as to form a network through which transmissions are routed between any two places. As far as possible the points are on hills so as to give the maximum line-of-sight distance over the horizon. Because of the high frequency of microwaves, and hence their large bandwidth, they are capable of transmitting without distortion.

Communication satellites

A weakness of microwave links is their demand for line-of-sight positioning. Since the horizon intervenes even between high points at distances exceeding a few tens of miles, microwaves cannot be used for transoceanic communications.

To overcome this problem repeaters are installed in satellites launched into geo-synchronous orbits several thousand miles above the earth. A geosynchronous orbit means that the satellite maintains a fixed position relative to the earth. The satellite repeater receives radio transmissions from earth and, after amplification, retransmits them to be picked up by receivers within its coverage. Only three satellites are necessary to cover the whole of the earth's surface, and so it is possible to link any two points on earth by means of communication satellites.

In practical terms an intercontinental data communications link uses most of the above media. The first stage would probably be a telephone line or coaxial cable, either private or rented from a common carrier, e.g. British Telecom. This would be linked to a microwave network, which takes the data to a radio station from where it is transmitted to a satellite. The satellite retransmits to a distant receiving station, after which the link is the reverse of the early stages. At various stages in the telephone system it is likely that optical fibres are alternatives to wires or coaxial cables.

Radio transmission

This involves sending radio waves from one computer to another with each computer being connected to some electronic circuitry which can send or receive the waves. This form of transmission is used where there might be difficulty in using cables, for example in remote parts of the world. It suffers from the fact that they are susceptible to static, radio interference and abnormal weather conditions.

Infra-red transmission

This involves the use of waves just below the visible light spectrum. This mode of transmission is used in local area networks where cabling is difficult to install.

Distributed processing

A distributed DP system has several interconnected points at which processing power, i.e. intelligence, and storage capacity are available. These points may on occasions act autonomously and at other times co-operate in handling a common problem. The locations of the processing points need not necessarily be physically remote from one another or from a central mainframe computer.

The main purpose of distributed processing is to give the end-users of computing facilities the control over and responsibility for their own data. The end-users become masters of their own destiny to a much greater degree than with batch processing carried out entirely within the IS department. In other words, they have considerable computing power under their control rather than delegating it all to the centralized computer.

Distributed processing presupposes that the user department automatically accepts responsibility for the correctness and completeness of its source data. The department in question is likely to be the only body aware of the source data in use and of the immediate results required from the system.

A distributed processing system may be composed of several processing points connected together in a wide variety of configurations. The points themselves can be minicomputers, PCs, workstations or mainframes.

Broadly, distributed processing systems fall into two approaches – hierarchical and lateral.

Hierarchical (vertical) systems

These have several levels, the most powerful of which consists of one or more mainframes forming the central complex of the hierarchy. This complex is capable of handling local batch processing, remote job entry, time-sharing, and the needs of the lower levels in the hierarchy. It is likely that the central complex is large and expensive.

The second level comprises a powerful, workstation-server minicomputer(s) acting as a satellite to the mainframe(s). This must be capable of administering a network protocol so that data and messages can be passed through it between the lower and higher levels. The minicomputers must also be able to handle local batchwork, interactive terminals and, possibly, communicate with other minicomputers. The third level consists of PCs or intelligent terminals dedicated to particular tasks such as point-of-sale processing. They are capable of controlling a number of keyboards and VDUs, and of communicating with the second-level minicomputers.

Lateral (horizontal) systems

These are similar to hierarchical systems except for the omission of the mainframe(s). The workstation servers/minicomputers in a lateral system are autonomous but are capable of communicating with one another. This intercommunication must be flexible in order that various arrangements can be set up. In some situations the servers co-operate in order to create a more powerful processing system, in others the communication is merely the interchange of messages or data.

Interco-operation of minicomputers in a lateral system infers that they act as standby and backup computers for each other. These requirements necessitate sophisticated software and, consequently, overheads in terms of storage, cost and time must be taken into consideration.

It is also possible to have stand-alone distributed processing in that the minicomputers are substantially autonomous with only occasional connection to other computers. This may occur when it is necessary to transfer fairly large amounts of data to or from the minicomputers. An example of data being sent from several minicomputers to one mainframe or minicomputer is where stock levels are transmitted from branches to head office. An example of the reverse is details of new products sent from head office to all the branches. These procedures are sometimes termed 'bottom up' and 'top down' respectively.

Client-server systems

Irrespective of the nature of the distributed system, many systems have a 'client-server' architecture. The server is a larger workstation or small mainframe which hosts the central resource, often a database but it may also include advanced peripherals, number crunching processors or any expensive resource which cannot be duplicated. Clients are PCs or workstations that are connected to the server via a network. Clients run processes on their own but rely on the server for central resources. Many applications originally written for mainframes are now being revised for client-server architectures, and this is referred to as 'downsizing'. The benefits are more local control of computing resources. Client-server systems are now dominating the distributed systems market place.

■ Data switching

Data is often transmitted from one point to another along a fixed, permanent communication link, such as between a terminal and a nearby computer. This is satisfactory when the points are in regular communication and no other, occasional, users are involved. When data has to be transmitted between various users and in an unpredictable fashion, a data-switching arrangement is necessary.

Circuit switching

Data can be switched by circuitry (hard wiring) in a similar way to telephone conversations; this method means that a physical circuit is set up and held in existence for the duration of the transmission. Circuit switching is an uneconomic method because the communication path is usually under-utilized during its existence.

Packet switching

A more efficient method of data switching is message switching, and particularly a variation of this, i.e. packet switching. The main concept of packet switching is that a piece of data, i.e. a message, is fragmented into packets of data of fixed length, e.g. 128 bytes (octets), and transmitted in this form together with its control data. Control data is essentially the identity of the sender and address of the recipient.

Packet switching is achieved through the employment of a packet switching system (PSS), also known as a packet switching exchange (PSE). This system accepts messages and directs them to their destination by examining an 'address header' incorporated into each packet. The address header is derived from that inserted into the original message.

A PSS takes the form of a computer network in which each computer redirects the packets it receives to the next computer along the appropriate route to the packet's destination. Although the original message is fragmented into packets, which may arrive at different times, the PSS ensures that they all arrive at their destination accurately and that they are reassembled correctly. Thus the recipient finishes up with an exact copy of the original message.

The high level of efficiency of packet switching is achieved by interleaving packets, that is to say, packets travel between various points along the same paths concurrently.

The computers employed for packet switching are called 'store-and-forward' computers since they have storage buffers in which the packets awaiting onward transmission are held. They are able to request transmission of packets found to contain errors on receipt, and retransmit packets at the request of the receiving point. It is also possible for the PSS to replicate packets for transmission to several recipients. Another facility is 'redirection'; this comes into play if the receiving point or a link is out of action. The packets are then redirected to a previously nominated address and the sender informed.

There are two main methods by which data is transmitted through a PSS. The first is a permanent virtual circuit otherwise known as a fixed-path protocol, in which all the packets comprising the one call (message) go by the same route. The other method, called a switched virtual circuit or path-independent protocol, transmits the packets of one call via separate routes.

It is also possible to transmit human speech via a PSS because speech is in bursts that can be packeted.

Virtual circuits

This is a facility that allows packets to be transmitted without them all containing the control data. The PSS makes a note of the first packet's control data and of its routeing so that subsequent packets can be treated accordingly. A similar arrangement is a 'permanent datacall'; this is provided by British Telecom for its users who wish to send large amounts of data between fixed points.

■ Local area networks (LANs)

A local area network (LAN) is based on the principle of several users sharing the IS facilities available in one organization by providing them with two-way access via a communications network. For instance, the common use of word processors, printers, PCs, databases, storage media, and so on; these are known as 'nodes' in the LAN. A LAN is generally confined to within one building or site but can be extended through the use of repeaters and gateways. The former allows longer communication paths; the latter enables one network to be linked to another.

There are many suppliers of LANs utilizing the various methods described below but typical characteristics are a data transfer rate of 30–40 megabits per second, a network length of up to a few kilometres, and a maximum of 100 nodes or workstations. As well as these characteristics there are three other important aspects of LAN: their access method, transmission mode and network topology. Access method, sometimes called protocol, is the way in which nodes are permitted to enter data into the network; two of these are described below.

Network topologies

The topology of a network is its physical layout of computers and other units, all known as nodes of the network. Topologies fall into four main types as described below, but in practice many networks are hybrids.

Tree networks

A tree or hierarchical network, as in Figure 3.3, inevitably means that nodes at the same level do not communicate directly with each other but via a node at the next higher level. The intention of a tree network is that the large-scale processing is handled by the mainframe, lower-level processing by the mini-computers, and the simplest processing by the intelligent terminals or PCs. Minicomputers may be sited at points intermediate between the mainframe and the terminals so as to reduce transmission costs. Tree networks tend to be used for large-scale, long-distance networking such as by international airline seat reservation systems.

Star networks

A star network, as depicted in Figure 3.4, has a mainframe or minicomputer as the central node. The lower-level nodes, i.e. PCs and/or terminals, are unable to communicate directly with one another but only through the central node. A star network is most suited to situations with many lower-level nodes wanting access to a central database controlled by the mainframe or minicomputer, such as with on-line banking.

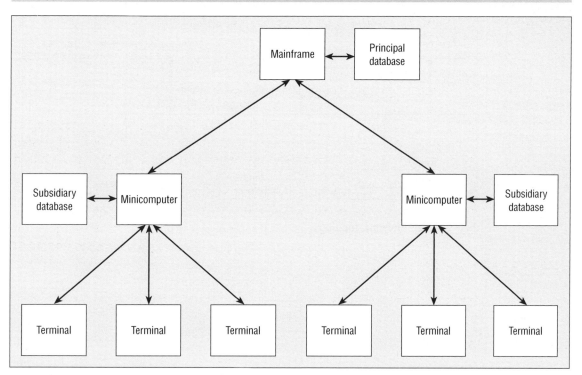

Fig. 3.3 Tree network topology

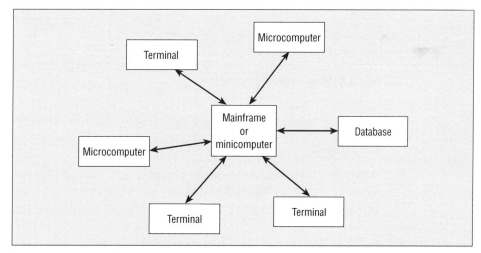

Fig. 3.4 Star network topology

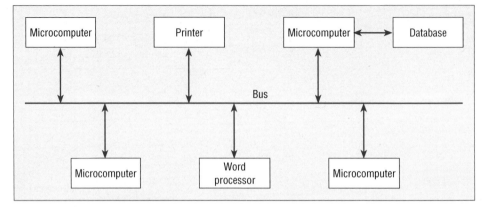

Fig. 3.5 Bus network topology

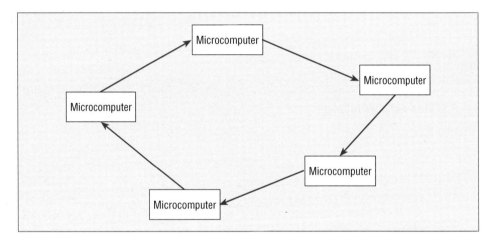

Fig. 3.6 Ring network topology

Bus networks

A bus is a main communication channel to which all the nodes are attached. Thus, as shown in Figure 3.5, they are all in direct contact with each other. This topology especially suits LANs, since nodes can be plugged in or out at will.

Ring network

As depicted in Figure 3.6, a ring network entails all nodes being linked together on an equal footing. Data is input to the ring by any node and transmitted round it. The appropriate node accepts data intended for it and other data is passed on through the ring to its destination.

Value added network services

Value added network services (VANS) are a means of simplifying the electronic exchange of data between users of the service. The concept behind VANS is that a user merely plugs into an interface provided by the VANS operating company and the network does all that is necessary thereafter. This service copes with different types of user companies by reconciling their protocols.

A particular VANS is aimed at a certain industry and is tailored to meet its individual needs, e.g. insurance, retailing or local authorities. A VANS may be utilized entirely within the one organization or between different organizations having the requirement for data exchange.

An example of the latter are estate agents and solicitors involved in property conveyancing having access to computer databases in local authorities via PCs in their own offices. A similar scheme is in operation for accountants and insurance brokers to have access to databases in insurance companies. It is intended that VANS will be extended to incorporate a worldwide network of companies and organizations.

Real-time and on-line systems

Real-time computing implies that the input messages and responses pass quickly enough between the environment and the computer to enable the latter to have some degree of control over the former. In the broad sense, real-time systems include plant and machinery control, and defence and space systems, but here we are concerned solely with real-time systems for business applications.

In a real-time business system input messages are transmitted from terminals, often in widespread locations. After the input messages have been processed by the computer, the resultant reply messages are transmitted back to the terminals sufficiently quickly for on-the-spot actions or decisions to be taken.

On-line systems are sometimes regarded as synonymous with real-time, but this is not truly the case; on-line means merely that the data is transmitted directly to or from the computer. The confusion arises because real-time systems are inevitably on-line; this must be so to attain the necessary short response time. To differentiate between scientific/engineering and business real-time systems the former are often called time critical systems, in the sense that if input messages are not processed in some short critical time period then the consequences can be dire. Hence these real-time systems are often safety critical as well; for instance, nuclear power station control and aircraft flight control systems. Two examples of business real-time systems are airline ticketing and automatic banking systems.

Airline reservation systems

All the world's major airlines now operate real-time seat reservation systems for their flights. These systems are sophisticated, expensive and heavily employed; moreover, airlines could not function without them.

The main intention behind an airline seat reservation system is to attract the extra seat bookings on each flight needed to make it profitable. There is a narrow margin between profit and loss in terms of seats occupied. Since prospective passengers turn to another airline if they do not get an immediate reservation, a rapid response is imperative.

The seat reservation system of a large airline typically has several thousand terminals situated in the airline's worldwide offices. These are linked to a central database holding up-to-the-minute records of the bookings for the next few months.

By this means and by using sophisticated dialogues (Section 8.5) passengers' enquiries and bookings are handled immediately. In other words, a passenger can reserve a seat on any flight in the foreseeable future from any point in the world.

Real-time banking/building societies

The concept behind real-time banking is that every clerk (teller) has immediate access to all the customers' accounts. This means that a customer is able to withdraw cash and enquire about his or her account at any of the branches. Also the bank or building society is able to keep a close check on the customer's current balance; thus the system has advantages for both parties.

In most real-time banking systems the terminal comprises a keyboard, a passbook printer and a display screen. The passbook printer is used to update the customer's passbook after every transaction. The display screen is the means by which the computer communicates with the clerk, and the keyboard enables him or her to transmit transactions and enquiries to the computer. Also familiar are automatic teller machines (ATMs) or 'hole in the wall' cash machines which accept the customer's cash card, validate the PIN (personal identification number) and then allow on-line access to account balances, cash withdrawals, etc.

System security is achieved in a real-time banking system by having more than one computer, by multiple data transmission paths, and by having storage equipment as part of the terminal. This latter facility allows transactions to continue on a local basis during periods of system failure. If this occurs the transactions are copied to the storage medium for subsequent transmission when the system again becomes operational.

Response time

An important aspect of all real-time systems is response time. This is the interval of time between the end of the input message and the receipt of the beginning of the reply message. It includes several contributory times such as the transmission times, message queuing and processing, and file accesses. These times have to be kept as short as is economic if a low response time is to be achieved. The response times acceptable for a real-time system call for careful study because of the many factors implicated – technical, economic and psychological. In broad terms, response times for good interactive working are in the order of 0.2 seconds.

When some delay may be expected (e.g. at the end of data entry sequence), a response time of 2–4 seconds is acceptable.

An important factor entering into response time is the 'traffic pattern' of the real-time system. The greater the number of messages handled by the computer in a period of time, the longer the response time tends to be. That is to say, a real-time system is designed to accept an average number of messages per minute, and if the actual volume of messages is above this, the response time increases.

3.4 THE INTERNET AND INTRANETS

The Internet is a collection of computer networks which themselves are connected via a variety of mechanisms ranging from advanced, broad-band satellite links to low-technology dial-up connections. As early as 1980 the US government had realized that, even in those early days, the hardware technology was available to connect computers together via communications links. In order to standardize the operation of any future communication networks it charged a body known as the Defence Advanced Research Projects Agency (DARPA) to develop a set of standards for the Internet.

In 1981 DARPA issued a document known as a Request for Comments which asked current and prospective users of networks for their views on the future standardization of a countrywide network. Central to this document was the use of two communications protocols known as TCP/IP. These were the dominant protocols being used on the relatively primitive networks which had been set up in the late 1970s. The adoption of TCP/IP for internetworking received two boosts in the 1980s. The first was the US Department of Defense making them mandatory for large projects and the second was the incorporation of TCP/IP within the hugely popular BSD 4.2 UNIX operating system.

■ The structure of the Internet

The Internet is not a single network but is a network of networks. These so-called sub-networks communicate via gateways. Figure 3.7 shows a small fragment of the Internet with six sub-networks connected via a number of gateways.

With TCP/IP all connections to the computers that make up a sub-network are through gateways. Gateways route all information in the Internet based on the sub-network name and not on the address of the computer to which the information is to be delivered.

The Internet can be regarded as a four-layer architecture. The sub-network layer consists of a number of computers connected together using a local area network. Above this layer is the Internetworking layer which handles the communication between computers through gateways. It is this layer that passes data from one gateway to another until the destination sub-network is found and the data delivered to that sub-network.

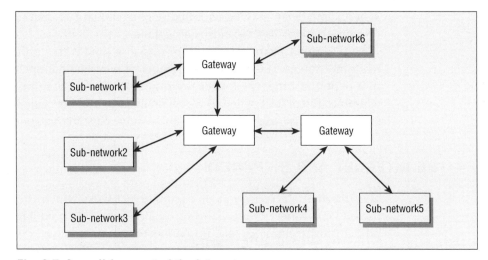

Fig. 3.7 A small fragment of the Internet

The Service Provider Protocol layer is responsible for the overall communication functions of the network and the Application Services layer is concerned with the communication between applications such as a Web browser or e-mail program with the Internet.

Each computer on the Internet has a unique address known as its IP address. This is a 32-bit address which is used to identify a computer and the sub-network in which it resides. If a network is isolated from the Internet and requires no connection with other computers on the Internet then any IP address can be assigned to a computer. An IP address is represented by four digits separated by dots. This form of address is known as dotted quad notation.

The IP address is used to route a message; for example, if a sub-network identified by 12.33 was asked to route a message to an IP address 12.33.45.78 then it would know that the computer for which that message was destined was situated on its own network and there was no need to send the message to a gateway.

Dotted quad notation is not the most convenient way to identify computers. From the early days of the Internet there was a perception that there was a requirement for sub-networks and computers to be identified using easy-to-remember names.

The current way that symbolic names are associated with Internet address is via the Domain Name System (DNS). It is organized hierarchically with the top level of the hierarchy being used to differentiate the type of organization administering the network; for example, *com* is used to designate a commercial organization, *edu* for an educational organization such as a school or a university and *gov* a governmental body. The next level is used to uniquely name an individual sub-network; this is known as a domain name. The administrator of a sub-network can partition his or her network into further small networks which are identified by sub-domain names. Thus, the name widgets.penelope.com designates the name of part of a sub-network belonging to a commercial company with the domain name of widget and a sub-domain name of penelope.

Each sub-network contains a *name resolver* which examines information in a domain name associated with some data to be sent through the Internet. If it finds the IP address then it sends the information associated with the address to a gateway. If it does not find the name then it asks a name server. This is a computer which contains up-to-date names of sub-networks. If this name server is not able to find the address then another name server is asked until the domain name and the address associated with this name are discovered.

Browsers and the World Wide Web

The most visible part of the Internet is the World Wide Web. It consists of four components: Web servers, Web documents, Web browsers and the HTTP protocol used to communicate between browsers and servers.

A Web document is a plain text document written in a language known as HTML (HyperText Markup Language) which is like a rudimentary textual markup language. The writer of an HTML document uses the various constructs in HTML to lay out entities such as headings, paragraphs and bullet points. These documents are stored on a computer known as a Web server and are accessed by a browser. When the user of a browser wishes to examine a Web document he or she specifies the address of the document on the Internet and the browser communicates with the Web server on which the document is stored. The server sends the page of the document requested back to the browser which then interprets the HTML on the page and displays the document; so, for example, if the browser encountered an HTML construct which indicated a large heading, that heading would be displayed in a large font and in bold.

An HTML document will usually contain visible links to other Web documents, either on the server from which it originated or on servers anywhere in the world; if the user clicks on one of the links then the page corresponding to the link is transferred to the browser and then displayed.

The protocol which mediates the transfer of Web pages to and from Web servers and Web browsers is known as HTTP (HyperText Transfer Protocol).

An overview of TCP/IP components

Because the acronym TCP/IP appears a large number of times in the Internet literature there is a perception that these are the only protocols and facilities offered by the Internet. Nothing could be further from the truth. The aim of this section is to describe the full collection of facilities:

87

- *Telnet.* This is a program which gives Internet users the facility to log on to other computers anywhere in the world. This assumes, of course, that the user has permission to do so.

- *HyperText Transfer Protocol.* This protocol, known widely by its acronym HTTP, is one of the most important components of the Internet. It provides the medium whereby text which makes up a World Wide Web document is sent to browsers.

- *File Transfer Protocol.* This protocol, often known by its acronym FTP, enables a user to copy a file on a computer anywhere in the world to his or her computer, provided that the user has permission to do so. This is probably the most venerable part of the Internet: the early documents which describe the net's growth mainly focused on it as a medium for transferring data and software.

- *Simple Mail Transfer Protocol.* The Simple Mail Transfer Protocol enables mail to be sent from a user to another user anywhere in the world. The protocol carries this out in the same way that FTP transfers files.

- *User Datagram Protocol.* This protocol, known by its acronym UDP, offers the user a fast transfer time for data transferred around the Internet. However, it suffers from one major problem: it is not very reliable; for example, data can be lost when sent via UDP. The major use for this protocol occurs in applications where reliability checking is built into the software that supports the application.

- *Domain Name Server.* This is described elsewhere in this book. It is a computer running a piece of software which allows numeric Internet addresses to be converted into symbolic names.

- *Transmission Control Protocol.* This protocol, known by its acronym TCP, is responsible for marshalling data produced by application software into packets which can then be sent using the Internet Protocol.

- *Internet Protocol.* This protocol, known by its acronym IP, is responsible for the bulk movement of collections or packets of data around a network. These packets have been assembled using either UDP or TCP.

- *Trivial File Transfer Protocol.* This is a very simple file transfer protocol which is built on top of UDP.

■ Ports and sockets

The final technical concepts in this section are those of port and socket. Applications that use TCP/IP are associated with a port number which identifies the application. Port numbers above 255 are for local use, port numbers less than or equal to 255 are reserved for fixed applications. For example, port number 21 is used for FTP, port number 69 is used for the Trivial File Transfer Protocol and port 25 is used for the Simple Mail Transfer Protocol.

Each communication loop associated with TCP is associated with a unique number pair making up the IP address of the computer and the socket number. This number pair is known as a socket; it provides the communication link between applications on different computers using TCP/IP.

The three ages of computer systems and the Internet

Before jumping into the technical content of this book it is worth revising the background description of the types of system that this book will describe and which were detailed in Chapter 1. However, it is worth revisiting these in the context of the Internet. Writers on management information systems have identified four types of information systems which have been developed since the mid-1960s.

The first type of systems were known as data processing systems. Such a system carries out some basic function which somehow improves the internal efficiency of a company. The classic example of a data processing system is one which processes the salary and wage records of staff and administers the payments associated with these records on either a weekly or monthly basis. In the 1960s this was one of the first areas that a company computerized, often dedicating a single computer to the task. The processing of a payroll is a routine operational matter; it has no *direct* effect on a business. There may be some indirect effect: savings made from computerizing the payroll might have been used for investment purposes; however, the main purpose of such a system is that of making some internal business process more efficient.

Next to be developed were systems commonly known as management information systems. The 1970s saw the rise of such systems which attempted to improve the effectiveness of the management of a company. A typical system which satisfies this description is the stock control system. A stock control system is used by a company which sells some physical product like computers, building materials or pharmaceuticals. It is used to keep track of the stock of items stored in some physical location such as a warehouse. The early stock control systems were data processing systems: they just kept a tally of the items in stock and alerted the company when a specific item was below some danger level such that there would be a good chance in the near future for an order for that item not to be satisfied. A management information system based on a stock control application would, of course, provide such functions but would also provide information about what items were popular and fast moving, what items such as sunglasses were subject to peaks and troughs in sales at particular times of the year and what items were subject to regional variations.

Such management information systems were often built upon successful data processing systems, the stock control system described in the previous paragraph being a good example. However, there was increasing evidence that in the 1980s – the golden age of information systems development – management information systems were being developed from scratch.

The third wave of application systems appeared in the early 1990s and were as different in intent from management information systems as management information systems were from data processing systems. Such systems are known as strategic systems. They are differentiated from information systems by virtue of the fact that they improve the competitiveness of a business by changing the way that they do business. A typical example of this type of system is one which is used to administer quotes for auto insurance.

If you wanted car insurance in the United Kingdom in the 1980s then you would normally phone or go to a local insurance broker. Such a broker would be an agent of a number of insurance companies and, given information about your circumstances, would attempt to find and sell an insurance package to you. Often this process would take some time: for example, the broker may need to ring a number of insurance companies if the companies were unwilling to trust the broker with a quote or, if they were willing, then calculate a quote. In the late 1990s a new breed of car insurance company emerged. Such companies still acted as an agent for other insurance companies; however, the important difference was that they computerized the process of offering an insurance quote, a five-minute phone call was usually all that was necessary to secure a competitive quote. Within two or three years such companies had captured a major slice of the car insurance market and had transformed the car insurance business; they had transformed it so much that the same model was used for other forms of insurance selling such as house contents insurance.

There is a fourth type of system known as a transformational system which transforms the functions of a business; however, the vast majority of systems fall into the first three categories and we shall concentrate on them here.

Ward and Griffiths (Ward and Griffiths 1996) have identified four types of strategic information system:

■ Those which share information using computer-based systems with either customers, consumers, other companies or suppliers and radically change the nature of the relationship between a company and some or all of these groups of people.

■ Those that effectively integrate disparate collections of information and use that information to either reduce costs or provide added value to a product.

■ Those that enable a company to develop new products or services based on information.

■ Those that provide strategic information to high-level management of a company.

Some examples of systems which fall into the first three categories and which use Internet technology are:

■ A system which shares booking and route information between two airlines. Since the early 1990s there has been a number of strategic alliances between airline companies which do not directly compete on many routes. Such an alliance enables customers of one company to be directly booked to destinations served by the other company as if the customer was dealing with only one company. Such companies will share route, ticketing and even marketing information as part of this alliance. A system which comprises the two computer systems used by the allied companies plus bridging software between these systems can be regarded as strategic; such a system would use TCP/IP and browsers to enable staff to carry out their jobs. This type of system falls into the first category.

- A system which provides financial advice for customers of a financial firm based on a browser-based interface. A good example of this is the Web pages of the American finance company Smith Barney which are concerned with educational financial planning. A major growth area in American financial services is that of investment products to provide funds to enable a customer's children to complete college. Smith Barney have Web pages which allow customers to input data such as the date of a child's first year in college and the name of the college that the child might attend and then receive a report which will detail information such as the annual tuition fees and the annual increase in these expenses over the last five years; it will then provide a financial plan for the customer to enable them to pay these expenses. This example falls into the third category of strategic system: those that involve the development of new services or products.

- A system for planning the production of an item such as a car which requires data on the stock levels of the components which make up that item. Such production planning systems were commonplace in the 1980s and are a good example of an information system. However, more advanced production planning systems which are strategic have emerged in the 1990s. Such systems allow the suppliers of components to interrogate the manufacturer's warehouse stock database in order to enable the supplier to make a decision about supplying components whose stock are low. The difference between this type of system and the sort of production planning and control system which was developed in the 1980s is that within specified stock limits the supplier makes the reordering decision rather than the manufacturer's staff and, consequently, has greater access to manufacturing data. This is an example of the first type of strategic system which changes the relationship between a company and its suppliers. Such systems are now highly distributed and involve a large number of computers with the communication between the computers being carried out by TCP/IP.

- A system which integrates data held by a company which sells agricultural products to farmers. Such a company might have two separate systems: one which carries out the ordering and another which administers stock with the ordering of items being processed by phone. Integrating both these systems and allowing sales staff, equipped with hand-held computers, to visit farmers to discuss their needs, predict delivery time and alert them to special offers changes one aspect of the business processes of the company. This is an example of the second type of strategic system: one which provides added value by integrating existing data. This type of system would not be possible without the technology available within the Internet.

■ A system for administering and obtaining information from loyalty cards. A loyalty card is a plastic card which is given to the customers of some retail business. When the customer buys something from a shop owned by the retailer the loyalty card is swiped through a special machine which logs the customer's identity, the items that have been bought and the total cost of the bill. Such a system has a number of functions: for example, it can be used to determine whether special offers should be made to the customers of a specific store. However, one use which places a loyalty card system into a strategic system of the fourth type, one that is used for high-level decision making, is that of determining the building of new supermarkets. Each customer using a loyalty card will have applied for it and given their address and hence a computer system which administers the cards is able to determine the distances that the customers have driven to do their shopping, consequently providing data which enables a new supermarket to be optimally sited. Again, this type of system is highly distributed, with a large number of computers communicating with each other over communication lines; without the Internet technology described above this type of system would not be possible.

These, then, are some examples of the type of systems built in the 1990s. Along with these types of systems there is also a general increase in e-commerce systems which enable customers to shop for items via a Web browser. One of the key features of these new systems is that of *sharing data*: a production planning system might share reordering data with suppliers, a system used by a finance company might share financial planning data with potential customers, two airlines which have formed a strategic alliance share their route and booking databases and a retailer might share stock availability with customers who are ordering items over the Web. The first decade of the third millennium will see a major increase in systems which distribute data across a number of enterprises; such systems are known as distributed systems.

Questions about which were the primary drivers to distributed systems, the trend towards strategic systems or the massive increase of computer power in the 1980s or 1990s, while interesting, are out of the scope of this book. What is clear, however, is that the first decade of the twenty-first century will see a massive number of these systems developed and that technologies normally associated with operating systems will become commonplace in application development.

What is also clear is that a very large number of applications will be developed using TCP/IP as a medium. This is due to a number of reasons:

- The ubiquity of TCP/IP. Virtually every operating system known includes these protocols.

- The cross platform nature of TCP/IP. Since virtually every operating system includes these protocols it is very easy to develop systems which have heterogeneous elements such as different computers.

- The cheapness of TCP/IP. A very large amount of software associated with these protocols is free or very cheap. For example, the base versions of Web browsers are now either bundled in with operating systems or given away free by their manufacturers.

Internets and Intranets

A term often used in connection with distributed systems is *Intranet*. This is a collection of computers which are private to a company, but which use Internet technology such as browsers and TCP/IP to communicate. Often these computers are not connected to the Internet. They are described by this term to show that they are similar in concept to the Internet, but differ in one instance: that they are not accessible to the normal users of the Internet. The system which sells agricultural products to farmers, described above, is a good example of an Intranet.

3.5 OFFICE AUTOMATION

The advent of networks and the parallel introduction of word processing were the principal spearheads of office automation. The nature of office work means that it will never be entirely automated in the sense that people will be eliminated from offices; there are too many discrepancies and variations for that to happen. Nevertheless the present generation of office workers are far more technology-orientated than those of previous generations.

Also referred to as 'electronic offices', office automation is a conglomerate of various technologies intended to improve the efficiency of office work and also to minimize the numbers of office staff. Office workers gradually increased their use of semi-automatic machines such as electric typewriters, calculators, photocopiers, tape recorders, etc., over a long period, but office automation brought more extensive changes to their work style. The principal areas of office automation are: word processing, PCs, intelligent terminals, COM and microforms, electronic mail (including facsimile), videotex, electronic funds transfer and desktop publishing.

■ Word processing (WP)

The increased need for typed documents combined with the high cost of manual labour and the reduced cost of microprocessors resulted in the development of WP on a large and expanding scale. Subsequent developments in microprocessor technology and its increasing cost have made it economic to incorporate a wide range of facilities into word processors. This has happened to the extent that a powerful word processor is nowadays available for the equivalent price of yesteryear's electric typewriter.

In its simplest form WP is an enhanced method of typing documents. Basic WP does not demand a high financial outlay nor necessarily the involvement of departments other than the one concerned. In its more sophisticated form, WP includes the composition, recording, transcription, editing and communication of text.

■ Desktop publishing (DTP)

DTP is the employment of a computer to create pages of mixed text and illustrations – as in most technical books. The software associated with DTP together with high-quality printers enable this to be achieved straightforwardly and to a high standard. The types of document likely to be prepared in this way include newsletters, notices, advertisements and catalogues.

By use of a mouse and monitor, frames are represented on the screen into which text is inserted via the keyboard or from disk storage. In the latter case the text may well have been generated and stored by means of word processing.

The text can be in several fonts and sizes on the one document. The illustrations are created on the screen by drawing with a mouse-controlled pointer or selecting from a pre-stored range. After a final check the document is copied exactly by the printer as many times as required or, alternatively, printed the once and then photocopied.

DTP facilities include the following:

- A complex of fonts (typefaces) and print sizes, with mathematical symbols and others such as arrows and stars.

- Columns of various widths into which the text is fitted automatically using sensible hyphenation. Reorganization of the columns is also automatic.

- Automatic spacing of characters and lines in order to fit the text into the available space.

- Scaling and cropping of illustrations so as to fit the available space.

- Page numbers inserted alternately left and right (as in books) in multi-page documents.

■ Electronic mail (EM)

Just as electronic funds transfer (EFT) has a natural connection with point-of-sale systems (Section 4.4), electronic mail (EM) ties in naturally with word processing.

EM means that letters and documents typed or scanned in one office are transmitted to another office very quickly irrespective of distance. This concept implies that a computer network using packet switching must be employed. The sender of a letter or document by EM includes an address code to designate the recipient(s). This accompanies the document along with the sender's code and a flag indicating the recipient's right to make amendments. This control is necessary because on some occasions a document is sent via several offices, each of which may contribute to it. For instance a report for head office might be created from sub-reports contributed by a series of branch offices. These contributions may need to be in a certain sequence because one office's sub-report depends on those of a previous contributing office(s).

It is sometimes desirable to have notification of delivery; this is achieved by the EM system automatically notifying the sender after delivery. In the case of confidential information, sight by an authorized person only is effected by inclusion of a password code. This must be matched by the recipient's password before the local computer will deliver the communication. Other features of EM include:

■ Automatic redirection of messages if the recipient has moved.

■ Authentication of sender for security purposes.

■ Notification to sender if a message cannot be delivered.

■ Multiple addressing, i.e. the same message being sent to several recipients.

■ Recording times of dispatch and delivery of messages.

■ Message filing and retrieval.

■ Automatic accounting and the billing of users.

Facsimile (fax)

Another communication requirement is for facsimile transmission, i.e. identical copies of documents, diagrams, handwriting, etc. It is obviously advantageous to be able to send documents rapidly between offices and companies, and it is sometimes important that these are *exact* copies of the originals, e.g. signatures. This latter requirement applies most particularly to legal documents.

Facsimile transmission is achieved by optically analysing the document's appearance at the sending end and converting it into a digital representation. The reverse process is carried out at the receiving end so that an exact copy can be produced. Facsimile is, in effect, the same as long-distance photocopying.

The main drawback of facsimile is its slowness of conversion, transmission and reconversion: an A4 sheet takes from 20 seconds to several minutes. This problem is caused by the huge amount of data involved. An A4 sheet is equivalent to about $\frac{3}{4}$ million bits, and this is thirty times as much as when it is represented by ASCII or EBCDIC coding.

Voice mail

A similar arrangement to EM is voice mail, this is the storing and forwarding of digitized speech. There is no need for the sender to be directly connected to the recipient but merely to speak his or her message, knowing that it will eventually reach the person(s) for whom it is intended, including multiple recipients.

Each participant has his or her own 'mailbox' from which messages are retrieved by calling on the phone; this may be up to a month after the origination of the message. He or she then replies in the same way as described above.

IBM's Speechfile is an example of voice mail; this handles not only speech but facsimile and encoded text.

▪ Electronic funds transfer (EFT)

The need to pay for goods and services has engendered a variety of methods of payment throughout history. These now include – in addition to cash – bank cheques, Giro transfers, credit cards, direct debit/credit and standing orders. The clearing banks deal with several million financial transactions per working day, many of which entail the preparation and handling of documents, such as cheques. This vast amount of work and immense cost has led the banks and other companies to move towards document-free methods of transferring money between bank accounts.

The concepts of EFT are, first, that details of financial transactions are represented by electronic or magnetic means instead of on documents, and, secondly, accounts are updated concurrently in computer storage by the corresponding debits and credits. The general idea is that the customer has a magnetized credit or debit card which is inserted into the point-of-sale (POS) terminal (Section 4.4). The amount of the bill is entered either via a keyboard or directly from the POS. In some systems the customer's balance is checked and his or her and the retailer's accounts are updated automatically.

EFT has now been extended, by means of Internet technology, into people's homes, enabling them to inspect their bank accounts, make transfers, settle bills and pay for goods purchased by using their domestic television. This is sometimes called 'home banking'.

A further aspect of EFT is the chip card. This is a card on which a predetermined amount of credit is magnetically recorded and the balance is reduced automatically each time it is used for a purchase. Usually the card is thrown away when the balance reaches zero but some types of card can be reloaded. An extension of this arrangement is a smart card holding the full details of a bank account. This could be used as above but also linked to a central computer for updating by other transactions such as direct credits and standing orders.

SWIFT and CHAPS

Another application of EFT is its employment by the clearing banks themselves in order to effect transfers of funds between their own accounts with other banks. Such a system, acronymed SWIFT (Society for Worldwide Interbank Financial Telecommunications), caters for interbank transactions at international level such as bank A instructing bank B to transfer funds to bank C's account with bank D. A development of SWIFT is CHAPS (Clearing Houses Automated Payments System): this is a bank clearing system operated through British Telecom's packet switching system and intended for the rapid transference of high-value transactions.

EXERCISES

Exercise 3.1   Office automation

An article in the February 1986 edition of *Management Accounting* stated: 'Within ten years or less, around 50 per cent of business transactions may be conducted electronically.'

You are required to answer *four* of the following, with reference to specific applications wherever possible:

(a) Explain what is meant by the term 'electronic office'.

(b) Describe the principal features of an electronic mail system either with a multi-access computer within an organization or using a national network.

(c) Explain what is meant by EFT.

(d) Describe what facilities would be expected within a terminal emulation package.

(e) A number of integrated software packages can be purchased, each of which combines terminal emulation, word processing, spreadsheets and a database. Explain how this integration may prove beneficial to an organization.

(CIMA, stage 2, Inf. Tech. Man., May 1988)

Exercise 3.2   Mainframe configurations

LM Ltd is a manufacturing company with many departments at a single site. It is considering the installation of a multi-user mainframe computer system.

You are required to describe the following hardware configurations for a multi-user mainframe computer system, identifying the advantages and disadvantages of each:

(a) a simple multi-user system;

(b) a host system with front-end processor; and

(c) a multi-processor system.

Illustrate your answers with diagrams.

(CIMA, stage 2, Inf. Tech. Man., May 1992)

97

Exercise 3.3   **Office automation**

Describe how the following information technology-based systems contribute to office automation:

(a) electronic mail;

(b) decision support systems; and

(c) tele-conferencing.

(ICSA, Info. Systems, June 1992)

Exercise 3.4   **Mainframe/distributed DP**

AB plc is a national freight distribution company with a head office, five regional offices and a hundred local depots spread throughout the country. It is planning a major computerization project. The options which are being considered are as follows:

(a) A central mainframe system with terminals at each depot.

(b) Distributed minicomputers at each regional office.

You are required to draft a report to the board of AB plc describing the ways in which each of the options would suit the company's structure and explaining two advantages and two disadvantages of each.

(CIMA, stage 2, Inf. Tech. Man., May 1992)

Exercise 3.5   **Data communications**

Describe the meanings of the following aspects of data communications:

(a) a modem;

(b) a multiplexor;

(c) baud rate; and

(d) a local area network.

Exercise 3.6   **Local area networks**

(a) Explain what a local area network is and the advantages accruing from its use.

(b) Describe briefly and diagrammatically each of the following:

(i)   a bus network;

(ii) a ring network;

(iii) a star network.

Exercise 3.7   **Data transmission**

(a) Explain, using diagrams, what you understand by the following data transmission techniques:

(i)   half duplex;

(ii) full duplex;

(iii) parallel transmission.

(b) Explain how data can be transmitted serially using synchronous and asynchronous techniques, clearly stating the advantage of each technique.

(BCS, GPII, April 1992)

Outline solutions to exercises

Solution 3.1   Refer to Section 3.5.

(a) The principal features are as follows:

- Office transactions represented and transmitted electronically.
- Office records held magnetically.
- Minimum of paper records and manual transcription.
- Facilities for word processing, extensive accounting and office control software, electronic mail, electronic funds transfer, facsimile.
- Personnel microcomputers linked to centralized and/or distributed databases.

(b) Refer to 'Electronic mail' in Section 3.5.

(c) Refer to 'Electronic funds transfer' in Section 3.5.

(d) A terminal emulation package enables a microcomputer to be used as a terminal linked to a mainframe or minicomputer. The package handles problems such as protocols, data coding and transmission line speeds.

(e) An integrated package combines together into the one package various types of applications software such as word processing, spreadsheets and graphics. This reduces the effort for certain work, e.g. preparing a cost variance analysis using a spreadsheet program, incorporating results into reports using word processing, and preparing a demonstration of costs using computer graphics.

Solution 3.2   (a) A simple multi-user system has the terminals connected directly to the CPU.

(b) The host system with a front-end processor is as Figure 3.2.

(c) A multi-processor system has more than one CPU to which all the terminals are linked, thus providing a high level of reliability and processing power. If other units, such as disks, are also duplicated and cross-connected, an even higher level of reliability is achieved. This is sometimes known as a polymorphic system.

Solution 3.3   Refer to Section 3.5 and to the Glossary.

Solution 3.4   (a) **Central mainframe**

Advantages:

- One database with only one level of control.
- Tighter control over system developments and computer operations.
- Greater security.
- Less overall floor space required.

Disadvantages:

- Cost of data communications.
- Less backup against failure.
- Users at depots more remote from the computer; this is largely psychological unless computing power is needed.

(b) **Distributed minicomputers**

Advantages:

- Fewer problems with failures.
- Users closer to computing power.
- Closer contact between regional office and depots.

Disadvantages:

- Greater cost of hardware.
- Less control over systems developments.
- More overall floor space necessary.
- More computer operators needed.

Solution 3.5
(a) Refer to 'Modems' in Section 3.3.

(b) Refer to 'Multiplexing' in Section 3.3.

(c) Refer to 'baud' in Glossary.

(d) Refer to 'Local area networks' in Section 3.3.

Solution 3.6
(a) Refer to 'Local area networks' in Section 3.3. The main advantages are:

- users have access to all hardware linked to the LAN, e.g. printers and backing storage;
- users can communicate with each other, and with more distant points via gateways;
- common use of applications software;
- common access to database records;
- transmission of data between offices.

(b) Refer to 'Network topologies' in Section 3.3.

Solution 3.7
(a) (i) Refer to Section 3.3.

(ii) Refer to Section 3.3.

(iii) Parallel transmission means that the data is transmitted along several channels concurrently thus minimizing the transmission time. This could take the form of the bits in a byte being transmitted along separate wires (circuits) simultaneously.

(b) Refer to Section 3.3.

References and further reading

3.1 Baron, R. and Higbie, L. *Computer Architecture* (Addison Wesley, 1992).

3.2 Brebner, G. *Computers in Communication* (McGraw-Hill, 1997).

3.3 Chalk, B.S. *Computer Organisation and Architecture* (MacMillan, 1996).

3.4 Coupille, P.A. *Computer Hardware and Data Communications* (Prentice Hall, 1993).

3.5 Dutton, W. *Information and Communication Technologies* (Oxford University Press, 1996).

3.6 Englander, I. *The Architecture of Computer Hardware and Systems Software* (John Wiley, 1996).

3.7 Fitzgerald, J. *Business Data Communication and Networking* (John Wiley, 1996).

3.8 Goodman, J.R. *Programmer's View of Computer Architecture* (W.B. Saunders, 1993).

3.9 Heath, S. *Microprocessor Architectures* (Butterworth, 1995).

3.10 Hennessey, J.L. *Computer Architecture* (Morgan Kauffman, 1995).

3.11 Hughes, L. *An Introduction to Data Communication* (Jones and Bartlett, 1996).

3.12 Hunter, P. *Network Operating Systems* (Addison Wesley, 1995).

3.13 Jaderstrom, S., Kruk, L., Miller, J. and Fenner, S. *Complete Office Handbook: the Definitive Reference for Today's Electronic Office* (Random House, 1996).

3.14 Leinwand, A. and Fang-Conway, K. *Network Management* (Addison Wesley, 1995).

3.15 Maitra, A. *Building a Corporate Internet Strategy: the IT Manager's Guide* (Van Nostrand, 1996).

3.16 McCabe, J.D. *Practical Computer Network Analysis and Design* (Morgan Kaufmann, 1997).

3.17 Miller, S.E. *Civilizing CyberSpace: Policy, Power and the Information Superhighway* (ACM Press, 1995).

3.18 O'Neil, S.L. *Office Information Systems: Concepts and Applications* (McGraw-Hill, 1989).

3.19 Ray, C.M., Palmer, J.J. and Wohl, A.D. *Office Automation: a Systems Approach* (South-Sestern, 1995).

3.20 Shulman, B., Adams, C. and O'Brien, K. *The Beginners Illustrated Internet Dictionary; From A to Zine* (Perspective, 1997).

3.21 Stallings, W. *Computer Organisation and Architecture* (Prentice Hall, 1996).

3.22 Terplan, K. *Applications for Distributed Systems and Network Management* (Van Nostrand, 1994).

3.23 Wilkinson, B. *Computer Architecture* (Prentice Hall, 1996).

4 Data capture and computer input/output

AIMS

After reading this chapter you should:

- be familiar with the various sources of data on a computer project;
- be familiar with the main categories of data capture hardware;
- be familiar with the main categories of data output hardware.

4.1 DATA CAPTURE

Data capture is a general term for the process whereby source data is collected and transformed into a form capable of being 'understood' by a computer. Computer input is the transference of the data from the computer-sensible medium into the main store of the computer. In some cases these two functions are one and the same, such as with on-line systems.

The collection and means of inputting data have undergone many developments in order to reduce the tremendous burden of data capture. This is just as well since it is a labour-intensive activity and consequently expensive. It is beneficial to capture data at its point of origin, thereby eliminating the need for keying, by constraining the originator to create it in a computer-compatible form.

The methods described in this chapter cover a wide range, and it is probable that developments of some of these methods rather than fundamentally new ideas, will be employed in the foreseeable future.

Before moving on to the methods it is useful first to consider a few points about the structure of business data. This nearly always takes the form of a group of related data items, called a record, that pertain to an activity or an entity, i.e. to something done or to something that exists. Thus if the activity took the form of depositing money with a building society, the record pertaining to this financial transaction would consist of data items such as the account number, date, amount and branch number. On the whole, the data items of an activity or entity are captured and input at the same time. This is obviously necessary because, for example, the amount mentioned above has meaning only if tied to a particular account.

Data capture is usually geared to the acceptance of a succession of similar records during a particular period and via a particular piece of equipment. This means that a key-operator or a source document user repeats the handling of a certain type of record many times during a certain period.

The principle of verifying is sometimes employed with key-operated systems. This entails keying the same source data, usually from documents, into the system twice, whereupon it is automatically checked for absolute similarity. Any discrepancies are thus detected and then corrected before the data goes for processing. As a general rule, the data is keyed by two different persons so as to minimize the chance of the same misreading occurring twice.

Because of the large volume of source data handled by many organizations, it is important to have an efficient arrangement for keying it into the system. The following are the three main aspects of data entry.

Source documents

Clear documents are important since it is possible that both the original and the verifier operators may misread a badly written figure in the same way, with a consequent error reaching the computer. The design of the source document is also of consequence in maintaining a high speed of keying. And although it is not always possible to arrange a document's layout purely to facilitate keying, this should be borne in mind in designing any document that holds source data.

Operator training

The training of keyboard operators has two main aims. First, the need for accuracy in keying the data. Second, achieving high keying speed while maintaining accuracy.

The former is promoted by showing the operators how their work fits into the overall system and pointing out the consequences of erroneous keying.

High-speed keying comes with proper training followed by the use of well-designed source documents. Operators can reach 18,000 key depressions per hour and, as this is an average of five key strokes per second, the operator obviously needs great dexterity and concentration. This activity has to be carefully designed to reduce operator fatigue and avoid repetitive strain injury (RSI).

Design of data entry equipment and dialogues

From the operator's point of view the keyboard and the display are the paramount factors and are likely to be the only 'window' through which he or she sees the whole information system.

The keyboard needs to be well laid-out, light to the touch and designed as far as possible to suit the type of data being keyed. Ideally the keyboard should have extra keys such as a numeric pad if a large amount of numeric data is involved. A numeric pad is a separate set of the keys 0 to 9 to one side of the main keyboard.

Data entry is a repetitive, boring job which requires high accuracy and attention to detail. Design of the data entry dialogue is therefore essential to reduce operator fatigue and errors. The main point to bear in mind is that a clear, consistent structure should be designed for the data entry sequence. The system should behave consistently and give feedback so that the operator can tell where they are in a sequence and whether anything has gone wrong. Furthermore breakpoints should be provided in the dialogue to allow the operator to have 'mini-rests' in a keying sequence (e.g. at the end of a record, or set of related fields). Breakpoints help combat fatigue by allowing a brief mental reset. Further points on the dialogue design are made in Section 8.5 on human–computer interaction (HCI).

Data entry may use a variety of equipment. Personal computers are often used as front-ends for data capture, with on-line validation to capture errors. Correct data is then up-loaded to the mainframe for updating the database and subsequent processing. This is usually achieved by a file transfer protocol (FTP) program which connects the PC with the host computer's data communications processor and then sends its files in chunks. The program checks that the data has been received correctly before sending the next chunk. Data transfer is dealt with in the section on data communications. It is also possible to transfer data to a host computer by diskettes, although this is a time-consuming and cumbersome procedure if a lot of diskettes are involved. Diskettes are robust enough to be handled and sent through the mail, this encourages dispersed recording. A further hazard may be the need to convert data formats; however, most host computers can read diskettes in PC operating system formats (e.g. Windows 98).

Alternatively, microprocessors may be used for data entry, or more specialized equipment. These are reviewed in the following section.

4.2 DATA ENTRY EQUIPMENT

Data entry is now performed on-line so that items can be checked (validated) as they are entered. The validation program may be resident on a front-end processor such as a PC, or it may be running on a mainframe.

4.3 BAR CODING AND POINT-OF-SALE SYSTEMS

Several media are grouped together under 'bar coding', although in some ways they are distinctly different from one another. Bar coding is the representation of code numbers or other data in the form of optical or magnetic bars on a data carrier. The data carrier may be a tag, ticket, label, plastic card or a container holding goods.

Fig. 4.1 Example of European Article Number bar code

Bar codes

The two types of bar codes are based on either the European Article Number (EAN) or the Uniform Product Code (UPC). These two systems are similar: the UPC is in use in the USA; the EAN is employed in various countries in Europe (see Figure 4.1) and comprises thirteen digits each of which is represented by two bars. The thickness and spacing of the bars is the means by which each digit is identified. Beneath the bars is printed a human-readable interpretation of the EAN or UPC in OCR–B font.

The thirteen digits of the EAN are usually made up of two digits to identify the country, five for the manufacturer, five for the product and one as a check digit. The UPC is similar except that only one digit is used for the country identifier, making twelve digits in all. By using five digits for the product, it is generally possible to identify uniquely all the products of a company and their variations such as colour, packaging form, weight, size, etc., bearing in mind that five digits cover 100,000 items.

The bar codes are printed on the containers at the time of manufacture and are read optically by either a hand-held reader (scanner or wand) or a static reader. Thus it is possible to identify entities automatically in a wide range of situations. These include goods in supermarkets, items in factory stores, library books and borrowers, and almost anything else.

Point-of-sale systems (POS/EPOS)

Point-of-sale (electronic point-of-sale) systems are used in supermarkets and other large shops in order to cope with a large throughput of customers. Typically a POS incorporates a scanner, i.e. a bar code reader, a printer and a keyboard built into each check-out point, of which there are generally between six and twenty-four in the one system.

105

The sales assistant passes each purchased item across a window in the counter so that the scanner can 'see' the bar code and identify the item. Another method is for the sales assistant to have a hand-held scanner which does the same thing. This latter arrangement is more suitable for large awkwardly shaped items that could not be passed easily over a static scanner.

When the scanner has read and recognized the bar code, it emits a short sound. If this is not heard the sales assistant repeats the scanning procedure. The price and the description of the item are then looked-up automatically by the computer, and the cost of each item is displayed and printed at the check-out point. The customer is thus provided with a printed receipt list showing the description, quantity, price and amount of all purchases.

The keyboard is for entering the prices of items not bar-coded, e.g. fresh vegetables. This operates in conjunction with electronic sales so that the price is computed automatically.

A further facility is for the terminal's printer to print all the entries onto the customer's blank cheque, leaving only the signature to be added. The change (if a cash transaction) is also computed and printed at the bottom of the receipt along with the date, time and reference data.

From the point of view of the supermarket's management, the POS enables stock levels to be updated immediately by each sale. This facilitates the prompt restocking of shelves, and also provides an analysis of sales so that sudden surges in demand are detected at an early stage.

Some problems with POS have arisen as a result of discrepancies between the prices marked on the shelves and those stored in the computer, bearing in mind that the individual goods themselves are not price-marked. Discrepancies tend to occur when prices have changed and have not yet been re-marked on the shelves. A possible way round this problem is to have electronically displayed prices, i.e. LCD displays, on the shelves linked directly to the computer so that all price changes are registered simultaneously.

Another significant advantage of POS is that the chore of price labelling is almost, if not entirely, eliminated. This is a large saving in terms of man-hours and, additionally, reduces errors that sometimes occur in labelling. There is also the advantage of preventing label swopping by dishonest customers.

4.4 VISUAL DISPLAY/KEYBOARDS

A visual display unit (VDU) is also sometimes termed a monitor. Although strictly speaking a VDU is an output device, it is often employed in conjunction with a keyboard as a means of input. That is to say, the keyboard is used to enter data which appears simultaneously on the VDU's screen, perhaps along with other data deriving from the computer. In any case the characteristics of a VDU apply equally well to its use for both input and output.

A VDU is basically the same as a domestic television set in that it contains a cathode ray tube scanned by a beam of electrons. This creates a set of horizontal lines (the raster) along which bright dots appear so as to give the outline of characters and shapes. Typically, characters are five dots wide by seven dots high. The most common display size is twenty-four lines of eighty characters each but variations occur between the different models of VDU. Another characteristic of a VDU is its character set (repertoire), i.e. the number of different digits, letters, symbols and shapes that can be represented on the screen. Often the repertoire consists of either 96 or 128 characters of the ASCII set.

VDUs may be able to display only characters and these are usually called character graphics array displays. The next step is to use the shapes which make up characters to build simple graphics as well as text and numbers; this capability is called extended character graphics. However, most VDUs now have a full graphics capability and can display pictures, drawings and text in a variety of different shapes and sizes.

The keyboard of a VDU reflects the repertoire to a large extent and, as already stated, has a similar layout to a typewriter but with additional control keys. Control keys do not necessarily cause characters to appear on the screen but send special signals to the computer for purposes such as text editing and cursor movement. A cursor is a symbol such as a line, square or arrow displayed on the screen to indicate relevant data or the next position to be used: it can be moved around the screen by means of the control keys or by a program.

Most keyboards are detachable from the screen and so can be interchanged and positioned to suit the operator's convenience.

Operational features of VDUs

The features described below are attained through a combination of hardware and software; the extent to which the VDU itself, as opposed to the processor to which it is linked, is capable of achieving them depends upon its level of intelligence, i.e. the amount of built-in processing power in the VDU.

Paging and scrolling: Paging is the displaying of a complete screen or page of information at a time and its immediate replacement by another page on request. This is analogous to turning the pages of a book, and so is suitable for inspecting sets of figures or pieces of text that fall into groups.

Scrolling is the movement of information up or down the screen a line at a time so that as one line appears at the bottom another disappears from the top and vice versa. This occurs rapidly and so there is no significant delay in inspecting long lists such as stock levels and account balances. Scrolling is only really suitable for lists that are in a known clear sequence, otherwise it is difficult to know which way to scroll to find a particular item.

Sideways scrolling is the movement of the display to the right or left across the screen. This is commonly utilized in spreadsheet programs but is otherwise not really suitable for lists or text.

Form filling: This is a changeable electronic or optical display of a document outline (entry form) together with the relevant headings and annotations. The keyed input data computed entries are displayed within the outline as though they are being entered manually. This method enables the operator always to have a clear picture of the point reached in the work.

An example of the practical use of an entry form is in the preparation of sales invoices from customer order data. The operator enters the account number, the commodity codes (or equivalent) and the quantities sold. At the appropriate points the computer inserts the name and address, descriptions, prices, amounts and other calculated information.

Double brightness: The various data items on the screen can be displayed at two alternative levels of brightness, these are sometimes called foreground and background data. The essential purpose of double brightness is to differentiate between data so that, for instance, data entered manually is clearly distinguishable from that emanating from the computer. It is also useful for drawing attention to certain data by making it brighter than the rest.

Highlighting: Certain data is given a solid bright background, i.e. the dots composing the character(s) are fully bright and the remaining dots in the character space are at half brightness.

Inverse video or reverse video: The characters are created by 'black' dots on a bright background, i.e. the reverse of normal. This is used mainly for headings in larger fonts.

Blinking or flashing: A data item or the cursor is made to flash on and off or from normal to inverse video. This is obviously useful for drawing attention to certain data.

Variable character size and shape: Text may be composed of larger characters than normal, e.g. double height and double width or different shape. Characters may be displayed in different shapes known as fonts, e.g. Times, Pica, Courier, Geneva. These are printing styles for the letters which most word processing packages allow to be set according to the user's choice.

Mouse: A mouse is a hand-held device which when run along the table causes a cursor to move correspondingly on the screen. When the cursor reaches a position of significance, the user presses a button on the mouse and the computer registers that this is of some interest. Thus a mouse is a convenient way of selecting parts of the display without having to use the keyboard. A mouse may either be attached to the VDU by a lead (its tail) or be unattached (tail-less) and operate by ultrasonics or by infra-red beams.

Touch screens: The concept of a touch screen is that the user is able to indicate items of interest by touching them on the screen. This is achieved mainly by a matrix of fine infra-red beams criss-crossing the screen and being broken by the finger or touching object. Other methods are also available or proposed such as conductive membranes, sound wave reflection, capacitative screens and piezo-electric crystal deformation.

Whatever method is employed, there can be only a limited number of touch points on the screen, otherwise their close proximity would cause confusion. A practical approach is to display a menu (Section 8.5) with a touch point alongside each item.

An extension of touch screens is hand-drawing whereby the user by moving his or her finger across the screen causes lines to appear. This is of more interest in education than business.

Windows: One of the problems with VDUs in business is the need to move frequently from one display to another in order to make cross-reference. For example, a customer's recent invoice, ledger account and payments history. This problem is largely overcome by having windows, i.e. separate small displays, on the screen simultaneously. Each window is capable of being scrolled, paged or edited individually as though it was on a separate VDU.

A process termed 'zooming' is a means of obtaining a finer degree of detail in the data shown via a window. Windows are frequently used on PCs in conjunction with a mouse and icons, often referred to by the acronym WIMP (windows- icon-mouse-pointing).

Graphics: Business VDUs are the raster type. As explained earlier in this section, the closeness of the dots on a high-resolution screen enables graphical shapes and coloured areas. This implies that a wide range of things can be represented, e.g. graphs, bar charts, pie charts, histograms, 'three-dimensional' representations, logos, icons and very large characters.

Colour: Some VDU manufacturers claim that their models are capable of reproducing hundreds and even thousands of colours. In so far as the three primary colours (red, green and blue) can be mixed at various brightnesses this is technically true. In practice the business user is unlikely to need more than a few dozen colours for graphical representations and production of presentation graphics. Judicious use of colours can make a display clearer and more interesting but overlavish employment is likely to be counter-productive by making the display too dazzling.

Audible output: An audible note, sometimes called a beep, is sounded by the VDU whenever circumstances warrant this. This is a convenient way of drawing attention to trouble or a special condition without altering the display. It must be backed up by facilities for detecting the problem and putting things right.

Panel display: This is a panel composed of a set of small tubes which are arranged to change their appearance so as to represent characters. A common use of small panel displays is in pocket calculators and in certain data capture-devices.

The tubes are either light-emitting diodes (LEDs) or, more likely, liquid crystal diodes (LCDs). The latter require so little electrical power that it is practical to incorporate a large number into battery-powered portable (lap-top) computers in order to create a twenty-four line by eighty-character display.

4.5 VOICE DATA ENTRY (VDE)

VDE, also termed 'voice input', 'speech recognition' and other names, differs from almost all other methods of data capture in that there is no digital input representing data. The input data is derived from a human voice and because of the wide range of voice sounds, it is very difficult for a computer to interpret that data with complete accuracy.

There are three aspects of VDE: isolated word recognition; connected speech recognition; and speech understanding systems. It is with the first aspect that we are concerned here, but a few words about the other two are worthwhile.

Connected speech recognition: This means that the computer must be capable of interpreting and making use of short strings of words, i.e. structured sentences. This implies the recognition and understanding of certain keywords in a string. Although the computer might not recognize all the words, there is sufficient understanding for certain actions to be initiated.

Speech understanding systems: These systems are of a higher level and necessitate the computer 'understanding' the grammatical and intellectual meanings of text and of continuous speech in real-time. These aims imply a high level of artificial intelligence, which in turn means the employment of a powerful computer. Inherent factors are speaker independence and differentiation between speech and background sounds. Speech understanding has progressed to the stage where good continuous recognition is possible with large vocabularies; these systems are, however, speaker dependent, i.e. they have to be trained to the voice of one particular individual. Once trained these systems have a high degree of accuracy and can deal with differences in intonation that we use when speaking. For multiple voice recognition (speaker independent), accuracies are lower and vocabulary sizes lower. From the business point of view, speech recognizing word processing packages are becoming readily available although these will remain solo voice for some time. Speech applications will become more common for simple situations in which the size of the vocabulary can be determined, e.g. order entry systems, ticket and financial transactions.

Isolated word recognition: This is a much less ambitious system in that it interprets only a limited set of words, e.g. up to 10,000 words in modern systems, and then only from known persons' voices. That is to say, the computer accepts input data only from persons whose voices it has been trained to interpret. It is trained by the person speaking each word a few times and concurrently inputting the word via a keyboard. This is known as 'training mode' and the computer is thus able to create and store a feature matrix of each word based on its pitch, tone and phenome characteristics. The feature matrix is, in effect, an average reference pattern for each particular word as spoken by a certain individual. If a person's pronunciation changes for one reason or another, it may well be necessary to retrain the computer.

The input of voice data for actual use is termed 'recognition mode'. If voice data is displayed immediately after being input, it is possible to amalgamate training and recognition modes. This occurs because corrections can be made to misinterpreted input and so the computer can modify the relevant feature matrix. During recognition mode the words are spoken into a small noise-cancelling microphone generally attached to a headband. There must be a short time interval (at least 0.1 seconds) between words in order to separate successive words; this is no problem in practice.

Applications of VDE

Until recently VDE was not generally employed for the input of a large amount of data as this could be handled by faster and more cost-effective methods. VDE at present is more suited to specialized data involving separate words or short strings of words spoken at intervals. This situation arises especially when the user is totally occupied with his or her hands or cannot divert his or her eyes from the task. Similarly VDE is useful for activating a computer or other machine to perform certain tasks.

Examples of VDE applications include the control of luggage and goods sorting/directing mechanisms, such as occur in airports, factories and postal centres. Another application is in slaughterhouses where the weights and qualities of carcasses are recorded by VDE.

The main area of development of VDE within information systems will be in the handling and input of large amounts of data and information. VDE technology has advanced massively over the last five years to the point where voice recognition packages can be found for sale in high street computer shops. This means that they can now be used in a wide variety of applications where direct data entry was previously the only option. These include:

- Word processing, where instead of keying in text the user speaks the text into a microphone.
- Data entry of forms such as applications for loans.

■ Applications where fast response is needed such as foreign exchange deal-ing, where a second or two can mean a loss or gain of thousands of pounds.

Voice recognition is also used in security systems, but in this application it is the person rather than the information that is of interest.

A final point worth mentioning is the challenge of language translation in real time; for instance translating Japanese into English over the phone while pre-serving the essential characteristics of the speakers' voices.

4.6 PRINTING

The printer is the workhorse of the computer. Computer printers come in a much wider range of capabilities than other peripherals, and this is reflected in their wide range of costs.

The descriptions that follow are intended to give the reader a broad and com-prehensive understanding of printer technology without, hopefully, confusing him or her with a mass of figures and technicalities. Some quantitative facts must, however, be taken into consideration otherwise we have no measurable bases of comparison between the different methods and models.

Computer printers fall into two main groups: serial printers and page printers.

■ Serial printers

Also known as character printers, serial printers print one character at a time across the line of print. This method of printing necessitates only one, or some-times two, print head(s) and operates in a similar way to an ordinary typewriter. In contrast to a typewriter, however, the print head of a serial printer moves across the paper while the paper remains stationary.

Serial printers fall into two categories: impact and non-impact. With the former the print head comes into contact with the print ribbon, generally with sufficient force to enable several copies to be made simultaneously. Non-impact printers inevitably mean only one copy at a time; this is a disadvantage because several copies are usually needed in business. The extra copies have to be pre-pared either by repeated printing or by photocopying off-line.

Non-impact printers are now most frequently found as ink jet printers. However, even these printers are being superceded by laser printers (see below).

Ink jet printers

Considerable technological development has gone into ink jet printers. A good feature of ink jet printers is their quietness of operation. They are at the top end of serial printer speeds and have great flexibility in their print style. Their capabilities include any font, simulated handwriting, variable size characters, diagrams and logos.

Another advantage of certain models is the ability to produce multi-coloured printing. The main disadvantage of an ink jet printer is its inability to print more than one copy at a time.

Thermal printers

The print head of a thermal (electrothermal) printer contains an array of heating elements that have electric currents switched through them so as to form the shape of the required character. When the print head comes into contact with the special paper, it burns away an aluminium coating to reveal a black core. Other models employ similar processes including the fusing together of colouring agents held in a surface coating on the paper.

The main drawbacks of thermal printing are the need for special paper and only one copy at a time. They are quiet in operation, however.

Characteristics of serial printers

Speeds: From 100 to 1,000 characters per second. The effective printing speed is increased by bi-directional printing and by 'look-ahead'. The latter means that the printer has sufficient intelligence to decide whether to print the subsequent line forwards or backwards in order to minimize the print head movement.

Character sets: The usual set is the ninety-six ASCII characters but a much greater number is possible with matrix and ink jet printers. The latter can produce cursive, i.e. imitation handwritten, fonts under software control. In effect, there is an almost unlimited number of fonts and characters available from certain models of serial printers.

Page printers

Page printers create a full page of print at a time using one of the techniques described below. They vary in size from relatively inexpensive table-top devices up to expensive but very powerful printing systems (see characteristics). There are four types of page printer: laser, ion-deposition, electrostatic and magnetic.

Laser printers

A laser printer functions by creating an image of a page of print in the form of microscopic dots on the photoconductive surface of a rotating drum by means of a controllable low-power laser beam. A special ink (toner) is then attracted to the laser-exposed areas of the drum. When the paper comes into contact with the drum, the image is transferred and then fused permanently onto it. The drum is then automatically erased before receiving the next page's image. This method uses ordinary blank computer stationery, either fan-folded or a continuous reel. The document's outlines and headings are created by flashing a photographic negative of the document onto the photoconductive drum while printing is in progress. Laser printers are now the most commonly encountered.

Since the mid-1990s laser printer costs have decreased substantially to the point where a good laser printer which prints at 600 dots per inch and six pages per minute can be purchased for under £300. Consequently a technology which used to be confined to large offices is finding its way into the home.

A recent development in laser printing has been that of the colour laser printer. This prints a colour each time; this means that a sheet of paper must be very precisely positioned. Because of this the hardware technology used in a colour laser printer is substantially more complex than that used in a black-and-white laser printer. This is reflected in their costs. A typical colour laser printer will cost around £3,000 to £6,000.

Ion-deposition printers

In contrast to the laser printer, this method is electrical rather than optical. Ions (charged electrical particles) are created in a cavity, and directed electrically through an orifice onto the dielectric surface of a rotating cylinder. By using an array of orifices and by switching the ions off and on, the required characters are formed as an electric charge image on the cylinder's surface. Toner is then applied to the charged image and transferred to the paper on which it is transfixed by pressure; this is known as cold fusion. This method enables logos, signatures, etc., to be printed as well as characters in a range of fonts. Ion-deposition printers are smaller and cheaper but have a lower speed than laser printers, typically forty pages per minute.

Electrostatic printers

Electrostatic printers can be regarded as page printers. The characters are created by dots at 200 per inch horizontally and vertically, thus giving a high quality that is not obviously matrix printing. The document outlines are software coded and then stored for printing concurrently with the data. Letterheads and logos are created electrostatically from a changeable metal cylinder.

Magnetic (magnetoelectric) printers

The basic principle of a magnetic printer is in some ways similar to a magnetic drum. A drum in the printer has a surface that can be coated with rows of tiny spots of magnetism by means of thousands of minute recording heads. As the drum rotates it becomes covered with these magnetic spots so as to form a latent image of the page to be printed. Following this, dry ink particles are brought into contact with the drum's surface and these adhere to the magnetized spots. The ink is then pressurized onto the surface and subsequently transferred onto paper. This method is claimed to be simpler and less troublesome than other methods, largely due to the use of a dry ink.

Characteristics of page printers

Speeds: Up to 2,000 pages per minute, this is equivalent to a maximum of about 120,000 lines per minute. The effective speed depends upon various factors such as the number of lines per document and options required, such as double-sided printing.

Character sets: Typically 128 but up to 256 characters per set with a range of fonts and sizes; the character fonts are stored in the printer's storage. The print pitch tends to be more compressed than other printers, e.g. fifteen characters per inch.

Copies: Produced by reprinting the document immediately as many times as necessary. This gives good quality on all copies, and permits differing data between them. Blanked-out areas and different names and addresses on standard letters are examples.

Intelligence: Built-in microprocessors and storage permit software control, e.g. a job control language for automatically selecting the print requirements for each type of document.

Output: The high rate of throughput of documents necessitates stationery handling facilities to be built into page printers. This means that it is possible to feed rolls of blank paper into one end of the system and for printed documents in their final form to come out the other end.

4.7 SPEECH OUTPUT

Speech output is also known as audio response, audio output, voice response, voice output and voice answer-back. It is a comparatively little-used method of outputting information or instructions from a computer. None the less it has been adopted for certain specialized applications, and in the long term may find a wider variety of uses.

The concept of speech output is that computers are capable of producing speech based on a stored digital representation of either words or sounds. In the former case, words spoken by a person are recorded, then digitized and stored for subsequent reproduction. Typically, the vocabulary would be a few hundred words. In order to save storage, certain commonly used words are stored separately from the sentence they are in and replaced by a code number after the style of an attribute value table. They are inserted into the sentence just prior to its output.

Storage of digitized speech is expensive because it is necessary to store the binary value of the sound's amplitude taken at frequent intervals of time. This is called pulse code modulation and demands something like 24,000 bits per stored word. The amount of storage can, however, be reduced significantly by storing the differences between the successive amplitudes rather than their absolute values.

115

It is also possible to store digitized sounds rather than complete words and therefore create the required words by combining their constituent sounds in the correct order. This is true speech synthesis but it requires more computer resource. The payoff is in a much more flexible system which can create speech for a wide variety of needs rather than a limited vocabulary which must be pre-recorded. However, synthesized speech still sounds metallic or artificial because it is hard to create natural human intonation, although systems are improving and starting to sound more natural. True speech synthesizers (e.g. DECTALK) are widely available and have been of considerable help to the blind. Speaking rates approximating to human capabilities are possible (200 words per minute) although less sophisticated systems are slower to ensure that word boundaries are clearly pronounced by leaving small inter-word gaps not present in continuous human speech. Commercial applications of synthesized speech are spreading, especially in environments where paper and VDU output may be inappropriate, e.g. factory and manufacturing applications.

Considerations regarding speech output

Before adopting speech output, there are several factors to be taken into consideration, as follows:

■ Speech cannot be scanned, i.e. if it is not heard or understood, it is necessary to have it repeated.

■ The amount of information, e.g. the number of words, that can be stored in human short-term memory is quite low. This is particularly true for numeric data (around seven digits is the maximum that is safely remembered).

■ Speech output must have a short response time and be continuous within a grammatical phrase or sentence. Pauses are acceptable but only at semantically appropriate places.

■ The stored vocabulary must be capable of being amended so as to meet the changing requirements of the business.

■ The quality of the speech must be acceptable to the listener, bearing in mind the environment and nature of his or her work.

Applications of speech output

Speech output is not a replacement for other methods when it comes to large volumes of output but in certain circumstances it has clear advantages. One of these is when the user is working in a situation demanding intensive effort or concentration of eyes or hands, e.g. highly concentrated work, disabled persons, visually impaired persons, or in darkness or poor light.

A popular business application of speech output is salespersons phoning in their customers' orders. The salesperson attaches a small keypad to a public telephone by means of an acoustic coupler, and then dials the computer. The computer is programmed to instruct him or her step by step in regard to the data to be keyed in, and these instructions are heard through the telephone. After each data item input, the computer responds by repeating verbally the data keyed-in so as to verify its accuracy.

This is a convenient, rapid and economic means of capturing sales orders, mainly because no humans are involved at the receiving end. A large amount of order data is automatically recorded from numerous distant sources and processed to initiate the dispatch of and accounting for customers' orders.

Current uses of speech output include the prompts and messages from electronic telephone systems, e.g. advice regarding changed telephone numbers, and credit card checking by means of a telephone link between a retailer and the bank's computer.

4.8 MULTI-MEDIA INPUT/OUTPUT

CD–ROMs are based on the same technology as the audio CDs that you purchase at record shops. CD–ROM stands for Compact Disc Read Only Memory. A CD–ROM has its data stored on its surface as a series of pits. To read data an optical head senses the pattern of pits which represent binary data and translates them to electrical signals. CD–ROMs have a number of characteristics:

- They have a very large capacity. A typical CD–ROM will contain a huge amount of data, currently this is around 700 megabytes, equivalent to 350,000 pages of text.

- Like their audio counterpart they are very durable: they can be bent or dropped without affecting the data stored.

- They are inexpensive to manufacture: a matter of a few pence.

- Most CD–ROMs are read-only and can only be used to distribute data, not store it. However, this is rapidly changing (see below).

Since 1997 there has been a huge expansion in technology which enables CD–ROMs to be created by the ordinary computer user. Currently on the market are a number of hardware writers which enable a computer to write data from its hard disk to a CD–ROM. These devices, which a few years ago were marketed at tens of thousands of pounds, can now be bought for under £350.

CD–ROMs have a number of uses:

- They are used to distribute software and data. In particular multi-media software and data are usually distributed using a CD–ROM since they can occupy large amounts of space.

- They are used for archival purposes. A bank, by law, has to keep data on its customers which stretches back many years. CD–ROMs are an ideal medium for such archival purposes.

117

They are not used for on-line storage since their speed of access is low and they can only be written to once.

■ Digital video interactive (DVI)

DVI allows digital computer storage of images in compressed form, and access mechanisms to retrieve and interact with digitized images. DVI therefore allows whole movies to be stored on hard disks and to be held in the computer memory. Images are manipulated by program and changed within certain limits (e.g. display colour, contrast, speed of animation, etc.). Ordinary analogue video can be input via a special video digitizer card and stored in DVI format. Movies and still pictures can then be processed under software control in desktop publishing and multi-media software such as Quicktime from Apple. This allows considerable flexibility: thus movies can be changed in size, speed of animation, and embedded in a word processor document so that the text speaks itself while showing a picture of the author doing the speaking. All this is now possible with multi-media input/output.

EXERCISES

Exercise 4.1   **Printers**

Describe the main characteristics of each of the following types of computer printer:

(a) daisy wheel;

(b) dot matrix;

(c) laser; and

(d) ink jet.

Exercise 4.2   **Display facilities**

Describe the characteristics and purposes of each of the following:

(a) a pointer;

(b) a graphics display;

(c) an icon;

(d) a pull-down menu; and

(e) a mouse.

Exercise 4.3   **Costing of data capture**

A company needs to capture data pertaining to monetary transactions. There are 5 million transactions per annum each involving the recording of fifteen digits. Alternative methods under consideration are:

(a) key-to-disk keyed from existing source documents; and

(b) OCR handprinting by the clerks who fill in the existing source documents, with ten transactions per OCR document.

Compute the annual costs of each of the two methods, taking into account the costs of the key-to-disk operators, the data media, the off-line equipment and the on-line computer peripherals. The costs shown in Table 4.1 can be assumed (do not take these as accurate for other purposes), overheads are included in these costs.

Table 4.1

Key-to-disk systems with 6 keystations	£800 per month
Key-to-disk systems with 8 keystations	£950 per month
Key-to-disk systems with 10 keystations	£1100 per month
Key-to-disk operators	£500 per month
Optical reader	£2000 per month
Magnetic tape drive	£125 per month
OCR documents	£12 per thousand
Existing stationery	£2000 per annum

Outline solutions to exercises

Solution 4.1   For technical characteristics of the printers see Section 4.6. Most suitable business applications are:

(a) *Daisy wheel*

- Reports demanding high-quality print but in only the one font and size.
- Correspondence needing a few simultaneous copies, e.g. word processing or desktop publishing involving one-font text but no graphics.
- Preparation of OCR documents.

(b) *Dot matrix*

- Routine reports needing several fonts but only low-quality print.
- Internal memoranda incorporating variable data extracted from the database, i.e. one copy only of each memo.

(c) *Laser*

- Desktop publishing involving graphics and multi-font text.
- Variable data correspondence including logos and headings.
- Reports containing graphics.

(d) *Ink jet*

- Reports demanding coloured graphics.
- Correspondence in imitation handwriting.
- Reports needing a range of fonts and symbols.

Solution 4.2   (a) A *pointer* is an arrow on a display screen that can be moved to indicate items of interest (see mouse).

(b) A *graphics display* is a display on a computer screen involving diagrammatic information such as graphs, charts, diagrams and pictures. These are in colour and usually associated with textual material and pointers, and can sometimes be modified by the user.

(c) An *icon* is a symbolic representation on a screen indicating the current position, e.g. a pen showing that it is possible to draw a line on the screen.

(d) A *pull-down menu* is a short list of options available to the user usually under a major heading, e.g. the editing facilities in word processing software. The option is often chosen by positioning a pointer alongside the requirement using a mouse and then pressing the mouse button.

(e) A *mouse* is a small hand-held device that is operated by running along the desk so as to move a pointer on the display screen. On reaching the desired point action is triggered by pressing a button on the mouse.

Solution 4.3   Table 4.2 shows the solution.

Table 4.2

(a) *Key-to-disk – annual costs*

5,000,000 transactions × 15 digits at 10,000 keystrokes
 per hour = 7,500 operator hours
7,500 × 2 (for verifying) ÷ 1,500 hours p.a. = 10 operators

10 operators at £500 per month	£60,000
Key-to-disk with 10 keystations	£13,200
Additional magnetic tape unit	£ 1,500
	Total £74,700

(b) *OCR hand printing – annual costs*

5,000,000 transactions at 10 per document = 500,000 documents

500,000 documents at £12 per thousand	£ 6,000
Optical reader at £2,000 per month	£24,000
	£30,000
Less cost of existing stationery	£ 2,000
	Total £28,000

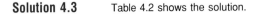

References and further reading

4.1   Angelides, M.C. *Multimedia Information Systems* (Kluwer, 1997).
4.2   Dutoit, T. *An Introduction to Text-to-speech Synthesis* (Kluwer, 1997).
4.3   Egerton, P. *Computer Graphics* (Prentice Hall, 1997).
4.4   Foley, J.D. *Computer Graphics* (Addison Wesley, 1990).
4.5   Hofstetter, F.T. *Multimedia Literacy* (McGraw-Hill, 1997).
4.6   Irwin, J.D. *Emerging Technologies in Multimedia Computer Communications* (Prentice Hall, 1998).
4.7   Miller, D. *Designing Web Multimedia* (New Riders, 1996).
4.8   Schindler, E. *The Computer Speech Book* (MacMillan, 1996).
4.9   Shuman, J.E. *Multimedia in Action* (Wadsworth, 1997).
4.10 Vaughan, T. *Multimedia* (Osborne, 1996).
4.11 Young, S. *Corpus-based Methods in Language and Speech Processing* (Kluwer, 1997).

5 Logical files and databases

AIMS

After reading this chapter you should:

- understand the concept of a file;
- understand the concept of a record;
- understand the concept of a data item;
- be familiar with the main categories of file;
- be able to perceive the differences between the main file types;
- have an understanding of the main processing that occurs on files;
- be familiar with the structure and properties of the main data structures;
- understand the defining features of a database and how it differs from a file;
- be familiar with the main functions of a database management system;
- have an understanding of the role of the data dictionary within a database management environment.

5.1 BUSINESS FILES

Business procedures, no matter how simple, call for the provision of up-to-date information in relation to the organization's suppliers, employees, products, customers, and so on. These are the 'entity sets' of the business, and each set has certain attributes associated with each and every entity in the set, e.g. a product has its selling price, labour cost and stock-in-hand. There is a host of terms associated with data structures, including many synonyms; Figure 5.1 shows the more common terms and the ways they interrelate.

In a non-computerized situation, each department in the organization has certain entity sets for which it is responsible. These are usually in documentary form and, as such, do not have the clear-cut structure associated with computer files. Nevertheless the same information about the entities is present, and is extracted and used in manual systems.

Reality	Information	System storage concept	Examples
The things with which the organization is concerned	Facts and figures about the things with which the organization is concerned	The way in which the facts and figures are held for use by the organization	A firm's customers, suppliers, employees, products, etc.
A range of entity sets	A range of entity record sets	Database or schema	All the data appertaining to the above
Entity set	Entity record set	File, logical file, record set or data store	A personnel file
Entity	Entity record	Logical record or record	An individual employee
Several closely associated properties of the entity	Group of attributes	Data item, group, data structure or data aggregate	Employee's work history
Property of the entity	Attribute	Data item, field or data element	Employee's grade
Value of property	Attribute value	Data item value	Engineer, grade 3

Fig. 5.1 Data structure concepts

■ Logical records and files

As stated previously, in business there is a natural tendency for entities to fall into sets and for each entity to have several different data items associated with it. This group of data items can be regarded as a logical record. That is to say, as far as the user is concerned the data items are always closely associated with the entity in question and can be identified by using the key of the logical record. The contrast between logical records and physical data, i.e. the way a database holds the data, is discussed in Section 5.3. For the moment we will remain with the logical aspect.

A logical record can consist of any number of different data items associated with the entity's attributes. It may also contain several repetitions of the one data item but with differing meanings, e.g. the amounts spent by a customer in each of the past six months.

Logical records appertaining to the entities within one set can be regarded as forming a logical file. The two principal types of logical record structures are described below. This section should be read in conjunction with the explanation of normalization and relational databases in Section 5.3.

File-record structures

Fixed length records

The simplest structured file is one which contains a separate record for each entity within a set, such as shown in Table 5.1. This example applies to an office furniture manufacturer's stock file and has a record for each and every commodity held in stock. Each record in the entity set is of the same length because it contains the same data items for each entity, and corresponding data items are allocated the same space within each record. This is known as a fixed length record and, if the structure or length of any one record has to be changed, then all the other records must be restructured in a similar way.

Table 5.1 Example A

Entity set	All commodities stocked, e.g. office furniture
Entity	A particular commodity, e.g. desk, 'Executive' style
Attributes of	Catalogue no., e.g. B263
entity	Selling price, e.g. £250.00
(data items)	Stock-in-hand, e.g. 12

In the example in Table 5.1 the actual data item values are shown and the record is arranged to hold the maximum value of each of these for any entity in the set. The space allowed in computer storage for the data item is usually termed a 'field', but from now on we shall refer only to data items.

SAQ 5.1 Accounts can be regarded as entity sets in a banking system. Can you think of a typical entity and attributes of that entity?

Solution

A current account is an example of an account, a savings account might be another one. For a current account typical attributes would be the number of the account, its balance and the overdraft limit.

Variable length records

The need for variable length records arises for two reasons: first, where there are differing numbers of data items associated with entities; and secondly, where the data items are allowed to occupy different amounts of space (variable length data items). These are shown in Tables 5.2 and 5.3 respectively. In Table 5.2 products do not all have the same number of constituent materials, and therefore the records vary in length accordingly. In Table 5.3, the names and addresses vary in length from one supplier to the next and, in order to avoid wasting space by allowing for the longest, the data items are allowed to occupy only the space they actually need.

A further but less likely reason for variable length records is where data items are nested within other data items. Table 5.4 makes the point, in that each product consists of several materials and each material comes from several suppliers.

If a systems analyst is confronted by logical records of this degree of complexity, it is worth making strenuous efforts to simplify them. In any event if the database is to be controlled by a database management system (DBMS), the logical files will be normalized (Section 5.3) as a matter of course.

Anticipating Section 5.3, normalizing means that, in effect, variable length records are converted into fixed length records by moving the repeated items into records of their own.

Table 5.2 Example B

Entity set	All products manufactured, e.g. gas cookers
Entity	A particular product, e.g. 4-burner cooker in white
Attributes of	⎧ Product no., e.g. 9CW 418
entity	⎨ Material no., e.g. 38051 ⎫ repeated for each raw material
(data items)	⎩ Material quantity, e.g. 15 ⎭ in the product

Table 5.3 Example C

Entity set	Suppliers, e.g. timber merchants
Entity	A particular supplier, e.g. Deal & Company
Attributes of	⎧ Supplier account no., e.g. 95026
entity	⎨ Supplier's registered name, e.g. A.G. Deal & Co.
(data items)	⎩ Supplier address, e.g. Fir Tree Ave., Woods End, Oakhampton

Table 5.4 Example D

Entity set	All products manufactured	
Entity	A particular product	
	⎧ Product no.	
Attributes of entity	Material no.	⎧ repeated for
(data items)	⎨ Material quantity	each raw
	Material supplier no. ⎫ repeated for each	material in
	⎩ Material supplier no. ⎭ supplier of the material	⎩ the product

---

SAQ 5.2 A tactical information system is used by a bank to track the loans made to individual companies. Can you think of a variable length record that can be used to store this information?

Solution

A simple variable length record would contain the name of the company together with details of each loan that has been made; for example, the amount of the loan and the date on which it was made.

■ Data items

A data item is a piece of meaningful data pertaining to an entity, i.e. it quantifies or specifies one of the entity's attributes. In this context an entity may be an artifact, person, company or other thing that is represented within the information system. A data item cannot sensibly be split into smaller pieces, and it is important to give careful thought to their structures as they form the 'bricks' of a database and might need to be specified accurately for use in a data dictionary.

As is seen from subsequent sections, data items are manipulated, structured, processed and stored in many different ways. And it is only through the processing of data items – from whatever source – that useful information can be output from a computer.

Categories of data items

It is useful to categorize data items into three main types: static, dynamic and changeable.

Static data items

These are non-changeable in their value and also largely in their layout (picture). Examples are account numbers, commodity codes and dates of birth. It would be only under exceptional circumstances that a static data item changes its value, most likely owing to an error having been discovered, e.g. a person's incorrect date of birth.

The layout of a static data item might be changed very occasionally, perhaps to bring it into line with other similar data items. An instance of this could be changing the layout of dates from the British format of DDMMYY to the American format of MMDDYY (DD = day, MM = month, YY = year).

Code numbers are generally static data items and also often are the keys of records (see 'Key data items' below).

Dynamic data items

These can expect to be updated at regular intervals as a consequence of activities taking place. A high proportion of data items are of this type and can be regarded as the lifeblood of business records. Examples include stock levels, earnings year-to-year, accounts balances and cumulative costs. In general, dynamic data items have a high level of activity although it is not impossible for a particular data item to remain the same for a long time, e.g. the stock level of a slow-moving stock item.

Changeable data items

A changeable data item generally remains the same for some time, i.e. it is not normally changed at regular intervals. Nevertheless it may be changed when

125

circumstances enforce this, e.g. a taxpayer's code when his or her allowances are changed. Other examples of changeable data items are VAT (sales tax) rates, sales prices and women's surnames on marriage.

SAQ 5.3 A bank keeps track of its customers and the current balances of their accounts. The system used for this keeps the following data items: the date of birth of the customer, the name of the customer, the current balance of an account and the overdraft limit of the customer. Which of these are static data items, dynamic data items or changeable data items?

Solution

The date of birth would be static, the name of the customer would be changeable (a female customer might marry and change her name), the current balance is a dynamic data item since it would change regularly and the overdraft limit would be a changeable data item.

Key data items (keys)

A key is an identifier of a group of data items, i.e. a logical record is identified by its key. Strictly speaking any data item can act as a key at a certain point in time. It is, however, generally code numbers, such as account numbers and commodity codes, that form keys.

A key may be formed from two or more concatenated data items if this is necessary for unique identification. Thus a stock record could need both the store number and the part number to give full identification. If only one of these two data items were present, only partial identification of the stock would be possible, i.e. either where it is or what it is.

A key might not necessarily need to provide absolute identification but only to the extent of the record being one of several or many similar ones. This would be true in the above example if we were interested solely in what stock we have regardless of its whereabouts; the part number would then suffice.

It is also possible for only a portion of a data item to form the whole or part of a key. For instance, the first letter of a UK vehicle's registration mark denotes its year of registration.

Designatory data items (designations)

A designation, otherwise known as a symbol or flag, serves to indicate the status or a characteristic of the entity, the record or the data item. Normally the designation needs only a small amount of space, e.g. one digit, and it is interpreted by the computer program so that the record or data item is processed in the appropriate way. See Examples 5.1–5.3.

Example 5.1   **An entity designation**

Commodities might either be manufactured by the company, purchased from a home-based supplier, or imported. These categories could be recognized by means of an entity designation in each record, with a value of 1, 2 or 3 respectively.

Example 5.2   **A record designation**

A record designation applies to the record itself and not to the entity. If we wish to know whether a record has been referred to during a processing run, the value of a record designation is changed when this happens, e.g. from 0 to 1, so that a subsequent check can be made. An example of this could be in order to detect which commodities have had their prices looked up during invoice printing.

Example 5.3   **A data item designation**

The quantity of a material might be measured in different units from one material to another. Consequently a 'unit of measure' designation is useful, e.g. 1 = units, 2 = metres, 3 = kg, and so on.

Example 5.4   **A data item status designation (field designation)**

Analogously to the record designation, the data item status designation applies to the field and not to the data item itself. Such a designation could, for instance, indicate whether the stock level of a commodity was changed during a stock updating process.

SAQ 5.4   A bank keeps details of each account it holds; accounts are associated with the name(s) of the account holders, the account number and the current balance of the account. Which of these can be used as a key data item?

Solution

The account number, assuming it was unique, would be used.

Characteristics of data items

An important aspect of a data item is its layout or 'picture'. The layout is its structure in terms of the constituent digits, letters or symbols. It is convenient to represent this in a similar way to that used in COBOL. Thus a digit is represented by a '9', a letter (alphabetic character) by an 'A', and either of these or a symbol, e.g. an asterisk, by an 'X'. So a recent UK vehicle registration mark has the picture A999AAA. To save space in writing, a long picture is abbreviated by the use of a bracketed number, e.g. A(30) means a name of up to 30 letters in length.

Another characteristic of a data item is the range of values within which it lies, and of course during processing it is the actual value of a data item that contributes to the result. In this context, value means any quantifiable or logical measure of its characteristic. Range means the span from the minimum possible to the maximum possible. For instance, a data item could be the commodity selling price, the values of which range from £10 to £50. Further explanation of data item characteristics is given under data dictionaries in Section 5.3.

Categories of IS files

There are several ways in which IS files can be categorized but we are here concerned with the usage aspect rather than the technological aspect. The four file usage categories are transaction (movement) files, working files, spool files and master files.

Transaction files

A transaction file is a logical file of records that relate to events or activities, e.g. orders received, jobs completed, goods dispatched. Such files are created at regular intervals geared to the organization's cycles of activities, e.g. payments received each day. Data such as this is known as source data because it derives from the source of the activity.

Quite often a transaction file is the output of a key-to-disk or other data capture system, and although by this stage it may have undergone a small amount of processing, it is still essentially the original source data.

One of the main purposes of a transaction file is to update or amend a master file, e.g. customers' payments update a sales (debtors') file. Another use for a transaction file is in order to analyse source data either entirely on its own or in conjunction with data read from another file, perhaps a master file. An example of the latter is a cost of sales analysis in which the sales quantities are in the transaction file and the unit costs are taken from a master file in order to compute the cost of each sales analysis total.

It is quite usual during the course of business activities for transaction data to stem from different sources or arrive at various times. This generally introduces a need for some storing and/or merging of several transaction files before the true processing commences. An example could be stock transactions arising from several stores needing to be combined before updating a common stock master file.

After being used in these ways, a transaction file has served its purpose and is no longer needed except for data security purposes.

Working files

Any temporary data which has to be stored is placed in a working file. This can be the results of a calculation, temporary pending files, partially processed data or when any data has to be stored while being exchanged between separate systems.

Spool (output) files

Spooling is the process of writing data to a magnetic medium so as to form a temporary file prior to printing or, alternatively, creating a copy of source data before processing it. These processes are necessary when the relevant units, e.g. printer or processor, are not in a position to accept the data. Spooling also means that a copy of the output or input data is available for future reference. Spool files are normally disposed of after a short time.

Master files

A master file is a set of records relating to things that exist, i.e. entities, and consequently is of a permanent nature. Its constituent records must be maintained in an up-to-date state so that they can be used to provide accurate data for processing.

It should be pointed out that a master file is a logical concept, and that within a database its physical structure may be entirely different from the way the user regards it. This is of no great consequence to the system designer, who, at this stage, should think of a master file as existing in logical form.

There are three main aspects that are applicable to master files: referencing (interrogating), updating and amending.

SAQ 5.5 A system for keeping track of the sales from a department store processes a file of sales made in a working day. This file is used to modify a file which contains details of sales made in the current year. Which of these files is the transaction file and which is the master file?

Solution

The first file is the transaction file and is used to update the master file which is the second file.

Referencing a master file

This process implies that the computer refers to certain records and makes a note of all or some of their contents without making any changes to the records. This is a usual procedure and referencing is the *raison d'être* of master file records. A given record may be referenced thousands of times during a single processing run, e.g. looking up the selling price of a popular item during invoicing. It is an obvious corollary that a master file needs to be up to date before being referenced.

Updating a master file

This means that each master record, as and when necessary, is brought into an up-to-date condition. In a payroll routine, for instance, each employee's record is updated as regards his or her 'earnings to date', 'tax paid to date' and so on, prior to being used for preparing his or her payslip.

With fixed-length records, updating entails altering the values of the data items in a record without changing its length. Updating variable length records tends to change their lengths because new data items are inserted and others are removed. This would be so for example if an additional material were introduced into a product. This change to the length of a logical record is more apparent than real if a DBMS is in use. The DBMS caters for the change but without necessarily needing actually to rearrange the record itself.

Activity ratios

Several ratios are measures of the amount of activity of a file. These ratios can be used to estimate the times taken to process the file. They are not defined absolutely, however, and various names have been applied to them. The reader should not be disturbed if he or she finds different names in other books.

The ratios defined below are normally in relation to one processing run. They would, however, be just as usefully applied over a period of time provided the comparisons cover periods of the same length.

A record is described as 'accessed' every time it is referenced or updated, i.e. transferred from backing store to main store.

The principal ratios, designated (A) to (C), are defined below:

$$\text{(A) Record hit ratio} = \frac{\text{Number of different records accessed}}{\text{Number of records in file}}$$

This ratio is never greater than one.

$$\text{(B) Hit records activity ratio} = \frac{\text{Total accesses to the file}}{\text{Number of different records accessed}}$$

This ratio is never less than one (assuming some activity).

$$\text{(C) Record activity ratio} = \frac{\text{Total accesses to the file}}{\text{Number of records in the file}}$$

This ratio can be any value above zero (assuming some activity).

- Ratio (A) gives a measure of the proportion of live records in the file. A low ratio indicates that many dead or dormant records exist.
- Ratio (B) provides a measure of the level of activity of the live records.
- Ratio (C) measures the average activity of the records. A high ratio means the file as a whole is busy but gives no indication regarding the pattern of this activity within the records.
- Ratio (A) × ratio (B) = ratio (C).

See Example 5.5 for examples of using the above ratios.

Example 5.5 | **Activity ratios: A, B and C**

A commodity file comprises 5,000 records. During an invoicing routine 4,000 of the commodities have their prices referenced and this occurs a total of 8,000 times.

(A) Record hit ratio = $4,000 \div 5,000 = 0.8$

(B) Hit record activity ratio = $8,000 \div 4,000 = 2.0$

(C) Record activity ratio = $8,000 \div 5,000 = 1.6$

Thus it is apparent that a fair proportion of the commodities are being sold but that the sales are thinly spread over the commodities within one invoicing routine, i.e. an average of only 1.6 sales per commodity.

Amending a master file

This means the insertion of new records, the removal of obsolete records, and changes to essentially static records. Amendment of a file is also referred to as file maintenance or housekeeping, but these terms also include the general tidying up of the file. The extent to which amendments are made depends largely upon the nature of the entities represented in the file.

Master files tend to fall into one of two categories: semi-static and volatile. Semi-static files are those that are really intended to remain in the same state for some considerable period of time. Volatile files are those that have a high volume of insertions and deletions. A typical semi-static master file is a commodity price file appertaining to a stable range of goods. A good example of a volatile file is a jobs-in-progress file; the jobs therein are perpetually starting and being completed.

It is useful in some situations to have ratios for amendments, e.g. a volatility ratio to give a measure of the amount by which a file is changing:

(D) Volatility ratio $= \dfrac{\text{Number of insertions} + \text{number of deletions}}{\text{Initial number of records in the file}}$

This ratio can have any value from zero upwards.

$$\text{(E) Expansion ratio} = \frac{\text{Number of insertions} - \text{number of deletions}}{\text{Initial number of records}}$$

See Example 5.6 for examples of using the above ratios.

The expansion ratio is a measure of growth of the file and may be any positive or negative value or zero.

Example 5.6 | **Activity ratios: D and E**

During a certain time period the commodity file in the previous example has another 750 commodities inserted into it and 250 obsolete commodities removed.

$$\text{(D) Volatility ratio} = \frac{750 + 250}{5,000} = 0.2$$

$$\text{(E) Expansion ratio} = \frac{750 - 250}{5,000} = 0.1$$

Thus the commodity range is quite volatile and tending to increase in size.

■ File processes

The main processes carried out with files are described below; these apply not only to master files but also to transaction and working files.

Sorting

During sorting, the records in a logical file are brought into a sequence as determined by a key in the records. A computer is capable of sorting records into a nested sequence, e.g. to employee number within department number within factory number. This entails multi-key sorting, i.e. the computer looks for several keys in each record and, in effect, treats them as a concatenated key.

In practice, sorting is often controlled by a 'sort generator'; this is part of the operating system software and comprises several sorting techniques called into use according to the parameters of the file and the sorting requirements.

The need for sorting has diminished with the demise of magnetic tape as backing storage. This is beneficial because sorting is time consuming and therefore expensive in terms of computer usage. Nevertheless there is always likely to be residual need for sorting owing to the wide variety of sequences demanded by business information. More is said about sorting in Section 6.3.

Merging

Merging implies that two or more files in the same sequence are combined into one file. There are two main aspects of merging.

File merging

Two or more separate files of similar records and in the same sequence are merged together so as to form one file. The records in the 'input' files need not necessarily be a one-for-one match, nevertheless the 'output' file holds all the input records. An example of file merging is the bringing together of sales records from the various branches of a firm, there being initially one file per branch and finally one file for the firm as a whole.

Record merging

The corresponding records from two or more 'input' files, usually in the same sequence, are combined into one record in the 'output' file. It is usually the case that the input records all match according to some key, and non-matching records are rejected. The input files need not necessarily contain similar records. An example is the combination of stock records with cost records on a one-for-one basis prior to preparing a stock valuation report.

An instance of non-matching files is stocks of parts in several stores; one part may be in all the stores, another in only one store.

Matching

Two or more input files, generally in the same sequence, are compared record against record to ensure that there is a complete set of records for each key. Mismatched records are highlighted for subsequent action.

An example is a file of purchase orders matched against goods-received notes, presorted into order number sequence.

Summarizing

Records with the same key in one file are accumulated together to form one record in the 'output' file. Summarizing usually applies to a file presorted into a certain sequence and the resultant file is in the same sequence. It is, however, possible to summarize a non-sequential, e.g. random, file if it is stored on a direct access device. Records to be summarized are generally of a similar type. An example is the labour costs of a work-in-progress file in which all the costs of each job are summarized together.

Searching

This entails looking for records with certain keys or holding certain data and in some way making a note of these. There may not, in fact, be extraction of data from the records but merely a count made of their characteristics. Several different counts or extractions may be made during the one search. An instance is a search for and count of all ledger records with a debt balance of above a certain amount and the totalling of these balances.

SAQ 5.6 In a purchasing system customers telephone their orders to a sales clerk. The sales clerk enters the details of their order. Each item that has been ordered is then checked against a file of items which are stored in a warehouse. If there are enough items in the warehouse the order is accepted. At the end of the day a file of orders is added to a file which contains orders for the current financial year. Which of the operations, merging, sorting, matching, summarizing and searching, are involved in this description?

Solution

The two operations are searching (the items are searched for in a file holding warehouse details) and merging (two files are merged together).

5.2 DIRECT ACCESS FILE ORGANIZATION

In this section we refer to a cylinder and a block. The former is a collection of data that can be accessed at the same time with the same physical operation, the latter is a collection of contiguous records.

■ Storage and access modes

There are three principal modes for storing and accessing records on a disk or in memory: serial, sequential and random.

Serial mode

The records are stored contiguously regardless of their keys (if any). The sole way of accessing serial records is to search through the complete file starting with the first record.

A serial file is normally of a temporary nature perhaps awaiting sorting and is likely to hold transaction records. A serial file may be keyless and there are a few instances of this in business.

Direct access sequential mode normally involves accessing sequentially a file that is stored sequentially. The search for a given record continues onwards from the previous record accessed, examining each record in turn. This method is efficient for high-activity but not low-activity files. This is because the transaction records need to be sorted, which is time consuming, and little advantage is gained from continuing the search onwards if there are large gaps between the records accessed.

Sequential mode is often associated with a master file held in a certain sequence and updated by a transaction file sorted into the same sequence, a typical example being the updating of stock records held in part-number sequence by stock issues and receipts.

SAQ 5.7 Would you use serial mode for the storage of the customer records in a bank?

Solution

No, this mode is normally used for temporary storage; details of bank customers are permanent.

Indexed-sequential mode

Indexed-sequential is a mode of storage in which records are held sequentially and accessed selectively, i.e. selective-sequential access. Groups of unrequired records are skipped past; this procedure is facilitated by the use of one or more indexes that are consulted when searching for a particular record (see 'Indexed-sequential searching').

If the accesses are known to be in the same sequence as the file, the search for a record can continue onwards from the previous record accessed, as with sequential mode. A separate smaller file is maintained as the index. This contains the record key and a pointer to its physical location. The index is searched sequentially or in some systems more intelligently as the index numbers are ordered so that the computer can guess roughly where an index record may be. Once the index record has been found the pointer is used to retrieve the data.

Problems arise when new records have to be added to the file, especially if the new record happens to be in the middle of a stored order. This means that the index has to be reorganized. The data record file may then be reorganized to reflect the new order by shuffling all the old records down one location and inserting the new record into the gap. Alternatively the old data file is left as it was and the new record is placed in an overflow area. The index points to the location of the record in the overflow area. The method, generally called the virtual sequential access method (VSAM), decouples the actual storage location from the order of records as reflected in the index. However, as more updates occur the index and the data file become progressively out of tune (i.e. overflows for overflows are required). File maintenance routines have to be run to reorder the file in line with its index.

Inverted files

These develop the indexing idea by storing more data about data. In other words not only is the record key stored on the index but also coded representation of

135

the real data itself. Inverted files can therefore be searched on values and properties of data items even if the key is not known. The fields on the inverted file usually contain a code to compress representation of the real data. For instance in a car sales recording system, the make of car may be given one code, e.g. F = Ford, M = Mercedes, R = Renault, the colour of the car another, and availability in each dealer location yet another. The coded data items, usually called facets, can be searched in a variety of ways to locate the record which matches the input search values. Inverted files are used extensively in databases where access to records has to be flexible. The penalty paid is the extra effort in coding the data itself into the inverted file index (see inverted lists and files later in this section).

SAQ 5.8 An airline receives a variety of enquiries about the flights that it schedules, for example 'Is there a flight to Amsterdam after 18:00 hrs. from Manchester?' Would an inverted file be a good choice for this type of application?

Solution

Yes, because the search occurs on a number of values and properties.

Random mode

This mode means that each record is stored in a location determined from the record's key by means of an address generation algorithm (AGA). There is no obvious relationship between the keys of adjacent records, and the only efficient way to find a record is to use the AGA. The essence of the method is to calculate the physical address of the record from one of its data items, usually the record key. The AGA takes the record key as input and outputs the record address. The main problem with this method is to ensure no duplicate addresses occur. Furthermore random access can be wasteful of storage as some disk locations (e.g. block, cylinder) may not be occupied.

Random mode is applicable to master files and is unlikely to be of any relevance to transaction files. Thus a master file stored randomly could be updated by transactions in no particular order. Random mode is useful with low-activity files because it is possible to go fairly directly to a wanted record.

The main advantages of random mode are:

■ No index is required, thus saving storage space.

■ It is a fast access method because little or no searching is involved.

■ Transactions do not need sorting, thus saving time.

■ New records are easily inserted into the random file provided they are not excessive in number.

The main problem with random mode is in achieving a uniform spread of records over the storage area allocated to the file. This problem is exacerbated if the file is volatile because the AGA tends to become less efficient when it has to deal with new patterns of keys (see 'Randomizing').

Processing and update modes

When amending or updating files held in a direct-access device, there are two fundamentally different ways that this is done: the reconstruction mode and the overlay mode.

Reconstruction mode

Reconstruction mode means that the complete original file after amendment or updating is written back to an entirely different storage area. The original records are thus preserved in their original locations and so at the end of the processing two complete sets of records exist.

Reconstruction mode is practical only with sequential or indexed sequential files for which all the updates or amendments to a record are done together. The new record is then written to the reconstruction area after it has been fully updated. If a master record is updated randomly, the next update must be applied to the record in the reconstruction area. This introduces the problem of finding it unless the reconstructed file is an exact facsimile of the original: this is not always possible if the records expand or increase in number.

At the end of the updating or amending run it is necessary to copy across the unaltered records so that finally a full set is in the reconstruction area. An alternative approach with sequential processing is to copy the unaltered records during the updating run. This means that wherever a gap occurs between two consecutive transaction records, all the intervening master records are copied as they stand.

Reconstruction mode has a security advantage in that, in the event of trouble, the processing run is easily repeated because the original file is still intact. Another advantage in the case of sequential and indexed sequential files is that if records expand during updating, either in number or size, they can be accommodated in the reconstruction area without difficulty, whereas this is not possible if the records have to be written back into their original storage area.

Overlay mode

This mode is usually applicable to a file of fixed length records that are being updated. The updated records are each written back to their original locations, and it is therefore quite possible that a record is overwritten many times during the course of an updating run.

This method has an inbuilt security hazard owing to the possibility of losing or damaging the original records. It also proves difficult if there is any expansion of records since they cannot then be accommodated in their original space, and so overflow methods are then needed. The security problem is minimized by either making a complete copy of the master file before updating or dumping copies at intervals during the updating run.

Reconstruction mode and overlay mode can, of course, be amalgamated within the one process – whatever arrangement gives best operational efficiency and data security.

■ Direct access addressing

The key of a record not only identifies it but also determines its storage location, i.e. address, either in memory or on disk. In some cases it is possible to form a unique address for each record. In other cases this is not practicable and so the key gives either a shared address or an approximate address for the record.

Self-addressing

Self-addressing is a straightforward method because a record's address is equal to its key's value, e.g. the record with key 458 is stored in address 458. Thus the file is inevitably stored in key sequence. The advantages of self-addressing are:

■ it leads directly to the wanted record;

■ no indexing or searching is required;

■ the key itself need not necessarily be held within the stored record, although it generally is.

The reason for the latter point is that the key is easily computed from the address, and so once a record has been found, its key is easily deduced. This need arises when a file of records without stored keys is being printed along with the key of each record.

The drawbacks to self-addressed storage are:

■ The records must all be of the same (fixed) length or, if they are of variable length, the storage space per record has to be sufficiently large to accommodate the longest record. Alternatively, variable length records can be split into sub-records chained together (see Chains later in this section).

■ Where records are missing from a set, i.e. those keys are not in use, the corresponding storage locations must be left empty in order to preserve the relationship between the key values and the records' locations.

SAQ 5.9 A bank keeps the details of all the transactions made by a customer during his or her life with the bank. Would self-addressing be used in such a file?

Solution

No, each record associated with a customer would be of variable length with some customers being associated with very large numbers of records.

Self-addressing with key conversion

This method is basically similar to self-addressing except that the key requires a little processing to turn it into the record's address. This leads to either a precise address, i.e. the record's position within a block, or an approximate address, i.e. the block only. Normally the former is possible as follows where the block no. is same physical location:

b = required block no.
p = record's position within the block
k = record's key value
k_0 = lowest key value in the entity set
r = number of records in each block
b_0 = number of first block in storage area in use.

Then $b = \left[\dfrac{k - k_0}{r}\right] + b_0$ where [ ] means the integer value of the quotient, and

p = the remainder from $\left[\dfrac{k - k_0}{r}\right]$ plus 1.

See Example 5.7 for an illustration of this method.

Example 5.7

Self-addressing with key conversion

Suppose we have 4,000 records with keys 12,000–15,999 held 10 to a block from block no. 1,500 onwards. The problem is to find the location of the record with key 13,874.

Then $b = \left[\dfrac{13{,}784 - 12.000}{10}\right] + 1{,}500 = 1{,}687$, with remainder 4, thus

$b = 4 + 1 = 5$

The required record is therefore the fifth in block no. 1,687.

Ordinal numbers

If gaps occur in a range of keys, as is quite usual owing to entities having become obsolete, the records may still be stored contiguously in order to save space. Each

139

key must first be converted into its ordinal number, and this is then employed to determine the record's address by using self-addressing. The ordinal number of a key is, in effect, the ordered position of the record in a sequence (see Example 5.8).

Example 5.8	**Ordinal numbers**

An entity set has keys 1,001–1,200, 1,351–1,600 and 1,870–1,950. Thus there are two gaps equivalent to 150 and 269 keys respectively, and an initial 'gap' of 1,000 keys.

To calculate the ordinal number of a key:

subtract 1,000 from keys 1,001–1,200 inclusive,
subtract 1,150, i.e. 1,000 + 150, from keys 1,351–1,600 inclusive,
subtract 1,419, i.e. 1,000 + 150 + 269, from keys 1,870–1,950 inclusive.

Thus key 1,915 becomes ordinal number 496, and its record's address is then found by using either self-addressing or self-addressing with key (ordinal number) conversion.

Matrix addressing

In some circumstances it is necessary to find the address of a record held within a multidimensional matrix of records, e.g. a table of sales records covering 400 products sold in twenty sales areas during each of twelve months. The above table therefore holds $400 \times 20 \times 12 = 96,000$ sales records. To find the address of a record, say, of product no. 1328, in area no. 13 for the seventh month, we must convert each of these to ordinal numbers, as described above.

If the product's ordinal number is P, the area's ordinal number is A, and the month's ordinal number is M, the record's position within the matrix (table) is determined from the equation:

$$(P - 1) \times 20 \times 12 + (A - 1) \times 12 + M$$

This equation assumes that the records are organized in months within areas within products. A different but basically similar equation is used if the table is organized in another way, e.g. areas within products within months.

Example 5.9	**Matrix addressing**

Suppose product no. 1328 turned out to be the 151st, i.e. its ordinal number $P = 151$, and $A = 13$, and $M = 7$. Then the required record is in position $(151 - 1) \times 20 \times 12 + (13 - 1) \times 12 + 7 = 36,151$ within the table.

This type of problem is quite common in business applications and may, in fact, extend to more than three dimensions.

140

Randomizing (hashing)

Randomizing applies to randomly stored files, as mentioned previously. The aim of randomizing is to obtain an even spread of records between the available blocks. This is attempted by using an algorithm (AGA), i.e. a randomizing technique, to operate on the keys to convert them into pseudo-random numbers. The pseudo-random number is then used to form the address of the record by one of the above methods. With randomizing it is probable that an address is shared because the AGA is bound to produce synonyms, i.e. identical pseudo-random numbers, for different keys. An address is normally that of a block and so several records are assigned to the same block.

A variety of randomizing techniques have been devised, just two of which are explained below. Since the object of randomizing is to obtain as even a spread of records as possible, two or more techniques may have to be amalgamated to achieve this. The designer should test the various methods allowing for the range of existent keys and number of available blocks.

If the spread of records is uneven there is the possibility of a block becoming too full and 'overflowing'. This is catered for by one of the overflow techniques explained later in this section.

Prime number division

The key is divided by the largest prime number below the number of available blocks. The remainder from this division is the relative block number, i.e. the number of blocks after the first. If the key values in a set have no bias in their end digits, non-prime numbers can be used as divisors, but the keys need careful checking first. Prime number division is a good method for use with sets of keys that have biased values but it is safer to analyse the remainders before adoption nevertheless.

Example 5.10	**Prime number division**

If 2,000 blocks are available, the prime number is 1,999. Key 27,859, for instance, when divided by 1,999 gives a remainder of 1,872, and so this is the relative block number for the record with this key.

Extraction

The digits of the key that have the most random values are extracted from it and, after adjustment, form the record's block number. The most random digits can be found by analysing the set of keys prior to setting up the digit extraction algorithm. If digits are taken from the end of the key, it is termed 'truncation'.

Example 5.11 ### Finding the block number by extraction

The most random-valued digits in a set of keys are found to be the second, third, fifth and seventh; 3,000 blocks are available.

Key 5,927,016 would therefore become 9,206; this is then multiplied by 0.3 in order to bring it within the number of blocks, i.e. $9,206 \times 0.3 =$ block no. 2,762.

■ Direct access searching

Whereas addressing determines the location of a record by using algorithmic methods, searching finds the record by scanning groups of records or an index, or both. The simplest method is to examine every record in turn until the required one is found (serial searching). This is prohibitively expensive of time for most files and so a short-cut is generally desirable.

Indexed-sequential searching

This method was mentioned previously and a more detailed explanation follows. Assume we have a disk split up into cylinders and blocks. A cylinder index is created (Figure 5.2) to hold the highest key in each cylinder – of which there are four in this example. Associated with each cylinder is a block index holding

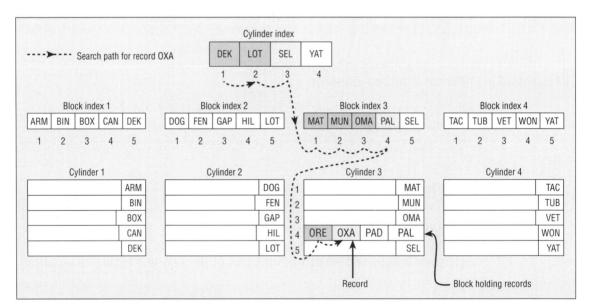

Fig. 5.2 Indexed-sequential searching

the highest key in each block within that cylinder – there are five blocks in each cylinder in the example in Figure 5.2.

When searching for a record's key in the index:

1 The cylinder index is examined key by key until one is found that is larger than or equal to the wanted key; this directs the search to the appropriate block index.

2 The block index is similarly examined, and the search directed to the appropriate block.

3 The block is searched record by record until the wanted record is found.

The records can be of variable length and need not necessarily be in sequence within a block since all records therein are examined in turn. In practice they tend to be in sequence and so other searching techniques, such as binary searching (see below), may be adopted for searching the block if it contains a large number of records.

The example in Figure 5.2 shows the search path for the record with key OXA.

SAQ 5.10 A company has a number of applications, some of which require reports on their customers to be produced in alphabetic order and some applications require details on an individual customer to be retrieved very quickly. Is an indexed sequential organization a good idea for a file which supports the company?

Solution

Yes, this form of organization enables a record-by-record form of processing to take place and also allows fast access to individual records.

Binary searching (binary chopping)

This technique may not always be suitable for searching through actual records since they are generally too widely spread in backing storage. It is, however, eminently suited to searching an index held in the main store. The indexes used for indexed-sequential are amenable to binary searching since in reality they are much longer than in the example of Figure 5.2.

The keys in the index to be binary searched must be in sequence and form a complete set. As shown in Figure 5.3, the search starts at the midpoint of the index and then moves half-way to the left or right (down or up) depending upon whether the wanted key is less than or greater than the mid-point key. The key at this half-way point is examined and each move from then on is to the half-way point of the remaining keys. The search ends either when the wanted key is found or when no more keys remain to be sensibly examined.

Figure 5.3 shows the search path for key 81 from among fifteen keys in the index. This example entails the maximum number of examinations of fifteen

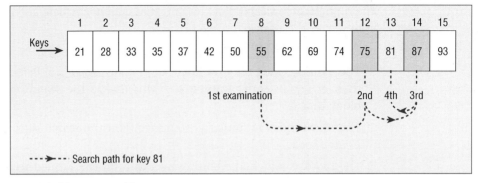

Fig. 5.3 **Binary searching**

keys, i.e. four, and it can be seen that each move is half the length of the previous one.

In practice, the index is unlikely to be as convenient as this example because it is not always possible exactly to halve each successive move. Complete exact halving is possible only when the total number of keys in the index is $2^n - 1$, where n is any integer (4 in the example). The way round this problem is either to insert dummy entries to make up the requisite number, or to simulate dummy entries in the binary search program.

The average number of examinations (comparisons) is $(\log_2 k) - 1$, where k is the number of keys in the index. This is immensely less than with a straight serial search of an index. For instance, if $k = 1,000$ the average number of binary comparisons is about nine as compared with 500 for serial searching.

SAQ 5.11 A manufacturing application keeps data on the number of different products manufactured by a company in random order. The details are stored in a file. Would a binary search be a good way of searching this data?

Solution

No, the data should be ordered and also should really be stored in main store.

Block searching

In this context a block is a sub-division of an index (not necessarily associated with a disk). A block is devised to contain roughly the square root of the number of keys in the whole index, e.g. an index of 900 keys would be sectionalized into thirty blocks of thirty keys each. It can be demonstrated mathematically that square-root blocks minimize the total number of index examinations needed to find a key.

The search is first through the block index to find the appropriate block and then through the records in the block. The average number of examinations is

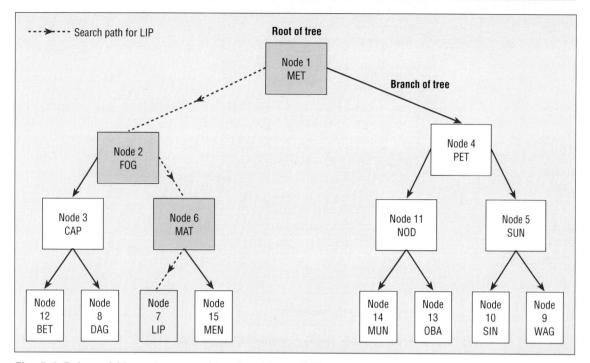

Fig. 5.4 Balanced binary tree searching (see Figure 5.5)

$\sqrt{k}$, where k is the total number of keys, i.e. $\sqrt{k}/2$ for the block index and the same for the block itself.

Balanced binary tree searching

A binary tree is a relationship of keys such that the examination of any key leads to one of two other keys. This process is illustrated in Figures 5.4 and 5.5 for a balanced tree, i.e. one with a balanced spread of its branches. The binary tree is actually in the form of an index (Figure 5.5) containing all the keys together with a directory showing the branches stemming left and right from each key.

Binary tree searching is suitable for an unordered, i.e. non-sequenced, file because the keys are chained together by the directory. The search is similar to binary searching in that each key examination halves the remaining keys on average. When creating the file and index, the records should be inserted haphazardly as this tends to create a balanced tree and hence minimizes the average number of examinations per search. That this is so is apparent if one constructs a tree for a string of keys in sequence.

Figures 5.4 and 5.5 show the search path for key LIP, i.e. via nodes 1, 2, 6 and 7. Having found the node containing the wanted key, the node number is easily converted into the record's address.

Node	Key	Left node	Right node
1	MET	2	4
2	FOG	3	6
3	CAP	12	8
4	PET	11	5
5	SUN	10	9
6	MAT	7	15
7	LIP		
8	DAG		
9	WAG		
10	SIN		
11	NOD	14	13
12	BET		
13	OBA		
14	MUN		
15	MEN		

Fig. 5.5 Binary tree index (corresponding to Figure 5.4)

An advantage of a tree index is that new keys do not necessitate the rearrangement of existing keys. Similarly, deletions can be made without leaving too many redundant nodes since these are eventually refilled with new keys. A new key is eligible for inserting into a node provided it is greater than all keys to its left and less than all keys to its right. Thus if the key MAT were deleted, it could be replaced by any key between LIQ and MEM inclusive.

A deleted key cannot be removed from the index without being replaced as this would destroy the continuity of the search. Instead a marker is inserted into the node to indicate its redundancy and invite later replacement by a qualifying new key.

SAQ 5.12 How should Figure 5.5 change if a left node is associated with node 12? This node will have the key YHI.

Solution

A new entry with a node number 16 is created. Its key would be YHI and the left node part of node 12 would contain the number 16.

Unbalanced tree searching

An unbalanced tree gives preference to the more commonly used keys by siting them in the nodes nearer to the tree's root. They are thus found more quickly during a search but the tree is inevitably unbalanced as a consequence. The simplest method of achieving this priority is simply to load the records into the file, and consequently the keys into the tree index, in descending order of frequency of usage.

Overflow methodology

The methods of addressing and searching for records explained in the previous sections merely mentioned 'blocks', the assumption being that a record is accommodated in its 'home block'. The home block of a record is the one in which it is normally housed. If, however, it cannot be accommodated therein, it is directed to an 'overflow block' by one of a number of overflow handling methods.

Causes of overflow

With a sequential file the reasons for overflow are:

- A variable length record expands during updating so that it can no longer be accommodated in its home block.
- A new (fixed or variable) record needs to be inserted into its home block but this has insufficient space.

With a random file, however, overflow occurs for another, fundamental, reason. When records are distributed by a randomizing technique it is a fact of statistical probability that, even though the algorithm is perfect, there is a maldistribution of records between blocks. This is analogous to a roulette wheel for which some compartments receive more balls than others. A disk's blocks are equivalent to the roulette wheel's compartments, and the records to the balls.

Handling overflow

Since overflow is unavoidable, we must be prepared to deal with it. The simplest method, if the overflow is minimal, is to accommodate an overflow record in the block following its home block. The search then starts at the home block and continues through successive blocks until the record is found. This method soon falls down when the amount of overflow increases because the blocks become jammed with each other's records, so aggravating their own overflow.

Overflow chaining

Specific storage areas are assigned as overflow blocks either within the same cylinder as the corresponding home blocks (cylinder overflow area) or as one area for all overflow (independent overflow area). The former method has the advantage of time saving because no arm movement is involved in going to the overflow area. The latter method gains by saving storage space because cylinder overflow areas tend to be left partially empty.

Whichever method is adopted, the computer must be directed to the overflow block. This is achieved by incorporating the number of the overflow blocks within each home block.

■ Data structures

There is a considerable range of data structures with many variations on the main themes. They tend to be used more in system software and database management systems than in user-designed systems. Nevertheless the IS student should be aware of the meanings and advantages of the main data structures.

Chains

A chain is a string of records each of which is connected to another record by means of a pointer. The pointer is held within the record and provides the address of the next record in the chain. From the application programmer's point of view the chained records are contiguous because the system software finds the whole group of logical records related to the first one demanded.

Chains are useful when logical records vary in length because they can then be made up of chained fixed-length physical records (as mentioned for self-addressing). Chains are also advantageous for linking logical records with similar characteristics, e.g. the 'hinges' in Figure 5.6.

The extent to which the chaining of logical records is viable depends largely upon the volatility of the file. Deletion of records poses problems owing to the need to change pointers, and this is particularly awkward when chains intersect, i.e. contain a record common to more than one chain. File amendment problems are eased if a chain is two-way so that both the preceding and succeeding records have a pointer in the record.

SAQ 5.13   Would a chain be a good choice to organize data about a customer of a bank and the transactions that he or she makes?

Solution

Yes, because chains can be made up of a variable number of fixed-length physical records they can be used for variable data.

Rings

A ring is a string of records linked by pointers in a similar way to a chain but with the final record pointing back to the head of the ring. Rings may be either one-way or two-way, and many intersect with other rings.

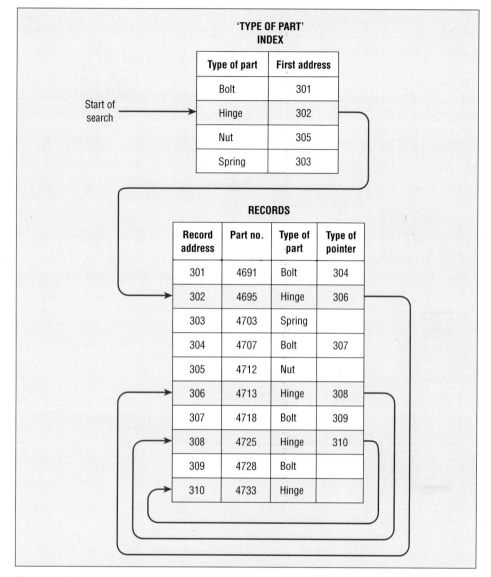

Fig. 5.6 Chained logical records showing linkage of 'hinge' records

The two-way ring in Figure 5.7 could apply to a ledger account, the head of the ring being the account identification and the linked records being transactions pertaining to that account. Although not shown, this ring could intersect with another ring applying to some other attribute of the transactions.

As compared with chains, two-way pointers are more secure since a break in the ring caused by a damaged pointer is overcome by going around the ring the other way. The records in a ring may also all be connected by pointers to the head of the ring. This is not shown in Figure 5.7 for the sake of clarity.

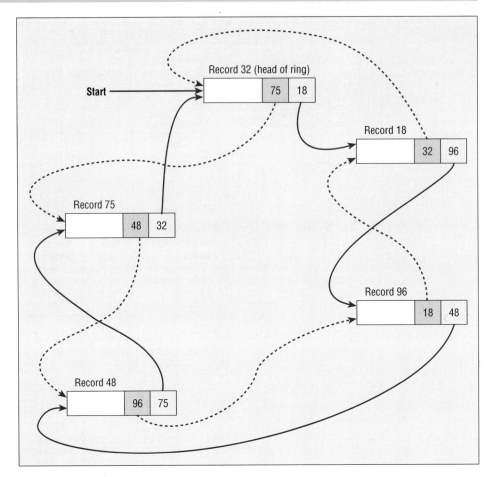

Fig. 5.7 Two-way ring

Trees

A tree consists of a hierarchy of nodes, the highest of which is the tree's root (the tree can be regarded as upside down). A node (element) may be a single record, a group of records or any other storage concept. Figure 5.4 is a simple tree structure. The main characteristics of a tree is that no 'child' (node at one level) has more than one 'parent' (linked node at the level immediately higher). A parent can, however, have any number of children – not just two as in the binary tree.

Network (plex) structures

A plex structure has its elements (nodes) linked to other elements at any level. A child can have more than one parent and be linked to its grandparent or higher

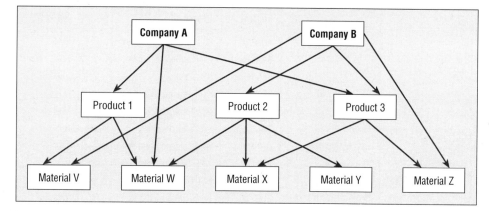

Fig. 5.8 Network or plex structure

level. The hierarchical aspect ceases to have meaning in the more complicated plex structures as it becomes unclear as to which level is which.

Figure 5.8 illustrates a situation where two companies supply products composed of materials, and also the materials themselves. The main advantage of a plex structure is that it is possible to link together all the characteristics of an entity without these being held in the entity's record. Thus, in the example in Figure 5.8, it is immediately apparent from company A's record that company A supplies products 1 and 3, and material W.

A network may also contain elements that are mutually linked, with the result that all elements are of equal status. For example, companies trading with each other could be linked, and so two such companies would have links in both directions.

Inverted lists and files

Whereas a conventional record contains an entity identifier (key) together with the attribute values of the entity, an inverted list shows the attribute value followed by a list of the identifiers. A group of inverted lists comprise an inverted file, and this may be completely or partially inverted depending whether all attributes are included as lists.

Figure 5.9 is a completely inverted file for students, each of whom has three attributes – course, year and sponsor. An inverted file is useful for rapidly finding the entities with a certain attribute value. If more than one attribute value is applicable, the computer selects the entities from one inverted list and matches them against those from the other lists. If looking for the names of unsponsored BA Business Studies students, for instance, BAKER, GREEN and LLOYD are extracted from the left-hand list and matched against COOK, FOX and GREEN from the right-hand list. This leaves GREEN as the only qualifier. Inverted lists are incorporated into certain types of databases.

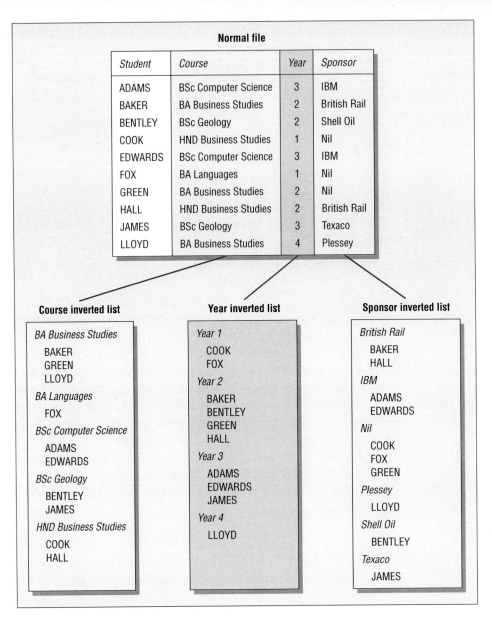

Normal file

Student	Course	Year	Sponsor
ADAMS	BSc Computer Science	3	IBM
BAKER	BA Business Studies	2	British Rail
BENTLEY	BSc Geology	2	Shell Oil
COOK	HND Business Studies	1	Nil
EDWARDS	BSc Computer Science	3	IBM
FOX	BA Languages	1	Nil
GREEN	BA Business Studies	2	Nil
HALL	HND Business Studies	2	British Rail
JAMES	BSc Geology	3	Texaco
LLOYD	BA Business Studies	4	Plessey

Course inverted list

BA Business Studies
  BAKER
  GREEN
  LLOYD
BA Languages
  FOX
BSc Computer Science
  ADAMS
  EDWARDS
BSc Geology
  BENTLEY
  JAMES
HND Business Studies
  COOK
  HALL

Year inverted list

Year 1
  COOK
  FOX
Year 2
  BAKER
  BENTLEY
  GREEN
  HALL
Year 3
  ADAMS
  EDWARDS
  JAMES
Year 4
  LLOYD

Sponsor inverted list

British Rail
  BAKER
  HALL
IBM
  ADAMS
  EDWARDS
Nil
  COOK
  FOX
  GREEN
Plessey
  LLOYD
Shell Oil
  BENTLEY
Texaco
  JAMES

Fig. 5.9 Completely inverted file

SAQ 5.14
A library is thinking of automating many of its facilities. One function which it is interested in automating is that of allowing users to search the main catalogue. Would an inverted file structure be suitable for this?

Solution

Yes, books would be stored with author, title, publication date, publisher and number of copies data, and users may wish to interrogate the system using one or a number of these attributes.

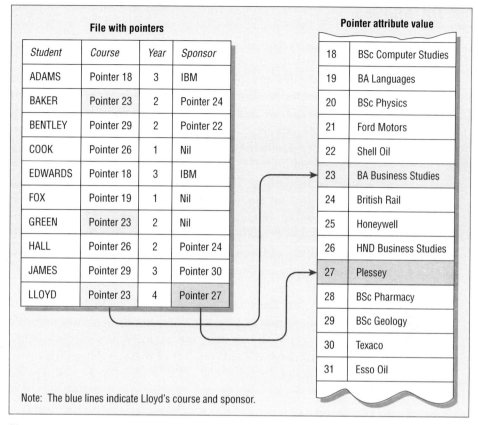

Student	Course	Year	Sponsor
ADAMS	Pointer 18	3	IBM
BAKER	Pointer 23	2	Pointer 24
BENTLEY	Pointer 29	2	Pointer 22
COOK	Pointer 26	1	Nil
EDWARDS	Pointer 18	3	IBM
FOX	Pointer 19	1	Nil
GREEN	Pointer 23	2	Nil
HALL	Pointer 26	2	Pointer 24
JAMES	Pointer 29	3	Pointer 30
LLOYD	Pointer 23	4	Pointer 27

File with pointers

18	BSc Computer Studies
19	BA Languages
20	BSc Physics
21	Ford Motors
22	Shell Oil
23	BA Business Studies
24	British Rail
25	Honeywell
26	HND Business Studies
27	Plessey
28	BSc Pharmacy
29	BSc Geology
30	Texaco
31	Esso Oil

Pointer attribute value

Note: The blue lines indicate Lloyd's course and sponsor.

Fig. 5.10 Attribute value table

Attribute value tables

It is apparent from even the brief file in Figure 5.9 that attribute values appear repeatedly in both the normal and the inverted file. To save storage space in either case, an attribute value table is employed. As is seen from Figure 5.10 the record does not hold the actual attribute value but a pointer instead. The pointer indicates where the value is to be found in the attribute value table.

Thus, many records may point to one entry in the table, and this need not necessarily be in any particular order; also it can hold values of different types of attributes, as in Figure 5.10 in which courses and sponsors are intermixed.

Short-length attribute values, such as 'IBM' and 'Nil', may not be worth putting in the table and so are contained in the record itself in the usual way. This is also done for infrequently occurring values.

153

5.3 DATABASES

In the early years of business computing the majority of computers were equipped with magnetic tape, and later disks, to hold files. Commonly, there were master files appertaining to payroll, sales, purchases, and so on. By and large, these files contained little or no redundant data, i.e. minimal duplication of entity values. As more applications were added to information systems, the likelihood of redundant data increased. Data items such as material costs, operation times, component descriptions and batch quantities appeared in more than one master file.

Another aspect of the development of business computing is the introduction of management information systems. Management information requirements may, on the one hand, be straightforward, being based on data derived from only one or two files. On the other hand, managers may demand information of a sophisticated nature calling for the complex integration of data from several files.

Generally when new routines are added to an information system the files have to be expanded or modified. This reorganization means that the existing application programs using these files have to be amended. This involves a considerable number of systems and much programming effort. Clearly there are advantages if the storage and maintenance of data is independent of the programs using the data. As a result of these problems the concept of a 'database' came into being.

■ Databases

A database is a collection of data supporting the operations of the organization. More specifically, a database entails creating and maintaining data in computer storage in such a way that it is usable for many purposes. In order to have an efficient database there are certain characteristics that must be met. A database must:

- be substantially non-redundant;
- be program independent (data independence);
- be usable by all the programs;
- include all the necessary structural interrelations of data; and
- have a common approach to the retrieval, insertion and amendment of data.

Let us consider these requirements individually.

Non-redundancy

Bearing in mind that redundancy is the duplication of identical data in the computer's storage, it causes certain problems. The three main dangers with redundant data are:

1 the probability of contradiction between the values of the data items in different files;

2 the waste of storage space; and

3 the problems of updating identical data items so that the master files are all equally valid.

Occasionally a minimal amount of redundancy is acceptable, such as when the need for data security or rapid access is paramount. If this is the case, the system must cope with multiple updating and thereby eliminate contradictory data values.

Data independence

This means that the data and the programs are mutually independent. That is to say, the data can be moved or restructured without the need to make alterations to the programs. Similarly an enforced program change does not call for rearrangement of the data.

These points are of great importance because if data independence is not achieved the programmers inevitably find themselves in a tail-chasing situation. Some IS staff have found themselves totally occupied in amending existing programs and file structures with a resultant delay in introducing new applications.

Program usage

A database needs to be usable not only by all the existing applications but also by all foreseeable applications. These are ambitious aims; nevertheless a database must be open-ended so as to accept new data items and changes to existing data items.

In practice this means that all possibilities of data usage must be considered from the outset. The compilation of data for the database needs to be far wider than that envisaged merely for the initial routines. In many cases it is impractical to store all the possible data initially, but it is advantageous if its existence is known from the start.

Data interrelationships

These are necessary owing to the fact that the various applications use data in different ways. One application may demand a link between an employee's name and his or her pension contribution, another between his or her tax payment and previous employer. Requirements such as these impose stringent demands upon the database's accuracy, security and flexibility, necessitating extensive use of the methods described in Section 5.2.

Common approach

This is in the interests of understanding and simplicity. Although application programmers are not concerned with the database's structure and techniques, a common approach simplifies the database control programs and facilitates the database administrator's work.

A helpful feature towards a common approach is the adoption of a data dictionary (see later in this section).

Logical and physical data

Logical data is the data as seen by the systems analyst and the applications programmer. In the course of their work these people put together entity attributes to form the logical records and files. The particular way in which this is done depends largely upon the future usages of the files and the need to maintain them.

Physical data is the actual structuring of data on the storage media. This is likely to be different from the logical aspect of the data. Physical data independence implies that the physical storage techniques and hardware can be changed without affecting the application programs.

The practical considerations of limited storage space and processing time impose restrictions on the extent of data interrelationships. Even sophisticated databases have limitations and it is therefore important for the systems designer to judge the priorities. If necessary, several independent databases are maintained as a means of simplifying the interrelationships.

In order to achieve these aims it is necessary to have a software interface between logical data and physical data. This go-between is called a database management system (DBMS).

SAQ 5.15　A company keeps data on the building products it stores. Each product contains data such as the name of the product which could be non-unique, the manufacturer of the product, the number in stock and the cost of the product. The systems that access the data regard the name of the product concatenated by the manufacturer as a key. Is this key an example of logical data?

Solution

Yes, the actual storage of these two data items is done separately, but at the systems level there is a concept known as a *key* formed by concatenating the items.

■ Database management systems (DBMSs)

A DBMS could be defined as software that organizes the storage of data to facilitate its retrieval for many different applications.

The characteristics of logical data are decided by the user's need for flexibility in processing data. In many business situations there are extensive demands upon the data; new problems and consequently new arrangements of data, occur frequently. The characteristics of physical data are decided by the need for high performance of the computer. These two requirements tend to come into conflict and it is the lot of the DBMS to reconcile them as far as possible.

It is apparent that DBMSs involve a high level of complexity in their design. The techniques described in the preceding section are employed by DBMSs, and in many cases at a higher level of sophistication.

Features of DBMSs

Bearing in mind that a DBMS on a mainframe or PC is required to handle a number of different systems concurrently, e.g. batch processing, on-line enquiries and transaction processing, it is obviously a complex piece of software. It needs to interface tightly with the operating system and the demarcation between these two pieces of systems software is blurred in some cases.

Another factor that imposes demands on a DBMS is a distributed database. The DBMS must then be capable of enabling a user at one location to access a database on a distant computer by using his or her own language and system procedures. That is to say, the distance and differences between the two systems must be transparent to the user.

DBMSs have a wider range of functions and capabilities. Some of these facilities are closely tied to the operating system. Examples of these facilities are:

- Screen formatting for ease of data entry (see entry forms in Section 8.5).
- Record and file locking to make multi-user systems secure.
- Sorting records into any sequence.
- Creation of an audit trail.
- Logging of transactions.
- Control of user passwords.
- Acceptance of user-written programs for enhancing the DBMS.
- Acceptance of high-level languages as a means of making changes to the contents of the database.
- Validation of data.
- Parameter-controlled report generation.
- Dynamic creation and maintenance of a data dictionary.

Structural types of databases and DBMSs

Databases and DBMSs consist of three main structural types: hierarchical, network and relational.

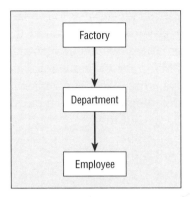

Fig. 5.11 Hierarchical data structure

Hierarchical databases

A hierarchically structured database is a top-down or branching tree arrangement. Each data item falls within a higher-level data item and so on to the top of the tree. Thus if the hierarchical database contained data pertaining to company employees, it might be organized into factories, departments and employees. This method facilitates the rapid retrieval of data provided and involves moving downwards through the hierarchy, e.g. if the employee's factory and department are known. If, on the other hand, higher-level details are not known, then a hierarchical database necessitates a long search, perhaps examining every record.

Figure 5.11 shows this simple hierarchical structure, and Figure 5.12 is the way in which the corresponding records appear. An instance of a hierarchical database is IBM's information management system (IMS). Although you can still find hierarchical databases in commerce and industry they are being superseded by relational database technology.

SAQ 5.16 A bank has a number of branches. Each branch is associated with a number of customers each of which is associated with one or more accounts. Is this a good example of a hierarchy? If so, what are the various levels?

Solution

Yes, it is. The hierarchy consists of bank, branches, customers and accounts.

Network databases

The above example is shown as a network structure in Figure 5.13; this is the same style as the structure in Figure 5.8.

There is a more direct connection between data items at the various levels than with hierarchical databases. This is brought about through the use of pointers linking data at different levels.

Factory–Department

Factory 1	Bristol
Dept. 3A	Metals stores
Dept. 3B	Liquids stores
Dept. 11	Purchasing
Factory 3	London
Dept. 2	Security
Dept. 10	Costing
Factory 5	Derby
Dept. 1	Machine shop
Dept. 8	Sales

Department–Employee

Dept. 1	Machine shop	
2573	Watkins	£156.00
2593	Bates	£123.75
Dept. 2	Security	
2588	Brown	£122.50
2655	Williams	£155.00
Dept. 3A	Metal stores	
2591	Porter	£150.00
2672	Cox	£171.00
Dept. 3B	Liquids stores	
2633	Moore	£168.75
2679	Forbes	£190.00
Dept. 8	Sales	
2569	Jones	£125.00
2580	Smith	£170.00
2618	Palmer	£178.30
Dept. 10	Costing	
2625	Ambler	£205.00
Dept. 11	Purchasing	
2548	Moss	£110.00
2703	Parker	£215.00

Fig. 5.12 Hierarchical database

In Figures 5.13 and 5.14 the directions of the arrows indicate the paths through which data is found. Thus in this network database it is easy to find an employee's factory and department from his or her employee number. It is also straightforward to find the departments in each factory and the factory that any

159

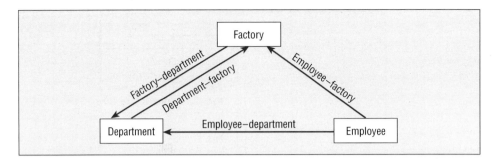

Fig. 5.13 Network data structure

department is in. If it was required to ascertain which employees are in a particular department or factory, more pointers would be necessary.

Although you can still find network databases in commerce and industry they are being superceded by relational technology.

Relational databases

A relational database can be regarded as a number of two-dimensional arrays of data items, i.e. otherwise known as flat files.

The theory of sets upon which the concepts of relational databases is founded introduces some new terminology that have not so far been mentioned. Whereas up to now we have referred to a record, this is known in this context as a 'tuple' and the data items are called 'domains'. The connection between domains within a tuple is called a 'relation'. The point about using these new terms is that they refer to the database structure and not to the logical structure as seen by the user. For the sake of simplicity it is the intention to stick to the terms 'data item' and 'record' in this section.

Figure 5.15 illustrates the flat files of the relational database of the above example. Although this diagram shows the department–factory relationship in department sequence, the DBMS could, if advantageous, also hold it in factory sequence. Much depends on the numbers of factories and departments and the consequent amount of searching and indexing involved. Examples of relational database systems are DB2, Oracle and Ingres.

Relational databases require special analysis techniques to be applied to data structures. These techniques, called 'normalization', are dealt with in Chapter 7. The idea behind this concept is to break the data down into small tables so that the fields are dependent only on the key and not linked to any other key field; this means that cross-referencing keys have to be embedded to maintain the associations between data (e.g. an order with many products will be split into order and product tables). A link table, or foreign key, has to be kept to trace the products in any one order. This means that relational databases consume more disk storage capacity and computer processing resources than other models.

Factory–Department

Factory pointer	Factory no.	Factory location	Department pointers
1	1	Bristol	3 4 7
2	3	London	2 6
3	5	Derby	1 5

Department–Factory

Department pointer	Department no.	Department name	Factory pointer
1	1	Machine shop	3
2	2	Security	2
3	3A	Metals stores	1
4	3B	Liquids stores	1
5	8	Sales	3
6	10	Costing	2
7	11	Purchasing	1

Employee–Factory–Department

Employee no.	Employee name	Employee wage	Factory pointer	Department pointer
2548	Moss	110.00	1	7
2569	Jones	125.00	3	5
2573	Watkins	156.00	3	1
2580	Smith	170.00	3	5
2588	Brown	122.50	2	2
2591	Porter	150.00	1	3
2593	Bates	123.75	3	1
2618	Palmer	178.30	3	5
2625	Ambler	205.00	2	6
2633	Moore	168.75	1	4
2655	Williams	155.00	2	2
2672	Cox	171.00	1	3
2679	Forbes	190.00	1	4
2703	Parker	215.00	1	7

Fig. 5.14 Network database

Factory no.	Factory location
1	Bristol
3	London
5	Derby

Department no.	Factory no.	Department name
1	5	Machine shop
2	3	Security
3A	1	Metals stores
3B	1	Liquids stores
8	5	Sales
10	3	Costing
11	1	Purchasing

Employee no.	Employee name	Department no.	Employee wage
2548	Moss	7	110.00
2569	Jones	5	125.00
2573	Watkins	1	156.00
2580	Smith	5	170.00
2588	Brown	2	122.50
2591	Porter	3	150.00
2593	Bates	1	123.75
2618	Palmer	5	178.30
2625	Ambler	6	205.00
2633	Moore	4	168.75
2655	Williams	2	155.00
2672	Cox	3	171.00
2679	Forbes	4	190.00
2703	Parker	7	215.00

Fig. 5.15 Relational database

Relational databases are now the overwhelming technology used in commerce and business. This technology has advanced to the point where there are large numbers of products available for both the small business and the large corporation. Typical products include SQL Server from Microsoft which is used in large distributed systems, ORACLE 8 from the ORACLE Corporation which, again, is used in large distributed applications, to ACCESS from Microsoft and paradox from Borland, both of which are small, cheap systems aimed at the small business or single user.

Object-oriented databases

This technology is comparatively recent. It is based on the concept of an object as an entity that has state: i.e. it contains data. This technology supports the storage of data which correspond to real-life entities such as a bank account or an invoice. The technology is somewhat immature with response times and storage demands which are inferior to those found in relational technology. However, object-oriented databases are beginning to be found useful in those applications where the underlying data is too complex for a relational system to handle. A typical application is in manufacturing where the explosion of a product such as a car into its components has to be stored in order to plan the manufacturing process.

SAQ 5.17   A travel agent wishes to computerize its holiday booking system. One of the tables that the developer wishes to construct contains customer details. What might you expect in this table?

Solution

Typically included would be the names of the customers, their addresses, their telephone numbers, the number of times they have booked a holiday and details of the current holiday that they have booked such as the name of the travel company and a booking reference number.

DBMS levels

As well as providing storage facilities, most databases come with a set of programs to help manage the data. These programs create, update and delete records, usually by a data manipulation language (DML). Most DBMSs are based on the ANSI/SPARC three-layer model of architecture which defines three levels of data description and the services associated with each layer.

Level 1 – end-user schema: This is data in the format in which the user wishes to present it and is not strictly defined in the model. An example would be data in tabular form, presented as a graph, with or without a sorted order, etc.

Level 2 – database sub-schema: This is a description of the data which is accessed by the user program. Sub-schema are a partial description of the whole database (assuming no one ever wants all the database at once) and are therefore a partial view of the data. A DBMS allows many sub-schema to access the same piece of physically stored data, thus implementing data independence.

Level 3 – physical storage schema, or the database schema: This defines how the data is held in the database itself as tables, networks or hierarchies. The database schema is not usually accessible by the end-user and is only changed by the database administrator when a significant change is necessary; for instance, when a new table is added to the database.

Associated with these three layers the DBMS provides a series of languages. The end-user schema may be supported by programs for report writing, table generation and graphical display. The sub-schema layer will be supported by a data retrieval language and access to the database schema is allowed by the DML. The database administrator has a data description language (DDL) to allow new records and tables to be defined.

SAQ 5.18

As part of the documentation for a project to computerize an air-traffic control system the developer constructs a document which describes how the plane arrivals data is chained in a file. What schema is this type of information associated with?

Solution

Since this deals with implementation matters it is associated with the physical storage schema.

DBMS languages and facilities

One service is to maintain data integrity. Databases are multi-user, hence when two or more users want to access, or worse still update, one record at the same time, a problem of access control arises. Most databases provide a locking mechanism so that updating clashes cannot occur. Several mechanisms exist but the most common is to use a wait-lock-commit command. The user program issues an update request to the DBMS. If the record is not being used it is locked so that no other user can access it. The record is then updated and unlocked for the next user program. Some systems lock just to reserve the record for use and then issue a commit to the DBMS when an update is ready. This gives a two-phase access control so a record may be locked and then released if no update was necessary.

Data retrieval and update languages

DBMSs also provide an access language to user programs to retrieve data. The languages are composed of data description languages and retrieval languages (DDL). The standard retrieval language is structured query language (SQL), which allows program-like phrases to be constructed from data retrieval and updating. SQL can be used either from within a user's program or as a stand-alone query language. The syntax in its simplest form is:

FIND <entity-name> WHERE <data items match constraints>

Entities are described in more detail in Chapter 7, but for our purposes we can equate them with records. An example of an SQL query is:

FIND Cars WHERE make = Ford AND year > 1990

This query will return all the makes of Ford contained in the Cars table which have a date of manufacture more recent than 1990 in the 'year' field. SQL has a rich syntax and mini-programs can be written for complex nested queries and update routines.

Failure and recovery control

DBMSs have to guard against the possibility that the computer may fail during an update. In this case the user's input could be lost and, more problematically, the user would be unsure as to whether the update had been completed before the computer crashed or not. To guard against these problems DBMSs maintain a transaction log of all input for updates and a trace of the status history of processing. This is effected by writing transactions to a temporary working file on disk as they are input or sent by a user's program. It is therefore possible to ascertain what stage the processing had reached when the crash happened. Two mechanisms are employed to correct the situation after the failure. One is to 'roll back' the database to the transaction before the failure occurred, the other is to 'roll forward' and complete the transaction using the data held in the log file. Which technique is used depends on the type of DBMS. Some use both, although the ability to roll forward is, of course, dependent on the permanent recording on disk of the input transaction just before the computer crash.

Update integrity control

Some DBMSs provide facilities for checking that updates are validated before they occur and for updating related data after one record has been changed. These operations are more generally the responsibility of the user or their application program; however, there is an increasing trend for DBMSs to provide more facilities for update processing. Simple update integrity control would be validation checks to ensure that a numeric field gets a numeric value, not a character, for updating. More complex checks are taken from the standard validation practices, e.g. type and range checks. More sophisticated update integrity links depend upon fields and records, so for example, in a scientific database, if a change to the temperature of a liquid also changed its density and this relationship was known, then the DBMS would perform the calculation and update the density field as well. In business applications update integrity control is more often involved with parts/sub-parts relationships; when a master part is deleted therefore, all its dependent sub-parts are also removed. Similarly, when a new master part is added to a component database, then the necessary sub-part records are also automatically created. These facilities all take some of the burden for basic data manipulation away from the end-user and application programs.

SAQ 5.19 Which of the following circumstances should be handled by update integrity control and which by failure and recovery control?

1 A computer malfunctioning and sending erroneous data to a database.

2 A computer failing and destroying data in a database.

3 A bank teller attempting to create details about a new customer who is in fact an existing customer.

4 A fire destroying the computer room containing a company database.

Solution

1 is handled by update integrity control, 2 is handled by failure and recovery control, 3 is handled by update and integrity control and 4 is handled by failure and recovery control.

■ Data dictionaries

A data dictionary (DD) is a store of information that describes and specifies the characteristics of each piece of data used in a system. In other words, it is an electronic glossary defining each of the data items incorporated in the whole system. DDs also contain the *metaschema*, the description of the data types and organization of the database structure, i.e. a description of the schema and permissible sub-schema. A DD may also include definitive descriptions of processes within the system, e.g. update and integrity process as described above.

The main point about having a DD is that experience has often shown that data easily become ambiguous. People are inclined to give different names to the same thing, and the same name to two or more different things. For example, one person may refer to the 'stock number' and another person calls it the 'commodity code'. Similarly, programmers are apt to employ different names for the same data item in their programs, e.g. STKNO and COMCDE.

A DD attempts to obviate these problems by pinning data down according to clear definitions. Each piece of data has the following characteristics ascribed to it:

■ *Name*: meaningful, standardized identification and preferably constructed in a program-defined form such as in COBOL, e.g. STOCK–NO.

■ *Description*: a brief explanation of the meaning of the data item.

■ *Aliases*: a list of alternative names that have been used for the data item, e.g. STKNO as an alias of STOCK–NO mentioned above.

■ *Related data*: it may be useful to draw attention to data that is closely connected with or has a similar name to the data item although it is not an alias, e.g. VAT status and VAT rate.

- *Range of values*: a data item may have a continuous range of values, in which case only its maximum and minimum values are included. Alternatively, the values may be discrete, e.g. discount rates of 0, 10 and 20 per cent only, or one fixed value, e.g. a fixed price. The other possibility is a coded value, e.g. 10 = Bristol, 11 = Southampton, and so on.

- *Layout*: this is the 'picture' of the data item as it exists outside the computer and so is conveniently specified and held in the form explained under 'picture' in Section 8.4.

- *Encoding*: an indication of the form in which the data is encoded, e.g. 1 = binary, 2 = ASCII, 3 = EBCDIC and so on.

- *Editing*: a specification of any editing or special checks that the data item must undergo, especially on input; these are discussed under data validation in Section 8.4.

Other characteristics of a data item that may be included in a DD are:

- *Ownership*: the department that initiates the data item.
- *Users*: the departments that refer to it.
- *Systems and programs*: those that reference and update it.
- *Security and privacy restraints*: those that are imposed upon its use.

These latter features take the DD within range of becoming a data encyclopaedia (see below).

Data dictionaries come in a variety of types: some are stand-alone, others relate to a certain DBMS. Small DDs of up to a few hundred items can be maintained manually by using a card index system. Larger dictionaries are automated, generally through the employment of a DD package. An automated dictionary has the advantage of being able to provide a variety of useful listings and reports. Examples are a full alphabetic list of all items, a selected list based on the first few letters of the name, and a search facility for letter groups, i.e. to find items whose names are not completely known. Most DBMSs have an automated DD incorporated into them.

The work of creating and maintaining the DD falls upon the data administrator – if such a post exists – and in any case it is important that all DD entries are channelled through one person. It is also possible that the DD is kept up to date by the software using it, this is called an active DD.

Example 5.12	**A data dictionary**

This example covers one data item taken from a stock control and sales analysis system for a jeweller's. In practice the information below would be entered on special forms that are part of the system design.

Name: article number, program form ART–NO.

Description: a number unique to each and every individual article. Even though two or more articles are physically identical, they still have their own individual article numbers.

Aliases: Article no., AN

Related data:

1 article description/description – a brief non-definitive descriptive name for general information only.

2 category – each article falls into one of nine categories (see layout).

3 manufacturer – each article in category 2 (watches) is from one of ten manufacturers including an 'others' group, i.e. from a smaller manufacturer not specifically nominated (see layout).

Range of values: minimum value 10000, maximum value 99999.

Layout: five numeric digits only, i.e. picture is 99999. The first digit denotes the broad category (numbered 1–9) into which the article falls. The second digit of category 2 (watches) denotes the manufacturer.

Encoding: ASCII.

Editing: limits (range) check in every input program. The limits are reassigned from time to time according to current usage.

Ownership: owned and allocated by the Goods Receiving Section.

Users: all departments.

Systems and programs: used in every process and program in the stock control and sales analysis processes.

Security and privacy: creating and allocation of article numbers is restricted to the Goods Receiving Section only.

■ Data encyclopaedias

An evolution of the DD is the data encyclopaedia (DE). This is a wider concept that includes additional information to that described above. This information is aimed at specifying the data item's usage, and so includes details of the processes, data flows and data stores (files) associated with it.

A DE is a combination of the following:

- Document specification form (Figure 7.3, Section 7.4).
- Data usage chart (Figure 7.4, Section 7.4).
- Output analysis chart (Figure 8.5, Section 8.3).
- Logical file record specification.
- Data item utilization.

Repositories

Encyclopaedias, combined with a database architecture and further processes for consistency checking and storage management make up a repository architecture. Repositories were primarily motivated by the need to provide comprehensive databases support for computer-aided software engineering tools (CASE). CASE tools need to store complex data structures which present models of the system (Section 8.10). The repository has to manage the storage of models and their components such as processes, data structures, events and relationships; furthermore it has to support graphical tools which present these models as diagrams to the user. Further services are required such as cross-checking between and within each model to ensure the models are consistent, and support to the development of the models themselves. Without going into the complexity of CASE tools and their support environment, it is sufficient to say that repository architecture requires many additional services beyond a standard DBMS.

Major manufacturers have used their proprietary relational DBMSs as a host for repositories, e.g. Digital Equipment Corporation RDB forms the basis for Corporate Data Dictionary (CDD), while IBM's DB–2 is the host for AD/Cycle. All repositories aim to support CASE tools for a variety of developments methods and diagramming styles. In addition repositories can manage fourth-generation reporting tools, report writers and graphical presentation tools.

Heterogeneous and object-oriented databases

Heterogeneous databases contain different types of data. Whereas the hierarchical, network and relational models are fine for stored tables of numeric data, they do not work so well for other data type such as text, graphics, sound and moving images. With the increasing use of multi-media DBMSs, manufacturers have moved towards providing storage facilities for different data types. Another influence has been problems encountered with relational DBMSs. Because these maintain all data in tables, a very 'flat structure' results. This means that it is hard to represent complex data in permanent storage because relationships such as hierarchies and networks are lost. This has promoted the idea of object-oriented databases which provide storage and manipulation facilities for complex data structures representing objects.

■ Text databases

These have existed for many years but have not been as commercially important for business as the relational and network models. Text databases exist to store text documents, e.g. books, articles, legal documents, design description, patents, and so on. The main difference between text DBMS and business DBMS is that text is not split into records, tables, fields, etc. The DBMS facilities are therefore more oriented towards access mechanisms. These are either by indexing, following concepts in index-sequential and inverted files. The text document is given one or more indexing terms. These are keywords which describe the document. Keywords may in turn be organized in facets, that is, sets of descriptive terms. Faceted descriptions allow more sensitive retrieval techniques to be applied by a combination of different keywords within and between facets. For example, a document describing company customer-service policy for major accounts may be indexed under the following facets: policy type = service; company division = customer support; and operational procedures = service rules. As with relational DBMS and SQL a query language is provided which allows queries to be submitted with a combination of Boolean operators (AND, OR, NOT).

An alternative to document indexing is to use free text search in which the text contents are searched for one or more keywords. The retrieval language may provide for combinations of words to be searched and in more sophisticated versions words may be searched for in proximity of each other. The penalty for free text searching is time. In long documents sequential search for one or more target words can be very time consuming, hence some systems restrict the search to abstracts and document summaries. However, intelligent techniques for free text search, such as skipping unlikely words, can speed up retrieval and many text-based DBMSs provide free text search within whole documents.

Text databases have a near relative in document management systems. These are DBMSs that manage large numbers of similar documents which have some text but the text is generally not long or complex. Document management systems use indexing techniques for describing and retrieving individual items and may be implemented in a standard relational database or in another technology.

■ Multi-media databases (MM–DBMSs)

Multi-media or heterogeneous databases are an emerging technology which attempts to put storage and management of different data types under one DBMS. As pictures, animation, sound, text and tables have radically different storage needs, MM–DBMSs generally use a mix of technologies, such as relational technology for tables and business data, text database for documents and image storage devices of graphics and animation. The MM–DBMS provides a suite of programs of data management as before. MM–DBMSs share many problems with object-oriented databases, the products generally referred to as OO–DBMSs (as described below). The main problem to be solved by MM–DBMSs is how to handle non-text or character-based material. Pictures, drawings, moving images and sound may be stored either in analogue or more commonly in digital form. However, the DBMS can only access those media items as a whole unless some access control is provided. In time-varying media, e.g. moving pictures and sound, access by frame is provided in DVI standards, and access to sound material in a tape recorder metaphor is usual. For more sophisticated access the images have to index with keywords in a similar manner to text-based documents. MM–DBMSs provide facilities for indexing and retrieval that are based on many text-based DBMSs.

Object-oriented databases (OO–DBMSs)

OO–DBMSs aim to provide facilities for persistent storage of items created by object-oriented programs. Object-orientation is surveyed in more depth in Chapter 8, but to introduce the concept here, an object is a complex collection of data items that relate to something in the outside world, such as a customer, product, ship, house, etc. Objects also exist in classification hierarchies: hence a ship object may be decomposed into sub-classes of cargo ship and passenger ship, passenger ship into ferry and cruise liner and so on. To complicate matters, objects are not just data, they also contain the programs which act upon the data. OO–DBMSs therefore have to provide storage facilities for large complex objects which are organized in hierarchies. Moreover the objects themselves may be composed of graphics, text, moving images, and so on.

This technology requires different storage and access procedures compared to relational DBMS. The unit of storage is larger, yet links must be maintained between objects in a hierarchy, and thus in this aspect OO–DBMSs have some similarity with the old hierarchical model. Access procedures, integrity controls, become part of the programs which are embedded in the object itself rather than being a separate DBMS facility, although recovery facilities are usually provided by the DBMS itself. No standard data retrieval language exists for OO–DBMSs, although SQL is being extended in this direction as OO–SQL. Data retrieval is therefore either by writing special programs with sub-schema in an object-oriented language such as C++, or sometimes limited query facilities are provided.

OO–DBMS technology is still in its early stages. Some examples of commercial products are Gemstone and O–2. It is unlikely that OO–DBMSs will replace relational technology from standard business applications because they do not offer the same advantages in data management. However, when data items are complex and heterogeneous items need storing as in CAD/CAM applications, OO–DBMSs may well come into their own.

EXERCISES

Exercise 5.1 **Binary tree searching**

(a) Create a binary tree index for the keys as under, entering them in the order shown. 58, 77, 94, 40, 49, 45, 63, 18, 72, 75, 25, 76.

(b) Delete keys 63 and 40, insert further keys 70, 46, 33 and 91, reconstruct the index.

(c) In what order should the 12 keys in (a) be entered to make the tree more balanced?

Exercise 5.2 **Database structures**

With the aid of diagrams, explain and contrast the principal types of database structure.

(ICSA, Info. Systems, Dec. 1992)

Exercise 5.3 **Searching techniques**

Describe and compare the following searching techniques by describing the algorithms used and their time and storage characteristics:

(a) serial;

(b) binary chopping;

(c) hashing; and

(d) binary tree.

Exercise 5.4 **Database management systems**

Describe the advantages and disadvantages of developing applications using a database management system (DBMS), as compared to a conventional file based system.
  Illustrate your description with an application of your choice.

(BCS, GPI, April 1992)

Outline solutions to exercises

Solution 5.1 (a) and (b) See Figure 5.16.

(c) There are numerous possible orders for the keys, such as 63, 76, 45, 25, 58, 75, 77, 94, 18, 40, 72, 49. This order gives four levels as against six in the original order. Construct a tree diagram to prove this.

Solution 5.2 Refer to Section 5.2 and Figures 5.12, 5.14 and 5.15.

Solution 5.3 (a) Serial searching means the same as linear searching, i.e. starting at the beginning and examining each item (key or record) in turn. If the items are wanted sequentially and they are in this sequence, the search can continue from the last item found. An unordered (no sequence) set must always be searched from its beginning.
  This method takes a large amount of time but no extra storage, only the records and keys themselves, i.e. no indexes.

(b)–(d) Refer to Section 5.2 for algorithms, the time and storage characteristics are shown in Figure 5.17.

Node no.	Key	Left node no.	Right node no.
1	58	4	2
2	77	7	3
3	94		
4	40	8	5
5	49	6	
6	45		
7	63		9
8	18		11
9	72		10
10	75		12
11	25		
12	76		

(a)

Node no.	Key	Left node no.	Right node no.
1	58	4	2
2	77	7	3
3	94	14	
4	33	8	5
5	49	6	
6	45		13
7	70		9
8	18		11
9	72		10
10	75		12
11	25		
12	76		
13	46		
14	91		

(b)

Fig. 5.16 Binary tree indexes for Exercise 5.1

Method	Time	Storage	Comments
Linear	Large	Small	No index
Binary chop-indexed	Small	Fairly large	
Binary chop-unindexed	Fairly large	Small	
Hash	Very small	Fairly large	Space needed for insertions and overflow
Binary tree	Small	Fairly large	Large index

Fig. 5.17 Characteristics of methods of searching

Solution 5.4 Refer to Section 5.3.

Applications lending themselves to DBMSs include stock control, payroll, sales accounting, production planning, and order processing.

References and further reading

5.1 Atzeni, P., Ceri, S., Parabosic, S., Torlon, R. *Database Systems: Concepts, Languages, Architectures* (McGraw-Hill, 1999).

5.2 Bestavros, A. *Real-time Database Systems* (Kluwer, 1997).

5.3 Carter, J. *The Relational Database* (International Thomson, 1995).

5.4 Connolly, T. and Begg, C. *Database Systems* (Addison Wesley, 1998).

5.5 Date, C.J. *An Introduction to Database Systems* (Addison Wesley, 1994).

5.6 Elmasri, R. *Fundamentals of Database Systems* (Addison Wesley, 1998).

5.7 Folk, M.J. *File Structures* (Addison Wesley, 1998).

5.8 Gorman, M. *Enterprise Database in a Client/Server Environment* (John Wiley, 1994).

5.9 Heinckiens, P. *Building Scalable Database Applications* (Addison Wesley, 1998).

5.10 Inmon, W.H. and Inmon, W.H. *Building the Data Warehouse* (John Wiley, 1996).

5.11 Ju, P. *Databases on the Web: Designing and Programming for Network Access* (MIS Press, 1997).

5.12 Larson, J. *Database Directions: from Relational to Distributed, Multimedia and Object-Oriented Database Systems* (Prentice Hall, 1994).

5.13 Purba, S. *Handbook of Data Management 1999* (Auerbach Publishing, 1998).

5.14 Steward, J.R., Scharle, J.J. and Greene, J.S. *Filing and Database Systems* (McGraw-Hill, 1989).

5.15 Ullman, J.D. *First Course in Database Systems* (Prentice Hall, 1997).

6 Programming and software

AIMS

After reading this chapter you should:

■ be familiar with the processes whereby a program is developed;

■ understand the main properties of third and fourth generation programming languages;

■ be familiar with the main categories of application software;

■ understand the nature of the Internet and the World Wide Web.

6.1 SOFTWARE

The term software describes the programs that are used by a computer. It can be taken as including application programs, utility programs, operating systems, translators, subroutine libraries and application packages.

Computers can do nothing without software. In fact, no other artefact relies so heavily on human instructions before it can function. In contrast, no other artefact gives such a close impression of having intelligence while accomplishing elaborate tasks in terms of repetitive logical processes and approach.

6.2 HIGH-LEVEL LANGUAGES

High-level languages were developed in order to ease the work of programmers by making the programming language more procedure-oriented. Whereas low-level language instructions tend to be machine-oriented, the 'statements' of a high-level language are closer to natural English or other natural languages.

A high-level source program is translated, i.e. converted, into an object program by means of a compiler or an interpreter.

Compilers and interpreters

The features of a compiler are:

■ It translates one statement in the source program into several, perhaps many, object program instructions.

- It checks for errors in the source program statements, such as invalid words and violation of the rules of syntax, i.e. the construction and relationship of statements.
- Open and closed subroutines, and macros are incorporated.
- The programmer ends up with a compiled program on magnetic storage and a printed copy, if required.

An interpreter is similar to a compiler except that the source program statements are converted one at a time into object program instructions immediately before execution, and so no object program is available as a complete entity. If there is an error in a statement detectable by the interpreter, an error message is returned on entering the statement.

There follow brief descriptions of a few of the better-known high-level business programming languages.

■ Third-generation languages (3GLs)

COBOL (Common Business-Oriented Language)

COBOL is an extensively used high-level language and since around 1960 several versions (dialects) have appeared. Although the original intention was that COBOL should be compiled and run on any model of computer, there are small differences between the various versions and so the corresponding compilers must also differ. While large amounts of software programmed in COBOL exist, its use is declining.

A COBOL program consists of four divisions:

1 *Identification division* – identifies the program.
2 *Environment division* –specifies the computer(s) to be used for compiling and processing.
3 *Data division* – specifies the format and characteristics of the files and data to be processed by the object program and relates these to the names used in the procedure division.
4 *Procedure division* – comprises the statements in the source program; this is the main part of a COBOL program.

The statements (sentences) of the procedure division are made up of 'verbs' and 'names'. The verbs are from a predetermined set of about twenty in all. Examples are ADD, MULTIPLY, MOVE, GOTO, READ and ALTER (see Example 6.1).

The names are created and assigned to data items by the programmer so as, in effect, to enable the compiler to find the data items in the computer's main store.

Example 6.1	**COBOL programming**

This example is a short piece of COBOL programming for computing the amount to be paid to a worker who receives time-and-a-half for hours worked above forty and normal rate up to forty hours.

The names adopted for this module are WORKED-HOURS, OVERTIME-HOURS, PREMIUM-HOURS, PAID-HOURS, RATE and AMOUNT-PAID.

```
25 SUBTRACT 40 FROM WORKED-HOURS GIVING OVERTIME-HOURS.
26 IF OVERTIME-HOURS IS NEGATIVE GO TO 30, OTHERWISE GO TO 27.
27 MULTIPLY OVERTIME-HOURS BY 1.5 GIVING PREMIUM-HOURS.
28 ADD PREMIUM-HOURS TO 40 GIVING PAID-HOURS.
29 GO TO 31.
30 MOVE WORKED-HOURS TO PAID-HOURS.
31 MULTIPLY PAID-HOURS BY RATE GIVING AMOUNT-PAID.
```

The most recent COBOL standard incorporates structured concepts with IF . . . END-IF for selections; PERFORM TEST BEFORE for iterations, and paragraphs and sections for modularity. For further information about COBOL, see References 6.7 to 6.12.

Pascal

Pascal is a high-level language named after the famous seventeenth-century French mathematician. It was expressly designed as a language to make programming more systematic and disciplined and, in these respects, lends itself to structured programming. Pascal's syntax is in a formal arrangement known as Extended Backus-Naus Form (EBNF). For example:

$$program = title\ statements\ \text{“.”}$$

means that a program is defined as a title followed by statements followed by a period; title and statements are further defined using the same form.

Pascal uses a range of special symbols in its vocabulary such as brackets and punctuation marks. It is efficient at handling sets, stacks, queues, lists and arrays, so making it powerful for programs that involve arranging and processing groups of data items (see Example 6.2).

Example 6.2	**Pascal programming**

In the following example the variables are the same as in Example 6.1.

```
IF wh > 40
THEN ph: = 40 + 1.5 * (wh – 40)
ELSE ph: = wh
ap = ph * rh
```

See References 6.15–6.21 for further details of Pascal.

PL/1 (Programming Language 1)

PL/1 was created by IBM in the early 1960s as a general-purpose, high-level programming language for both business and scientific applications. It is designed so that simple programs can be constructed without a knowledge of the full language. The concept is for the programmer to utilize a subset of PL/1 with which he or she is familiar. This is referred to as 'modularity' although it has no direct connection with structured programming.

A PL/1 program is composed of blocks of statements called 'procedures'; these may form part of larger blocks after the style of subroutines. Each procedure is compiled separately starting with the lowest, and they can thus be used in several different programs.

Example 6.3   **PL/1 programming**

Below is the short piece of program used previously, the meanings of the words are the same as in Example 6.1.

```
IF WH > 40
THEN
   PH = 40 + 1.5 * (WH – 40);
ELSE
   PH = WH;
AP = PH * RH;
```

C

The C programming language was introduced as early as 1972 but did not become well known until the 1980s. Its rise was linked to the increasing use of the UNIX operating system whose software is mainly written in C.

C is a very powerful language since it gives the programmer access to the low-level architecture of a computer, and it also has found use when software is required which can be moved from one computer to another. In order to obtain the full power of C it is necessary to have UNIX as the associated operating system.

An error checking program known as Lint is often applied to a C program before compilation, which is just as well, since C is a difficult language for non-specialists. C allows the programmer low-level control over input and output while providing the structured programming constructs found in Pascal. C can be regarded as the precursor of the object-oriented programming languages C++ and Java discussed below, as much of the syntax of C can be found embedded in their syntax. Example 6.4 shows the fragment of program code which was reproduced as Example 6.1, 6.2 and 6.3, but written in C.

Example 6.4   **C programming**

In the following example the variables are the same as used in Example 6.1.

```
IF (WH > 40)
   PH = 40*1.5*(WH – 40);
ELSE
   PH = WH;
AP = PH*RH;
```

■ Object-oriented programming languages

Since the mid-1990s there has been an explosion in the deployment of object-oriented programming languages. Such languages operate on objects, an object being a collection of data which mirrors an entity in an application. The data associated with an object is known as its state. As well as having a state an object is also associated with methods. These are chunks of code – very much like subroutines – which can be used to carry out operations on an object.

In a seat reservation system for an airline a typical object would be a flight. The data associated with the flight – its state – would be items such as the flight number, the number of seats on the flight, the type of plane that is being used, the current number of passengers booked and so on. Typical methods which would be associated with this type of object would be code which adjusts the number of passengers when a booking is made or a seat is cancelled.

Object-oriented programming languages have been around for more than twenty-five years. The first language which had object-oriented features was Simula, a language which was invented in the late 1960s and was used for simulation applications. Another early object-oriented programming language was Smalltalk which originated from the Xerox PARC research centre.

C++

It is true to say that these early programming languages made little impression on software developers until the mid-1990s. The first industrial strength object-oriented programming language which was deployed in some numbers was C++. The main reason for the rise of such languages in the 1990s was the increasing availability of libraries of code which reduced the amount of programming required to develop a system.

C++ is a superset of C: it contains all the facilities of C but also includes new facilities which enable objects to be defined and manipulated. There are now a large number of versions of the language and increasing evidence that developers who formerly used C are increasingly switching their software development to C++.

Java

Java is a programming language which is based on C and C++. It has had an explosive growth since it was announced in the mid-1990s. There are a number of reasons for the popularity of Java:

- It is based on the programming languages C and C++. There are many programmers who are fluent in these languages and it is not a difficult process retraining them.

- Programs written in Java are very easy to transfer from one operating system to another: many Java programs require no change when transferring them from an operating system such as Windows 95 to a dissimilar operating system such as UNIX.

- Java is an Internet system programming language. It enables the programmer to carry out tasks such as connecting to another computer on the Internet and loading down a file to another computer easily. Tasks which, in the past, required hundreds of lines of code can now be done in tens of lines of code. Java's popularity coincided with the huge growth of the Internet.

- Java was the first programming language which properly interfaced with the World Wide Web. The World Wide Web consists of a series of documents, some of which may be geographically many thousands of miles from each other linked together by means of textual links. The user of the Web reads these documents with a program known as a browser. Until the advent of Java, Web documents were very static. Java enabled programmers to include such things as animation programming and forms processing on the pages of a Web document; its ability to do this coincided with a period when the users of the World Wide Web started asking questions about its limitations. A programmer using Java can now embed entities known as applets into Web documents. These applets are executable Java programs.

- Java is a portable language. Java produces an internal code known as a byte code. This byte code can be executed on any computer that has the Java system implemented on it. This means that anyone can move a Java program from one computer to another with no extra work at all, over and above that of transferring the code using some magnetic medium or transmission mechanism.

- Java has extensive facilities for distributed processing. It enables programmers to easily develop programs which communicate with other programs over some network, usually a network which has TCP/IP implemented on it. Previously this type of programming involved very extensive and complicated programming. Java has reduced this by an order of magnitude.

180

A fragment of code written in Java is shown as Example 6.5.

Example 6.5   **Java programming**

```
class Employee {
String name;
int currentSalary, worksNo, currentBonus;
. .
    public int calculateThisMonthSalary ( ) {
    return (int) currentSalary/12+currentBonus;
    }
. .
}
```

The code describes objects which represent the staff whose details are processed in a personnel system. An object is given the generic name Employee in the first line, the next two lines introduce some of the data associated with an employee: name, current yearly salary, works number which uniquely identifies him or her and the current bonus which they are to be paid. The fifth to seventh lines define a method which calculates the current monthly salary of an employee. This is calculated by dividing the yearly salary by twelve and adding the bonus to it. The keyword return means that this value is returned to any software that uses Employee objects.

Fourth-generation languages (4GLs)

A fourth-generation language (4GL) is a higher-level language than third-generation languages such as COBOL and C. In other words it demands fewer lines of manual coding to achieve a given task and is less verbose in its written form. It is difficult to be definitive about 4GLs because this is largely a matter of opinion.

Nevertheless it is generally accepted that a 4GL must be easy to learn, have a programming productivity of at least 10:1 over COBOL and incorporate facilities for prototyping (Section 7.9). 4GLs are slow in executing the job in hand and need powerful processors; however, they have brought a measure of programming capability to end-users. 4GLs have become divided into two classes: those which are more of a professional programmer's tool, known as application generators; and other types of 4GL which lend themselves to end-user application.

The characteristics of 4GLs vary, some are non-procedural, i.e. the programmer states what needs to be done as a set of IF . . . THEN rules rather than how it is done. This implies that the 4GL software decides the procedure, whereas the programmer merely declares the requirements. Others have procedural syntax with the standard components of sequence, selection and iteration. Most 4GLs incorporate data description, report writing and screen painting facilities, so that creating files, query screens and formatting reports becomes a simple task.

4GLs are usually programmed interactively; this means that programming errors are detected at an early stage. This is reinforced by the fact that 4GLs are generally interpretive. Interactive programming reduces, or even eliminates, the need to remember a set of mnemonic statements as these are replaced by selection from menus and by use of semi-natural language. Similarly, screen graphics in which the programmer fills in the blanks and extensive use of WIMPs contribute to ease of programming.

4GLs in the broad sense consist of the following types:

■ Spreadsheet-based, e.g. Excel. These include concise syntax and powerful commands, making them of a higher level than earlier spreadsheet software. There is, however, a limit to what can be achieved through the spreadsheet approach.

■ Database-based, e.g. dBASE IV and Oracle. These are based on the management and interrogation of databases.

■ Application (code) generators, e.g. Ideal, Telon and Natural. This type utilizes form-filling on screens related to systems design. They are then compiled to generate COBOL coding.

■ Information centre-based, e.g. Focus, Ramis II and Nomad II. These are capable of producing complex reports, handling sophisticated queries and generating intricate graphics.

An extensive range of 4GLs is now available which with time will no doubt shrink considerably. This large range is in some ways detrimental as it reduces portability and dilutes programming expertise.

SAQ 6.1   Which of the following statements are true and which are false? If you think a statement is false explain why.

1 C has achieved its popularity by virtue of the fact that it has extensive facilities for accessing databases.

2 One reason why Java has achieved its popularity is because of its Internet programming facilities.

3 C++ and Java are quite similar.

Solution

1 is false, C achieved its popularity because of its connection with UNIX and via the fact that it provides quite a lot of low-level processing facilities. 2 is true. 3 is true since they are both based on C.

SAQ 6.2 What programming language would you use for the following three applications?

1 A banking system which uses a large number of screens and accesses a relational database and also produces a large number of reports.

2 A program which accesses financial information over the Internet.

3 A system where some degree of reuse is required.

Solution

1 would use a fourth-generation language. 2 would almost certainly use Java and 3 would use an object-oriented programming language, either Java or C++ would do there.

■ Software development environments

One of the most startling programming developments since the mid-1990s has been the rise of the software development environments. Such environments have produced major productivity gains by automating a large number of functions that the programmer has to carry out. A typical software development environment is Symantec Café which is used for the development of Java programs. A typical environment offers the following facilities:

■ Fast compilation: thousands of lines of source code statements can be compiled in a fraction of a second.

■ The ability quickly to build up windows-based (direct manipulation) interfaces to a system. Using a software development environment a programmer is quite capable of developing interfaces which contain a large number of widgets involving buttons, menus, scrolling areas and menu items in less than an hour – a task which in the past would be programmed by hand and would often take days.

■ The ability easily to program the interactions between elements in a direct manipulation interface and the underlying software. Many software development environments provide facilities whereby all the programmer needs to do is to select a particular widget and is then prompted by the environment as to what programming actions are required when an interaction with that widget occurs.

■ The ability to support the development of a system which contains a large number of chunks of source code. Keeping track of the elements of such systems is a really difficult task and modern software development environments organize the code into entities known as projects and provide screens which enable the programmer to see the complex relationships between the chunks.

■ Object-oriented development environments often contain component libraries which enable the software developer to integrate chunks of software with a specific functionality into a system that is being developed; for example, some development environments offer the facility to add a simple spreadsheet to a user interface.

6.3 SYSTEMS SOFTWARE

■ Utility software

Certain computing requirements are common to a high proportion of computer users and so generalized utility software is available to meet this need. Utility software is intended to be sufficiently flexible to meet most users' requirements and is tailored to meet their precise needs by parameters entered prior to use.

Some or all of the utility programs described below may be incorporated into the operating system that is used with a particular computer.

File conversion

This covers the transference of data from any medium to any other, e.g. magnetic tape to magnetic disk. This may be done either as an exact copy or with simultaneous editing and validation.

File copying

An exact copy of the file is made and written to the same storage medium, e.g. a replica of data records on disk is made on another disk area.

File reorganization

As explained in Section 5.2, direct access files overflow and consequently the overflow records are stored in designated blocks. This is acceptable up to a point but from time to time it is necessary to reorganize the file so as to remove the overflow. This entails reorganizing the cylinder and block indexes, and transferring overflow records back into their home blocks.

Sorting

Quite often it is necessary to arrange records into a certain sequence based on the values of their keys, as described in Section 5.1. A number of factors enter into the choice of the most suitable sorting method. These are:

■ the number and average size of the records to be sorted;

■ the desired sequence (ascending or descending);

■ the degree of sequentiality already present in the set of records; and

■ the hardware units available, e.g. the number of magnetic tape decks.

The above parameters are used by a sort-generator to set up the most suitable type of sorting program.

Dumping routines

These transfer the program and its working data to the backing storage at regular intervals. A dump routine is used in conjunction with a restart program, which reloads the main store with the program and working data.

Housekeeping operations

These are programs or parts of programs not directly concerned with the solution of the problem in hand. Examples are clearing areas of storage now redundant, writing magnetic tape labels, updating common data in records, e.g. the current date, and so on.

Trace routines

These entail the dumping, display or printing of the program or other contents of the main store during program testing to facilitate error detection.

Program control

The concept of multiprogramming, i.e. the ability to run several programs at once, was introduced in Chapter 3. To accomplish multiprogramming, an 'executive' or 'supervisor' program is employed to control the application programs. The executive program has the power to interrupt a running program and pass control to another program. This happens when, for instance, a printer signals that it is ready for more data. The executive program interrupts the current program, loads the print buffer with data waiting to be printed and then returns control to the original program. The printer then prints from the buffer autonomously while processing of current data continues.

When two running programs both require the same printer, the supervisor allows them both to proceed concurrently. One program is allowed to output data to the printer as it runs. The other's data is loaded to disk, i.e. spooled, for subsequent output when the printer becomes free.

It is necessary to hold the executive program permanently in main store, i.e. it is main store resident. The application programs, on the other hand, are moved in and out of partitions of the main store to suit the jobs being currently carried out.

Virtual storage

The concept of virtual storage is to enable a program to address more storage than is currently in the main store. Without virtual storage a program needing, say, 3 megabytes of main storage cannot be run on a computer with only 2 megabytes.

Virtual storage operates on the principle of holding programs on disk and transferring segments (pages) of the programs into main store for execution. This is a complicated process, known as 'paging' or 'segmentation', and is controlled by a special program in the operating system (see below). The pages of main store assigned to each program are adjusted dynamically in order to maximize the efficiency of the multiprogramming.

Operating systems

As mainframes and minicomputers became more powerful, their ability to run jobs in multiprogramming mode outran the abilities of their human operators. The set-up times of the jobs became proportionally greater, with the result that the computer remained idle during these times. An operating system precludes this by allowing the operator to stack jobs for subsequent continuous processing.

It should be remembered that in some IS departments the computer carries out a wide range of activities, e.g. in educational institutions. These include the translation of source programs into object programs, especially compiling, and the running of a wide range of jobs. Often the jobs can be segregated into two main types – background and foreground. Background jobs are regular and of known duration and operational requirements. Foreground jobs are occasional and of less-known characteristics. Operating systems endeavour to maintain a balance between the requirements of these various jobs.

An operating system consists of a suite of programs, one of which, the master, kernel or executive program, remains resident in the main store. This program controls the other operating system programs in the suite and between them they control the application programs.

Personal computer operating systems are often integrated with user interface management software, i.e. the operating system includes routines for window, icons, mouse and menus (WIMPs) in its suite of programs. This results in a common look and feel to all the applications which use the operating system, i.e. the interfaces all have the same appearance in terms of window and menu bars (see HCI section in Chapter 8) and they interact in the same way by a combination of mouse and menu commands. Another trend is to bundle databases and networking facilities with the operating system, so Windows NT from Microsoft comes with a database and network communications software which allows LAN (local area network) and other connections. In some cases a range of applications software is controlled by the operating system and supplied automatically with the PC. Examples of such software include word processing, electronic mail, networking, spreadsheets, graphics and file handling.

When considering a PC and its operating system, a number of factors should be taken into account:

■ The amount of memory occupied by the operating system.

■ Does it permit fast running of the application programs?

■ How many processes and VDUs can it control simultaneously?

■ What choice of programming languages does it allow?

Well-known microcomputer operating systems are:

■ **System 7**   An operating system for the Apple Macintosh.

■ **OS–2**   A 32-bit multi-tasking multi-user operating system produced for IBM by Microsoft.

■ **PS–2**   The IBM version of OS–2 comes with DB–2 database and SAA communication network control and CUA (IBM's GUI).

■ **Microsoft Windows 98**   A WIMP-based operating system for microcomputers.

■ **Microsoft Windows NT**   The successor to Windows as a 32-bit OS.

■ **UNIX**   A 32-bit multi-user system with file-handling capabilities, based on the structured language C.

■ **XENIX**   An offshoot of UNIX.

■ **LINUX**   A free version of UNIX.

Currently, the microprocessor market is dominated by Windows 95 or 98 for single microprocessors. The minicomputer market is dominated by Windows NT and UNIX. The Windows 2000 operating system which will be launched by Microsoft in 2000 will replace Windows 98 by Windows NT technology.

Functions of an operating system

■ **Priority assignment**: jobs awaiting execution are scheduled according to either a predetermined or a dynamic assignment plan.

■ **Control of multiprogramming**: as described above.

■ **Spooling**: the control of input/output peripherals in order to achieve their best utilization.

■ **Communication**: control of data transmission between terminals and the computer, and computer to computer.

■ **Dynamic allocation**: of main and backing storage, including virtual storage.

■ **Database**: control of the database management system.

■ **Software control**: of assemblers, compilers, utility software and subroutines, so that these are immediately available when required.

■ **Operator communication**: via the console printer or VDU.

■ **Operations log**: recording of details of jobs carried out by the computer.

■ **Debugging and editing**: new programs, in conjunction with the compiler, and passing error messages to the user.

■ **Application package control**: especially with PCs, as described above.

Job control languages

Each job coming under the control of a mainframe or minicomputer's operating system is specified by its 'job description'. The operating system is told how to carry out a job by a job control language. This comprises a number of control commands that are retrieved from disk storage along with the job data.

Below is a selection of typical job control commands:

EXECUTE – carry out the current program.
SORT – sort a disk file.
ABORT – abandon the current job.
DELETE – removes a program or file from storage.

6.4 APPLICATION SOFTWARE

Application software comprises the programs that are written specifically to achieve results appertaining to the company's activities. In other words, application software is user-oriented as opposed to systems software which is computer-oriented.

Application software comes from two sources, i.e. the user-company's own staff or from external agencies. It is necessary to compare the cost and staffing problems of maintaining the user-company's staff against the cost and risk of relying upon software suppliers.

Purchased software may be found to be too restrictive, unreliable, badly documented, and expensive to update so there has been a reluctance to rely on third-party (externally sound) software. More recently there has been an increased use of third-party software, especially as applied to PCs. Firms using PCs do not generally have IS staff and, since other staff are fully engaged with their normal work, it is necessary to purchase application programs.

■ Application packages

A business application package is a complete suite of programs together with the associated documentation. It covers a business routine, and is usually supplied by a computer manufacturer or software house, on lease or purchase.

A package is normally intended to meet the needs of a number of different user companies. In order to achieve this aim, most packages are of modular design and so can be constructed on a building-block principle to cater for the needs of the individual user. A package often also contains a number of options, these are selected by the user by the insertion of parameters before use.

Of the many different applications covered by packages, those prominent are financial accounting, auditing, payroll and stock control, and in most cases it is important that these applications are integrated with each other. Thus the sales accounting, purchase accounting, nominal accounting and stock control should be integrated so that the final accounts, i.e. balance sheet, profit and loss statement, etc., can be produced automatically.

In addition to general packages there are those directed to specific types of user companies. Examples of these are stockbrokers, insurance brokers, estate agents and travel agents.

Advantages of application packages

The following benefits should accrue from the adoption of an application package:

- *Implementation* is quicker and possibly cheaper.
- *System design, programming and system testing* are minimized.
- *System documentation* is provided with the package.
- *Portability* from the existing system to a new system.
- *Efficiency* in terms of speed, storage requirements and accuracy.

Considerations regarding application packages

Definition of requirements

The user company cannot abandon the study of company objectives, systems investigation and consequent definition of IS requirements. In these respects the approach is the same as when designing an IS for in-house programming.

Study of range

A range of packages should be examined in depth before a choice is made, and existing users of the packages consulted for their practical experiences and opinions. It should be remembered that the more commonly used packages result in pressure on the suppliers to keep them up to date.

Interfacing

How easily does a package interface with the user's own routines, both existing and future? How well does the package run on the company's platform – this means the computer, its operating system and operating environment, e.g. the database and data dictionary? Only too often software vendors claim a system is supported when in fact the application package has not been run on the particular operating environment.

Amendments

At least one member of the user-company staff must be completely conversant with the operation and capabilities of the package adopted. This facilitates in-house tuning and subsequent amendments. Moreover, it is important that a package's output is fully meaningful to the end-users.

The package supplier must be able to provide all the necessary ongoing amendments, especially those related to legislation such as taxation.

Performance

How efficient is the package in terms of its average and maximum run times on the computer? What resources does it demand such as peripherals and amounts of main and backing storage?

Contract terms

The contract should embrace factors such as the terms of payment, supplier's assistance with implementation, extent of documentation, and future maintenance. It is important to have some measure of security should the supplier go out of business. Most application packages are supplied in executable form only. This means the source (program) code is not available to the purchaser. If the vendor goes out of business no further modifications will be possible. Some agreement such as depositing a copy of the source code with a bank as an insurance is advisable.

■ Types of software

As mentioned above, PCs have engendered the development of a large amount of software. This has tended to fall into the main types explained below. Versions of these types of packages are also available for larger computers.

Application wrappers

This is a collection of software tools which allows you to enclose existing software with an interface that enables more modern software to interact with it. A typical use of a wrapper is where a developer might have hundreds of thousands of lines of old COBOL code and has started to use an object-oriented programming language such as C++: an object wrapper tool surrounds this code with a layer of software which enables the C++ language to view the code as a collection of objects and to interact with this software. Such tools are becoming very popular as companies have invested so many resources in their existing software that they find it impossible to redevelop it in a more modern language and yet want to take advantage of the facilities offered by languages such as C++ and Java.

Groupware

This is the term given to a collection of software which carries out the following functions:

- electronic mail
- keeping diaries and meeting dates
- scheduling meetings
- administering large collections of data such as spreadsheets
- co-ordinating the actions of a number of staff, for example ensuring that all participants in a meeting have agreed to a decision before releasing documentation from that meeting.

Groupware is often used as a component of an information system, for example it might be used to keep and co-ordinate sales records in a purchasing system. It is also gaining use as a tool for administering software projects; probably the most well-known groupware system is Lotus Notes. One of the major features of groupware is that it contains interfaces which allow companies to write software which interact with the groupware in order to carry out application-specific tasks which the groupware cannot do. Most groupware products have interfaces to languages such as C, C++ and Java or have some special-purpose scripting language associated with them.

Online Analytical Processing tools

Online Analytical Processing Tools or OLAPs are programs which enable managers to query a set of databases with complicated questions whose answers are needed for the efficient functioning of their business. For example, an OLAP enables a question such as:

What sales were below the liquidity ratio we adopt for fast-moving items over the period when the sales of our medium-speed items were running at a 40% below market share proportion?

OLAPs normally translate such queries into the form of a standard query language such as SQL which then interrogates a database.

Workflow software

Consider a company that issues motor insurance policies and also processes claims on those policies. This type of company is normally a heavily paper-intensive organization, where items of paper often flow between a large number of staff. Workflow software enables handling processes to be highly automated. A typical workflow tool will store items of paper which require processing, provide screens which enable workers who have to carry out such tasks to enter data, schedule tasks among staff and report on the progress made with individual items.

Business process capture tools

A business process is an individual task which is carried out in an application. For example, evaluating a customer's request for a loan is an example of a business process that makes up part of the business processes in a bank. Business process capture tools enable analysts to write the various steps and rules that make up a business process and translate those rules into program code which carries out standard software processing such as updating a database. Increasingly companies who use information systems codify their business processes using formal or semi-formal languages; business process tools are able to take advantage of this by producing much of the routine processing code which, in the past, would have been hand crafted by programmers.

Data mining tools

These are a collection of tools which allow the information systems customer to discern trends and common information in what could be a large number of databases which may have evolved in a company with little thought for integration – with some of the databases being very unstructured. A data mining tool would examine the data stored in the databases and discern whether there are any interesting relationships between entities which are stored in them. For example, data mining tools have been used to predict which customers would be likely to move from an existing company to another one, to detect which financial transactions are fraudulent and decide which customers should be sent which particular mailings for a range of products which they might buy.

Spreadsheet software

A spreadsheet, also known as a worksheet, is a multi-purpose software tool that is usable for a variety of planning, modelling and forecasting purposes, e.g. budgeting sales analysis and break-even analysis.

The principle of a spreadsheet is that it simulates a large matrix of cells within each of which a data item or formula can be held. Once in a cell, the data item or formula can be replicated, moved, sorted, filed, printed and so on, at will. These facilities provide the means of easily and rapidly processing the data without the user needing to write a program. The spreadsheet is controlled by a set of user commands in conjunction with a small menu that appears at the bottom of the screen.

A display of cells, i.e. a window, is scrolled up or down and right or left so that any cell may be inspected at will. The display shows around 160 cells at a time out of many thousands that exist on one spreadsheet. Some versions allow different areas of the spreadsheet to be displayed and scrolled simultaneously but independently, i.e. several windows. It is also possible to transfer data from one spreadsheet to another so as to build up the final file or document.

To take a simple example of using a spreadsheet, suppose we wish to create a table of mortgage repayment accounts for various amounts loaned over various repayment periods. The usual procedure would be to enter manually the appropriate formula into one cell and then replicate it in modified form in certain other cells. There would be one cell per amount/period combination and the modified formula in each cell would represent an actual amount/period combination. Having done this, all the formulae are automatically evaluated so giving the repayment amounts. These are then printed along with headings and annotations to form the required table.

It must be emphasized that this simple example is by no means the limit of spreadsheet capabilities. An almost limitless range of work can be performed to give a wide variety of results. Among the better-known spreadsheet programs are Excel and Lotus 1, 2, 3.

Word processing software

The concept and features of word processing (WP) are explained in Section 3.5. These are so commonly required as to permit the development and utilization of standard packages. A feature of WP software is its ability to interface with other text that automatically incorporates the relevant names, descriptions and numeric data.

The well-known WP software packages include WordPerfect and Microsoft Word.

Database management software

Databases and DBMSs are described in Section 5.3. DBMS software for mainframes and minicomputers are closely related to the computer itself and supplied mostly via the computer manufacturer. These packages are largely geared for use by professional IS staff and are completely isolated from the end-user.

At the PC level progress has been quite different. A considerable number of DBMS packages are available, some of which do not really justify the name. The salient features of PC DBMSs are their user-friendliness and breadth of capabilities.

Accounting and other software

This includes a wide selection of software for business applications such as payroll, sales, purchase and nominal ledgers, stock control, financial modelling, survey and statistical analyses, operational research and graphics.

■ Middleware

This is the term used to describe software which often sits between application software and system software. Rarely can it be categorized directly as either system software or application software and so it is increasingly being placed in this category by software vendors. The term is also used to describe software which does not readily fit into any of the established categories.

Code converters

These are software tools which take an existing set of programs and convert them into programs in a different programming language. There are a whole series of programming language converters which range from those which just convert from one dialect of a language to another dialect, to those which convert between widely differing languages. Such tools are normally used when a customer moves from an old, perhaps obsolescent language to a newer one.

Database tuning tools

These are tools which optimize a company's use of a database. They work by monitoring the activity that occurs over a period of accesses to the database and fit in between application programs and the system software. The more sophisticated tools will then automatically reorganize the database in order to increase the speed of response to user requests. The less sophisticated tools will provide a report on where the performance bottlenecks are in a database and make suggestions for improving it; however, they leave the reorganization of the database to a human database administrator. A typical action that a reorganization tool would suggest is to move very frequently accessed data to the main memory of a computer.

Compatibility software

These are tools which are aimed at an information systems user who has a variety of computing platforms. Consider such a customer who keeps all his or her data on a large mainframe computer running an operating system which is specific to that computer, but yet also has a number of client computers running a variety of operating systems ranging from UNIX to the Macintosh System 7 which update and query the database. Compatibility software are those programs which interface between a database and client computers; they format any requests and updates from the clients into a form that is recognized by the mainframe computer.

Security software

An increasing problem with modern information systems is keeping their data private. Even in the early days of information systems a computer's file store would contain sensitive data such as an employee's salary which could be used by outside agencies. Today, where the computer is often used as a strategic business tool, there is a huge imperative on the information systems developer to include a high degree of security. Operating systems provide some degree of security, for example by providing login facilities whereby every user has to provide a password. However, often this is not enough, and further software has to be used. Some typical security tools are described below:

- Security monitors. These are programs which sit between the user and the computer's operating system. They monitor the usage made by members of staff and build up a profile of the ways in which they interact with the computer: what time they normally log on, what time they log off, how many times they log on and log off, what files they access, what programs they use, etc. These monitor programs, as well as building up this profile, will also notify systems administration staff when a non-typical access occurs, for example when a member of staff logs on at night. In this way an intruder can often be caught.

- Hostile software detectors. One way for a system to be compromised is for a program to be inserted into a computer system which carries out some gathering of data and then, for example, e-mails the data found back to a criminal. Hostile software detectors continually monitor a computer system for programs being executed which have a non-standard behaviour and immediately terminate them and provide a report back to a system administrator.

- Password dispensers. One of the problems with passwords is that most users of a computer system choose passwords which are easy to detect. Typical choices that users make include their name, their dog's name or the name of their spouse. A password dispenser provides a password which is not so easy to detect but, at the same time, can be remembered. A typical password produced by such a tool is *plipType23*.

SAQ 6.3 Under which of the categories, system software or application software, do the following items of software fall?

1 A logging program which keeps track of the usage of a computer by users.

2 A program which allocates memory to individual programs.

3 A program which carries out salary calculations.

4 A piece of software which administers a warehouse of stock items.

5 A program which asks for a user password before he or she is allowed to use a computer.

Solution

1 and 2 are examples of system software, 3 and 4 are examples of application software and 5 is an example of a piece of system software.

■ Internet and Intranet software

Since the late 1980s there has been a huge increase in the use of the network of computers known as the Internet. The term *Internet* is given to a collection of networks of computers which link companies, governmental organizations and educational establishments throughout the world. Companies are now building their own networks which mimic the Internet and to which the term Intranet is given. The next ten years should see a large deployment of company Intranets as organizations attempt to make their computing more distributed. Both the Internet and Intranets depend on major items of software; this section describes them.

The Internet

The key part of the Internet is the network. This consists of a number of computers physically connected together using cables or wireless links. Each component of the Internet usually consists of a number of sub-networks associated with a single company. Computers in the Internet send messages using a mechanism known as TCP/IP (Transmission Control Protocol/Internet Protocol). This enables a message to be accurately sent from one computer to another computer, even across insecure and unreliable communication lines. The Internet supports a number of components – both hardware and software; this section concentrates on the software.

E-mail

This is probably the most popular use of the Internet and predated it as companies which used their own networks often used electronic mail to communicate over networks which might just be connected by a simple local area configuration. E-mail offers the facility for someone to send messages, attached files and documents to another user of the Internet. E-mail software allows users to send messages, receive messages, place them in folders, filter messages so that, for example, you can ignore messages from another user, recover files which have been sent using e-mail and set up mailing lists to which you can send messages directly.

Web documents

One part of the Internet which has experienced explosive growth has been the World Wide Web. It was initially developed as an internal information system at CERN, the European Nuclear Research Establishment, by the computer scientist Tim Berners-Lee. The World Wide Web consists of a number of documents written using a simple language known as HTML. You will see very little that is remarkable when looking at a World Wide Web document. It will contain paragraphs of text; headings; sub-headings and sub-sub headings; figures; bulleted

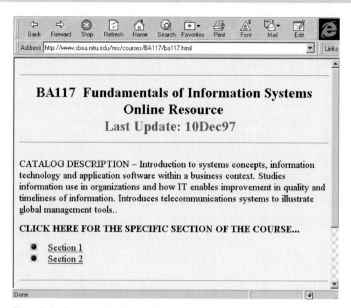

Fig. 6.1 A page displayed by a browser

lists and quotations. Effectively they look like word processed documents. A typical Web document is shown in Figure 6.1.

The one thing which distinguishes a Web document, however, is the presence of links. Figure 6.1 shows two such links: the first to another part of the document and the second to a document held on a computer at another location. What these links represent are the addresses of another Web document. These documents may be on the same computer on which the Web document is found or might be found on a computer in another continent. It is this linking which gives the World Wide Web its name: that it resembles a spider's web of connections.

The connections in a Web document allow the users of the document to access related information. For example, a Web document which contains information about, say, educational courses on information systems might contain pointers to university departments which present such courses.

Web browsers

A Web browser is a program which reads Web documents and displays them on a computer. It has two functions: the first is to display the documents properly formatted, it will do this by interpreting the HTML code that the document is written in; the second is to establish the communication between the computer which contains the browser and the Web document and whatever other documents the user references by following links to other Web documents. There are a number of browsers in existence, almost certainly the most well known are *Netscape Navigator* and *Internet Explorer*. They are now very sophisticated items of software and typically offer facilities such as:

- The ability to browse through pages in the World Wide Web.
- The ability to move from one page that has just been read to the previous page that was being read.
- The ability to store Web pages on the computer where the browser is being used.
- The ability to transfer large files across the Internet.
- The ability to keep track of popular Web documents and connect to them quickly.

Figure 6.1 shows a page being displayed by the Internet Explorer browser. The top bar contains icons which allow the user to navigate through a Web document in order to search for particular pages and alter the way in which the browser displays these pages. The page displayed is the first part of a syllabus for a course in Information systems at Michigan Technical University. The underlined parts are links to other pages.

Web servers

A server is a computer in a network which usually carries out a dedicated task: mail servers administer the transfer of e-mail; file servers administer the access to files in a system; print servers allow users access to limited printing facilities and Web servers are computers on which Web documents are contained and which provide browsers with access to these documents.

Mail lists

A mail list is a file which contains a number of e-mail addresses of computer users who have some common characteristic. For example, on the Internet there are e-mail lists for academics who have common research interests, users who want latest product announcements and staff such as operating system admin-istrators who carry out the same tasks.

In a company, mailing lists will normally be formed by virtue of the fact that the members of the list are in the same department or are currently working on the same project.

Newsgroups

A newsgroup is very similar to a mailing list. It is a group of computer users who have some common interest such as a particular programming language or some form of music. Where newsgroups differ from mailing lists is the fact that they are a little more interactive. A member of a newsgroup might send a message to the group, for example asking for help on a particular topic. This request would be displayed to all the members of the newsgroup who might then respond. Any responses are automatically posted to all the members of the group.

Search software

One of the major problems faced by users of the Internet is of finding information. When you consider that the Internet consists of tens of millions of computers, each storing tens of millions of words of information, it is quite surprising that any information is found. Some of the most useful tools found on the Internet are search engines. These are programs which ask the user what type of information is required and then search for it. Normally these tools prompt the user to provide keywords which are then searched for over the Internet. For example, you may want to know if any documents are relevant to your interest in the programming language C and the use of the language in manufacturing applications within the auto industry. Such a query would be communicated to a search engine by typing in

C + Manufacturing + Auto

The addition sign means that the user wishes to search for all the Web documents which contain *C* and *Auto* and *Manufacturing*. If the query was for all the documents which contained the word C or the word Manufacturing or the word Auto, then the user would type

C Manufacturing Auto

Figure 6.2 shows part of the output from a search engine which was asked to find all the Web documents containing the terms Information and System (only summaries of the first three documents found are displayed). Each entry contains a link to the document and a summary of its content.

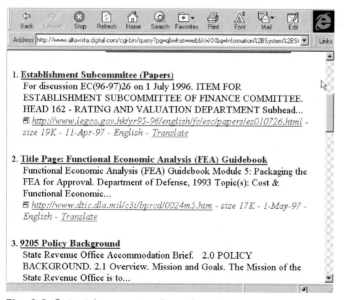

Fig. 6.2 Output from a search engine

■ Software components and reuse

As more applications are developed it becomes increasingly obvious that the same thing is being written time and time again. The idea of software reuse has developed to try to stop the endless cycle of reinventing the 'software wheel'. Reuse has been practised on a modest scale for many years. For instance it is commonplace to keep data definitions in COBOL in a data dictionary and then import them ready-made into the program. Similarly many algorithms for graphical routines and scientific calculation are well known and publicly available in libraries such as the NAG (Numerical Algorithms Group). Subroutines can be imported from these libraries into the application program.

The advantages of software reuse are the saving of coding effort and use of reliable tested code. However, problems have arisen when reuse was scaled up from small algorithms and data structures. One of the motivations for object-oriented design was to improve software reuse (see next chapter); however, this too has failed to deliver much improvement. Against this somewhat dismal picture there are signs of increasing reuse in practice. Partly this has been a management problem. Programmers do not trust other people's code so there is a tendency to program anew each time. Also there has been a problem of critical mass; until there is a large library of reusable software components in a repository, there is insufficient choice for reuse. In some companies these problems are being solved and reuse of up to 90 per cent of the modules in a new application have been quoted. This is a considerable saving in development manpower.

The increase in popularity of object-oriented programming languages has given rise to an increasing number of component libraries. There are two categories of such libraries:

■ Portable libraries. Information systems users often have to put up with a number of different computing platforms; for example, a company may use both Windows 95 and UNIX as their main operating systems. There are major problems in porting an application developed for one platform to others. One solution to this is to use cross-platform libraries of software. A number of libraries provide facilities for direct manipulation interfaces across a wide variety of platforms: such libraries ensure that applications have a common look and feel across a variety of implementations.

■ Object component libraries. The last three years have seen a large increase in the availability of libraries of objects which can be used directly in a program. Such objects range from simple spreadsheets to individual HCI components such as buttons and text fields and also include complicated components such as animators which allow moving pictures to be displayed on a computer screen. Such libraries contribute to major reductions in programming cost and move software development towards the day when the process of constructing an information system will mainly consist of assembling components.

Reuse has promoted development of 'vertical markets in software', whereby a software vendor develops not just one application package but a whole series of modules to tackle problems in one particular market sector, e.g. banking, shipping, insurance, etc. The software purchaser can then buy a library of modules and design by composition of building blocks. The vendor may sell a template as an application frame with a set of more specialized modules that purchasers then fit into the template according to their needs. Alternatively a library of modules may be purchased for an application area and the purchaser designs new applications using these as building blocks.

Thus there has been a successive change in the software market from buying specific application packages to tailorable application packages, and, in the future, to buying a library of modules for development by reuse. Application packages are already tailorable; for instance, most accounting packages allow the user to specify report formats, load different cost centre and account codes, and change cross-posting routines. The vision for the future is in flexible development by reuse. Nevertheless there are certain problems which stand in the way. One is the legal issue: who is responsible for the reliability and maintainability of a reused system, the original author or the purchaser? Secondly, there is a problem of ownership: could the purchaser then sell a composed but reused system? In spite of these problems software reuse will probably continue to increase in economic importance and increase the market for third-party software.

Software and re-engineering

Over the years software systems have been developed and then, in response to changes in requirements, have been modified time and time again. This has led to systems which while satisfying their requirements are baroque and unstructured. These so-called legacy systems have been written using old technologies such as hierarchical or network database systems and have become a nightmare to keep going. Typical areas where systems which were written years ago are still in existence include payroll and stock control systems.

There comes a point, however, when the cost of maintaining a system becomes excessive, and the only option left is to re-engineer it: that is, re-implement the system using a newer technology and reduce the system's complexity by partitioning it into smaller and smaller modules. Increasingly the cost of this process is driving software developers into considering techniques where software can be reused.

Exercise 6.1   **Microcomputer software**

CD Ltd is a large finance house with many microcomputers that are used for a wide variety of administrative tasks.

You are required:

(a) to describe briefly the major categories of software which CD Ltd is likely to have on its stand-alone microcomputers, giving an example from each category; and

(b) to explain the criteria which CD Ltd should use in selecting its software.

(CIMA, stage 2, Inf. Tech. Man., May 1992)

Exercise 6.2   **Fourth-generation languages**

What is a fourth-generation language? Explain the main features and usefulness of one such language.

Exercise 6.3   **Software quality**

Organizations that are introducing IT to help them perform more effectively will be concerned about the 'quality' of the software to be purchased. Explain what you understand by 'quality' of software.

What techniques would you expect your supplier to be using to ensure software quality?

(BCS, 2E, April 1992)

Outline solutions to exercises

Solution 6.1
(a) Refer to Sections 6.3 and 6.4.

(b) The main criteria are as follows:

- *clarity* – easy to understand, well documented, clear guidance displayed;
- *reliability* – prevention against loss of data, detection of input errors, no latent bugs;
- *portability* – compatibility with other PCs;
- *maintenance* – regular updating by supplier with good documentation;
- *support* – training and on-tap advice from supplier;
- *efficiency* – speed of operation, economy of memory and disk storage if extensive software;
- *interfacing* – with other systems and applications where relevant; and
- *adaptability* – to cope with changes, especially increased volumes of data items.

Solution 6.2
Refer to 'Fourth-generation languages' in Section 6.2. For example, the main features of FOCUS are as follows:

- mainframe and microcomputer versions;
- own relational database;
- modular form of purchase, i.e. input, dialogues, displays, report generation, graphs, financial modelling, statistics;

- huge gain in programming productivity over 3GLs;

- sophisticated database interrogation including non-FOCUS files;

- interactive involvement during processing.

Solution 6.3   Refer to Solution 6.1(b).
Techniques to be expected are:

- structured programming;

- WIMPs (Glossary);

- program error-checking methods;

- human–computer interface (Section 8.5); and

- low-level programming language.

References and further reading

C++

6.1   Ammeraal, L. *C++ for Programmers* (John Wiley, 1995).
6.2   Dattari, K. *C++: Effective Object Orientated Construction Concepts, Principles, Industrial Strategies and Practices* (Prentice Hall, 1997).
6.3   Horton, I. *Beginners Guide to Visual C++* (WROX Press, 1996).

Java

6.4   Cornell, G. *Core Java* (Prentice Hall, 1996).
6.5   Horton, I. *Beginning Java* (WROX Press, 1997).
6.6   Tittell, E. *Discover Java* (IDG Books, 1997).

COBOL

6.7   Arranga, E.C. and Coyle, F.P. *Object-Oriented COBOL* (Prentice Hall, 1996).
6.8   Griffith, A. *COBOL for Dummies* (IDG, 1997).
6.9   Parkin, A. *COBOL for Students* (Edward Arnold, 1996).
6.10 SAMS Development Team, *COBOL Unleashed* (Howard Sams & Co., 1998).
6.11 Stern, N.B. *Structured COBOL Programming* (John Wiley, 1997).
6.12 Wessler, J. *COBOL Unleashed* (SAMS, 1998).

BASIC

6.13 Cornell, G. *Core Visual BASIC 5* (Prentice Hall, 1997).
6.14 Eddon, L. *Active Visual BASIC* (Microsoft Press, 1997).

Pascal

6.15 Buchanan, W. *Mastering Pascal and Delphi Programming* (MacMillan, 1998).
6.16 Findlay, W. and Watt, D.A. *Pascal* (UCL Press, 1994).
6.17 Hahn, B.D. *Pascal for Students* (Edward Arnold, 1996).
6.18 Holmes. B.J. *Computing: Pascal Programming* (Letts Educational, 1991).
6.19 Sargent, D. *Coding in Turbo Pascal* (McGraw-Hill, 1994).
6.20 Shaffer, D. *Abstractions and Programming in Turbo Pascal* (W.B. Saunders, 1994).
6.21 van den Weyer, R. *Pascal in a Nutshell* (Hodder & Stoughton, 1997).

C and C++

6.22 Bronson, L. *A First Book of ANSI C* (West, 1995).

6.23 Deitel, P.J. *C++* (Prentice Hall, 1998).

6.24 Horstmann, C. *Computing Concepts with C++ Essentials* (John Wiley, 1998).

6.25 Irvine, K.R. *C++ and Object-oriented Programming for Alphabetic Reasons* (Collier MacMillan, 1996).

6.26 Kelley, A. and Pohl, I. *C by Dissection* (Benjamin Cummings, 1996).

6.27 King, M., Pardoe, J. and Vickers, P. *A First Course in Programming Using C* (McGraw-Hill, 1994)

6.28 Meyers, S. *More Effective C++* (Addison Wesley, 1995).

6.29 Perry, G.M. *Borland C++ by Example* (Prentice Hall, 1994).

6.30 Perry, J. *Advanced C Programming by Example* (PWS, 1998).

6.31 Schildt, H. *C++* (McGraw-Hill, 1992).

6.32 Sedgewick, R. *Algorithms in C* (Addison Wesley, 1990).

6.33 Skansholm, J. *C++ from the Beginning* (Addison Wesley, 1997).

Fourth-generation languages

6.34 Bernknopf, J. *Practical Guide to Fourth Generation Programming Languages* (McGraw-Hill, 1989).

6.35 Crumlish, C. *ABCs of the Internet* (Sybex, 1996).

6.36 Eager, W. *Using the World Wide Web* (QUE, 1994).

6.37 Frost, A. *Exploiting the Internet* (John Wiley, 1997).

6.38 Garfinkel, M. *Web Security and Commerce* (O'Reilly, 1997).

6.39 Hahn, H. *Harley Hahn's the Internet Complete Reference* (Osborne, 1996).

6.40 Holloway, S. *Fourth Generation Systems: Their Scope, Application and Methods of Evaluation* (Chapman and Hall, 1990).

Internet

6.41 McKinley, T. *From Paper to Web* (Hayden, 1997).

6.42 Meehan, R. *An Introduction to Fourth Generation Languages* (Stanley Thornes, 1991).

6.43 Minoli, D. *Web Commerce Engineering* (McGraw-Hill, 1997).

6.44 Nichols, S. *Inside the World Wide Web* (New Riders, 1996).

6.45 Pitter, K. *Every Student's Guide to the Internet* (McGraw-Hill, 1995).

6.46 Ross, J. *Discover the World Wide Web* (IDG Books, 1997).

6.47 Watts, R. *Application Generators Using Fourth Generation Languages* (Blackwell, 1987).

Operating systems

6.48 Deitel, H.M. *Operating Systems* (Addison Wesley, 1990).

6.49 Ericksen, L. *Quick, Simple Windows 98* (Prentice Hall, 1999).

6.50 Kantaris, N. *A Concise Introduction to UNIX* (Bernard Babani, 1988).

6.51 Nutt, G.J. *Operating Systems: a Modern Perspective* (Benjamin Cummings, 1997).

6.52 Silberschatz, A. and Galvin, P. *Operating System Concepts* (Addison Wesley 1998).

6.53 Tanenbaum, A.S. *Modern Operating Systems* (Prentice Hall, 1992).

Spreadsheets

6.54 O'Neill, S. *Essential Spreadsheet Skills* (Stanley Thornes, 1994).

6.55 Richardson, R. *Professional's Guide to Robust Spreadsheets: Using Examples in Lotus 1 2 3 and Microsoft Excel* (Prentice Hall, 1996).

6.56 Robson, A.J. *Designing and Building Business Models Using Microsoft Excel* (McGraw-Hill, 1995).

7 Systems investigation

AIMS

After reading this chapter you should:

- understand the nature of systems investigation;
- be familiar with the human aspects of any system;
- be familiar with the duties, attributes and personal qualities of a systems analyst;
- be familiar with the processes that comprise the activity of planning a systems investigation;
- understand the nature of users' information requirements;
- understand how analysis is carried out in terms of the flow of information within a company;
- be familiar with the main notations used to document information in an application;
- be familiar with the ways in which business processes are documented in an application;
- understand the way in which entity modelling is used by the systems analyst;
- understand the differences between the four normal forms;
- be familiar with the various techniques used for fact finding during systems investigation;
- understand the nature of prototyping during systems investigation.

7.1 GENERAL ASPECTS

Systems investigation, sometimes referred to as systems analysis, is not a precise science. It is more of an art augmented by a logical approach to the definition and recording of facts. We cannot always precisely define the objective, in other words the aims of a systems investigation are often rather broad. Although the objectives have been defined by management in clear terms, it does not always follow that these are immediately translatable into investigation procedures. In fact, the systems investigation inevitably starts as a reconnaissance in its early stages. As the picture unfolds, it becomes more apparent where to look next, eventually what to do and how to do it.

It is better if the systems analyst in an investigation avoids having preconceived intentions. Because a method solved most of the problems on previous occasions it does not follow that this will again be the case. Often it is more beneficial to look for differences in the present situation than to reach back for a previous solution.

Each circumstance and problem should be seen as part of a wider situation. Consequently the systems analyst is well advised to keep 'why' in mind as much as 'what, when, who, where and how'. There is obviously no one answer to all the differing situations but there is usually plenty of scope for innovations and lateral thinking in system development. As a general rule, try not to meet a problem head on, i.e. don't use the sledgehammer until the nutcrackers have been considered.

It is also interesting to remember that the person closest to a problem is often the least able to spot a solution. Although it is realized that procedures are not as efficient as they could be, he or she cannot always see past a psychological barrier to the essential ideas beyond. And thus, in course of time, he or she has come to accept the situation as being inevitable and unalterable, the methods as inviolate, and the documents as sacred. The systems analyst must guard against falling into this trap during the course of the systems investigation.

The advent of fourth-generation techniques (4GTs) including fourth-generation languages (4GLs) has introduced the likelihood of new approaches to systems investigation and design, such as prototyping. More is written about prototyping in Section 7.9; for now, prototyping can be regarded as a number of methods that enable systems analysts and end-users to 'talk the same language'. That is to say, the end-user is drawn more closely into specifying what is required by being enabled to see the outcome of a system at a much earlier stage. This facilitates making decisions, and alterations, thereby moving more rapidly towards the final specification.

Figure 7.1 shows diagrammatically a number of stages in systems investigation from the initial objectives through to the detailed fact finding. These stages are explained in full in the ensuing sections of this chapter.

■ Assignment briefs

As shown in Figure 7.1, an assignment brief is a directive from either management or the steering committee to the systems department. Where a steering committee (Section 1.6) exists, the assignment is drawn up by its members. If there is no steering committee, an assignment brief is prepared by the senior manager responsible for management services or, alternatively, by the manager of the department most affected by the assignment.

The significance of an assignment brief depends a great deal upon the extent of the company's present information system. Where a computer system is already in operation, it is likely that the assignment will lead to an extension of its use. It must not, however, be assumed that expanding a computer's utilization is

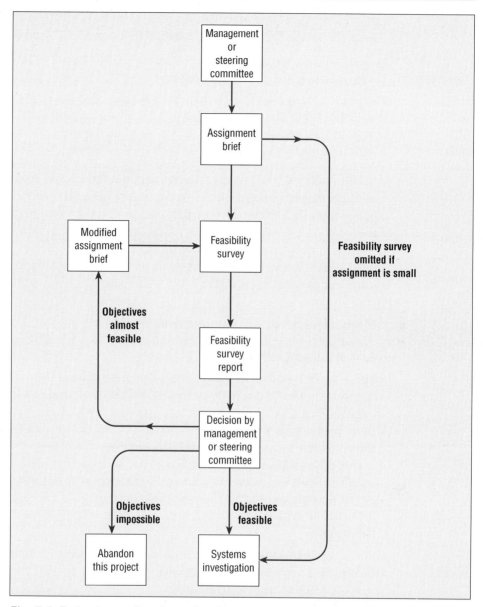

Fig. 7.1 Early stages of systems development

necessarily a trivial exercise. If the existing applications and database are not both open-ended, there may be a considerable amount of work in integrating a new application. This type of assignment sometimes arises as a result of a manager's need for further information usually related to that already coming from the computer system.

If the need for further information is accepted by the steering committee after taking into account the systems work involved, it is generally possible to draw up an assignment brief in entirely precise terms. Because the manager knows what he or she wants, the aims of the assignment are clear and can be specified exactly.

Example 7.1	**Procedure for a works order**
	'Design a system to provide a stockholding report of bought-out parts based on the updated month-end stock positions. The contents and layout of the report are to be shown on the attached specimen.'

In the case of a company whose previous computer use has been either minimal or non-existent, the assignment brief usually covers a much larger area of activities. It is inevitably couched in less specific terms because the aims are wider and the possibilities greater. Although it is likely in this situation that the steering committee has computer utilization in mind, it is better if this is not put forward too strongly. The systems analyst will do a better job if he or she starts out with an open mind.

Main aspects of an assignment brief

Aims of the assignment

These are in broad terms in the case of a general feasibility survey but in more specific terms for a detailed systems investigation. Certain of the aims are liable to be amended as a result of the feasibility survey. A revised assignment brief is then prepared before commencement of the systems investigation. This is shown as the iteration in Figure 7.1, which may be repeated several times.

This process of moving towards the precise requirements is abetted by utilizing prototyped results (Section 7.9) if the manager or steering committee is unsure of the possible outputs of the new systems.

Authorization

The assignment brief acts as an authorization as well as a directive, and should be regarded as a request from top management for all staff to co-operate in the proposed survey or investigation.

In this respect the staff should be made aware of the existence and broad purpose of the assignment so that they do not regard it as an undercover operation by the systems/computer people.

Scope and limitations

The scope of the proposed systems work is indicated in the assignment brief, including any limitations or dictates imposed upon it. These may be absolute and unalterable, such as a finite starting date for the new system or a top limit to the expenditure incurred. As far as possible constraints should be avoided as they tend to restrict the true potential of a new system.

Assistance

Information which may be of use in the investigation or design of a system is provided. This includes factors such as proposed changes in the company's policies and/or structure, reference to previous work of a similar nature, names of persons who can be of immediate assistance, and the company's relationship to outside bodies.

Human aspects of systems

The initiation of a systems investigation often causes disquiet in the minds of the staff in the departments concerned. An investigation, probably to be closely followed by changes within the company, is very disturbing to some people.

The systems analyst, by being aware of the possible consequences for people, is able to allay apprehension and encourage interest and participation. It is therefore worth considering the main reasons for apprehension and the ways in which it can be alleviated.

Career prospects

Middle-level staff, having reached a position a few rungs up the promotion ladder, are inclined to believe that a new system will damage their career prospects. It cannot be denied that a few careers have been damaged by computerization but, on the other hand, many have been greatly enhanced.

For the most part, middle-level staff are able to play an important role in the development and functioning of computer systems. They must, however, adopt a flexible attitude towards their workrole and position, and be willing to learn and operate new methods.

Job security

The possibility of redundancy tends to apply more to low-grade clerical staff since their jobs are more vulnerable to computerization. This is particularly the case where work is of a mundane nature requiring no great level of intellect. Also computers are often installed in a company for the express purpose of reducing clerical costs. In many situations, however, it turns out that there are no redundancies in spite of the original intention to cut costs.

As with middle-level staff, clerical employees can enhance their career prospects by accepting changes and taking advantage of the resultant opportunities. These include a willingness to be retrained into IS jobs such as computer operators, data preparation clerks, keyboard operators and computer programmers. These possibilities should be borne in mind by the systems analyst when assessing future staff requirements.

Explanations

Where extensive changes are to be introduced in a company, the departmental staff may find themselves feeling totally confused. They cannot see how their work skills and jobs can be accommodated within the new system. This is largely because they do not as yet understand the new system or even the general concepts of computerization.

Starting at the investigation stage, it is beneficial if the systems analyst endeavours to allay these fears by explaining how the staff may fit into the new system. This is no easy task in the early stages because the staff's relevance to the new system has yet to become apparent.

Nevertheless staff are encouraged and reassured by seeing input screens or computer-produced results that they recognize and understand. This is a point at which early employment of prototyping can be advantageous.

Involvement

If the departmental staff are made to feel that they are in some way contributing towards a new system, they are less likely to be apprehensive and more likely to be enthusiastic towards changes. They then believe that, to some degree, they are in control of their own destinies. One way in which involvement is brought about is by project teams (Section 7.2), another is by encouraging suggestions and ideas. Much has been written elsewhere about user involvement but in practice only too often the users are frozen out of the development. The consequence is a system developed according to the technical viewpoint of IS staff and not a system that the users want. One method is to encourage participative design in which users lead the system steering committee and IS staff act as advisers. In practice this is difficult as users are not always acquainted with the technical possibilities and limitation of computer systems. A weaker version of participative design is to have users as active members of the design team. This can pay dividends as requirements are transmitted and refined continuously rather than at periodic reviews or, worse still, after the system is delivered. However, the planning of mixed IS and user staff in development teams is tricky and considerable effort must be made to 'speak the user's language' and avoid technological jargon. Another suggestion is to develop the expertise of one or more users as local experts who can explain the technicalities to other users as necessary. These 'local experts' then act as a bridge between the users and the technical IS developers. The danger here is that local experts can end up being seen as one of 'them', i.e. staff, rather than one of 'us'. Clearly user involvement is difficult but every effort should be made to encourage it.

During the early stages of the investigation, the departmental staff should be shown films and given short talks to stimulate their interest in new methods. This approach is especially relevant to computer technology and the usage of PCs, computer terminals and word processing in offices.

■ Systems analysts

The work of most systems analysts is in establishing and maintaining computer-based systems in the fields of business and administration. Their employers are industrial and commercial companies, national and local government departments, financial institutions, computer manufacturers, hospital boards, consultancy firms, and others.

It is evident from the above that the range of work undertaken by systems analysts as a whole is very broad. The tasks of an individual systems analyst are, of course, much narrower during one period of time, but nevertheless provide plenty of scope for innovation.

Duties of systems analysts

What are the specific duties of the systems analyst?

1 Investigating the existing information usage, systems and procedures of the company with a view to discovering inefficiencies and problems. Of significant interest are areas of work incurring high costs, long throughput times and considerable clerical effort.

2 Analysing the findings of the investigation so that they can be used effectively in designing a new system.

3 Designing a new system that increases efficiency, minimizes problems and achieves the objectives set by the steering committee.

4 Testing and implementing the new system, including its appraisal before and after implementation, its documentation and continuing maintenance.

SAQ 7.1 The following is a list of tasks which are carried out on a software project. Which of these would you normally expect to be done by a systems analyst?

1 Designing the detailed structures of the database.

2 Checking that the system meets the requirements of the customer by testing it.

3 Talking to the customer in order to find out about what the functions of the required system are.

4 Examining an existing system in order to find out what the current problems with it are.

5 Programming the system.

6 Managing the work of the programmers.

Solution

The tasks associated with an analyst are 2, 3 and 4.

Programmer/analysts

A programmer/analyst is a combination of the jobs of systems analyst and programmer. The following factors influence the desirability of this arrangement.

The size of the company

If the company is small, there may not be scope for the two distinct occupations. The dual role of programmer/analyst provides more flexibility in assigning work since on some occasions the work load demands analysts and on other programmers.

Communication between analyst and programmer

Two people working on the same problem do not always communicate with one another all that well. They perhaps fail to appreciate each other's capabilities, function and job problems, and this lack of empathy creates a communicative gap.

The communication problem obviously disappears with the dual role, and so a number of larger companies have adopted the programmer/analyst policy for this reason.

Conflict of personal attributes

Systems work calls for a fair amount of extroversion and ability to communicate orally, especially during the investigation stage. The programmer's job, on the other hand, demands great concentration whilst working individually. It is unusual for one person to have the qualities necessary for these two dissimilar tasks.

The increasing use of 4GLs will alleviate this dichotomy. As it becomes simpler and quicker to create computer programs, there are greater prospects of more programmer/analysts.

SAQ 7.2   A programmer has good communication skills with other programmers in that he or she can easily communicate the technical problems and features of a system. Would this member of staff automatically be a good analyst?

Solution

Not automatically. Analysts need to be able to discuss systems at an application level with customers and their staff. There is no implication that someone who can communicate technical detail is able to transfer these skills to the user domain.

The attributes of the systems analyst

As is the case for many other occupations, an extensive list of desirable attributes is compilable for systems analysts. And, although it is unlikely that any individual possesses all the qualities described below, they indicate the most suitable type of person. They also help the aspiring systems analyst to know the qualities to cultivate and the training to acquire.

Education and training

Nowadays a high proportion of systems analysts have degrees, diplomas or professional qualifications. The disciplines of most immediate application are business studies, data processing, accounting, information systems, computer science and management studies.

The more desirable areas of knowledge and expertise are:

- An understanding of the aims and purpose of the company from the point of view of its management.
- The procedures, techniques and problems entering into these areas:
  - financial and management accounting;
  - stock and stores control;
  - personnel administration;
  - production planning and control;
  - sales and marketing; and
  - operational research and statistical methods.

A systems analyst is clearly not expected to be fully conversant with all these areas but some knowledge of each is desirable.

- The methodologies of systems analysis and computer-based IS. Some knowledge of computer programming strategy is useful but there is no need for great expertise.
- The capabilities and purposes of computers, allied equipment and software. The systems analyst must, however, guard against becoming a computer addict as this tends to produce a machine-oriented instead of a problem-oriented approach.
- The sociological, psychological and legislative factors relating to computerization and automation.

Personal qualities

The qualities below are ways of behaving and thinking that facilitate the work of the systems analyst.

- **Confidence**: a self-assured manner but not in any way supercilious.
- **Responsibility**: a willingness to accept the consequences of his or her decisions.
- **Creativity**: the ability to innovate and so avoid automatically applying old solutions to new problems.
- **Patience**: a readiness to listen and to explain, repeatedly, if need be.
- **Communicability**: oral and written, so as to transmit information about and enthusiasm for new methods.
- **Logicality**: the ability to comprehend both the overall concept and the detail of a problem, and to think it through to a logical solution.

■ **Persistence**: to obtain information not easily procured and gain results not easily attained.

■ **Empathy**: the ability to self-project into the position of the person with problems.

■ **Diplomacy**: in dealing with staff whose jobs are in jeopardy or who are inclined to be uncooperative.

Other techniques

These include document analysis which is covered elsewhere (see Section 7.4) and fact acquisition techniques which may be used as an alternative to interviews. The following are some of these techniques.

Protocol analysis: In this case the user is asked to think aloud while performing a task. The protocol is tape recorded and video recorded. This technique yields a large amount of detailed information about the user's mental activity and reasoning which may not be apparent in interviews. This technique is consequently more popular in knowledge elicitation when rules and details of an expert's reasoning are required to create an expert system. Protocols are useful when combined with a retrospective interview to follow up with questions about why the user did a particular activity. Protocols are expensive in terms of analysis time as the verbal reports have to be analysed.

Activity groups: A group of users are asked to converse about the system and its problems while the analyst acts as a facilitator but otherwise takes a low-profile role. Group discussion can work well when there is a group of vocal users, but it can suffer from lack of verbal expression of tacit knowledge.

Card sorting and scaling techniques: Users are asked to sort a series of cards which have a set of related facts written on them. The users are asked to sort cards in order of importance or into related sub-groups, etc. Alternatively the analyst may ask the user to rate facts in terms of their relatedness using a scale. These techniques are useful for narrow problems such as deciding the categories of objects and ranking and ordering priorities.

Analysts often use a combination of techniques, although interviews remain the most popular and cost-effective means of fact gathering.

■ Present situation and system

It is an unwise systems analyst who attempts to introduce a new system without first closely researching the relevant existing systems and situations, and the company's aims and problems. This would be 'shooting in the dark' at an 'undefined target', and so the chance of success would be negligible. It is impossible to imagine a situation where no information whatsoever is needed before a new system is designed. Even when there is no existing system, or for that

214

matter no existing organization, there must still be some aims and purpose for future systems. Without at least some indication of these, a new system cannot even be conceptualized.

This is still the case even when a prototyping approach is adopted. It is sometimes suggested that systems investigation is unnecessary when using prototyping because new versions of the system can be prepared so rapidly that a heuristic approach is viable. This philosophy is risky and is best shunned except for the simplest of applications. It is more pragmatic to employ prototyping to augment the conventional systems development life cycle (SDLC) (see Sections 7.9 and 8.1).

It is all very well to say 'define the problems'; this is simple only if the problem definitions are expressed in broad terms. Aims such as increasing the company's profitability, reducing overheads, keeping within the departmental budgets, and satisfying customers' demands are all major problems that spawn a host of minor problems. All aspects of problem definition must therefore be investigated, and this entails discovering as much as possible about the present and future situations.

It is vital to find the facts appertaining to the present situation, and especially when a computer-based system is to be introduced or extended. Computers do not like vagueness and are adept at turning it into chaos. Their programs, input/output and database must be defined precisely if they are to be properly utilized. Inaccuracies, particularly in business programs, may cause the computer to run amok and so make a fool of its programmer and an enemy of its user. A user never quite trusts the computer again after erroneous results have been received from it.

Ideas for improvements

As well as actual facts about the existing situation, there is room for the consideration of people's ideas. Provided these are clearly recognized as subjective judgements and are differentiated from facts, they make a valuable contribution to the design of a new system. The analyst should therefore be on the lookout for well-founded ideas and should encourage the staff to put forward their suggestions during the investigation stage.

At a later stage these ideas can be encapsulated into specimens of input/output in the course of prototyping, enabling the user to confirm that the ideas are understood. These ideas will become the 'requirements' of the new system and usually fall into the following two sections.

Enhancements

Some of these will come from the users as a wish list of desirable features that the new system should have. The analyst may also be able to suggest improvements. Some wishes may be clearly impossible, others will be expensive but possible to implement, and others quite easy. The wish list has to be analysed in terms of the consequences of change on the system organization (how it will change people's working practices) and the likely cost of change. Wish lists can then be prioritized as changes which are essential, desirable, optional, etc.

Problems

These will be statements of what is wrong with the current system. These may be volunteered by the users, but more often it will be the analyst's responsibility to uncover deficiencies, inaccuracies, inconsistencies, etc. Problems should be analysed for their causality, i.e. why they occur. Design improvements can then be suggested to cure the problem, but again the consequence of change and their cost should be estimated and explained to the user.

Checklists

It would be convenient to have a comprehensive checklist showing all facts to be collected and questions to be asked during the course of a systems investigation, and merely to fill in the answers. This method is not recommended, however, unless the systems analyst has in-depth knowledge of the applications in question. This is because working from a predecided list does not bring to the surface the more subtle information and latent ideas. An individualistic approach does this more satisfactorily. None the less, within a given situation a simple checklist acts as a useful *aide-mémoire*, especially for the less experienced systems analyst.

Information required

It has been suggested that a systems investigation should be carried out with no thought for the future of the facts gathered. The concept behind this idea is that the investigation is then completely unbiased in its findings and so all possibilities are given equal consideration. Interesting as this approach may be, it is likely to result in the accumulation of many more facts than are truly relevant. In practice, most investigations are carried out with certain possible outcomes in mind.

One type of information that could be usefully gathered almost without limit are entity attributes for inclusion in the data dictionary. If there is any possibility of an entity being involved in a system, it is wise to collect information about its attributes.

7.2 PLANNING SYSTEMS INVESTIGATIONS

It is unwise to attempt to work to a rigid timetable in planning a systems investigation, first, because the amount of work involved cannot be predicted precisely, and secondly, because the investigators are dependent upon the availability of the departmental staff whom they wish to interview.

As regards the first point, approximations of the time needed will have been made as a result of the feasibility survey. This means that a rough timetable can be prepared showing the weeks during which each project team or individual will be engaged in each project.

The employment of a Gantt chart is beneficial in that it gives a clear indication of the arrangements, and actual progress can be marked thereon. It is probably not worthwhile using network analysis at this stage owing to the imponderables involved.

The following points are to be remembered when planning a systems investigation:

- Investigate the applications or departments in the order of their natural sequence of activities, e.g. customers' orders before accounts receivable.

- This implies that normally the first applications investigated are those involving the most source data.

- Work from the senior staff downwards, seeking the permission of all managers and supervisors before investigating their departments' activities or interviewing their staff.

- Make arrangements well beforehand for visiting the departments so that the departmental staff can make preparations.

- Target interviews at the correct people: senior managers for policy (why type questions); middle and junior management for more immediate goals and some operation detail (why and what questions); clerical and operational staff for procedural detail of how the system works.

- Arrange for the investigation to be officially announced so that everyone concerned is aware of its imminence and legitimacy.

SAQ 7.3 Janet is an analyst who is in charge of the initial stages of a system that keeps track of the large amounts of paper which are generated by a motor insurance company. She carried out the following activities in the order shown below. Can you point out where she went wrong?

- She interviewed the clerks about the paper handling process and when they thought the system should be implemented.

- She interviewed the managing director of the company in order to find out what was the best way to handle the flow of paper within the office.

- She interviewed the manager of the department responsible for setting up customer details when an insurance policy is taken out.

Can you criticize what she did?

Solution

She did not seek permission to interview junior staff. She did not have the project officially announced. She asked junior staff an inappropriate question about when the system should be implemented. She asked a very senior member of staff a question about document flow which he or she would not be concerned about. She would have been better employed dealing with the departments in reverse order: the new policy department first followed by the claims department.

■ Project teams

The concept behind project teams is that the future users of a system should be involved in its design and implementation. They also take part in the investigation work because users have a deep knowledge of existing systems.

In a large company several project teams are formed at an early stage of the investigation, each team comprising a mixture of systems analysts and user department staff. Project teams are organized in a variety of ways dependent upon the particular circumstances, and are assigned to the areas under investigation. Each team consists of three or four members who work closely together.

Organizing project teams

- ■ Each application area or department to be investigated is regarded as a project. Projects are planned to occupy a pre-estimated period of time, usually a few weeks. The senior systems analyst administers the project teams and defines the boundaries of each project.

- ■ A team includes at least one member who is familiar with the day-to-day working of the department or area under investigation. This person is assigned to the team on a full-time basis throughout the duration of the project.

- ■ The members of project teams drawn from user departments probably need some training in systems investigation methods. They then work with an experienced systems analyst for a few weeks before undertaking investigations on their own.

- ■ In some circumstances project teams are composed of a mixture of business analysts and technical analysts. The former are responsible for defining the business problems precisely, and the latter for arriving at technical solutions.

- ■ Each project team has a leader who is the most experienced systems analyst in the team. This appointment is temporary in that it lasts for only as long as the arrangement is advantageous. It is important to maintain flexibility and so teams need to be reorganized from time to time.

- ■ Regular meetings, say fortnightly, take place between project team leaders in order to ensure that no gaps or overlaps occur in the areas under investigation. A record is kept of the progress made by the date of each meeting.

■ Feasibility surveys

A feasibility survey (study) acts as a reconnaissance for the systems investigation coming later. As is seen from the aims stated below, it provides information to facilitate a later in-depth investigation. In situations where the extent of an investigation is strictly limited or where the background information is already well known, the feasibility survey can be omitted.

The main sources of information gathered during a feasibility survey are top and middle (line) management. It is not necessary to unearth a large volume of detailed facts at this stage, and so lower levels of staff are not usually involved.

The following are the main purposes of a feasibility survey.

Realism of assignment

To decide whether the aims stated in the assignment brief are realistic within the imposed constraints. Most aims are attainable if unlimited time and money are available, but this is never the case. If the assignment's aims are found to be unrealistic, it is necessary either to revise them or to remove the constraints.

Principal work areas

To determine the principal work areas relating to the aims stated in the assignment brief. This information permits planning of the full investigation, including the optimum deployment of systems analysts and project teams between the departments and applications involved. It is also possible to arrive at estimates of the times needed for the various parts of the investigation, and hence their costs.

Scope for improvements

To discover the applications and departments where scope exists for saving money, time and effort. It is possible that the problem areas identified during the feasibility survey do not coincide with the aims stated in the assignment brief because they were not then recognized as problem areas. Nevertheless wherever there are large amounts of office work, administrative procedures or active information files, there lies potential for savings. Similarly, in work areas involving high levels of stock, sales or production, it is likely that more efficient information could yield benefits.

Specialist assistance

To determine whether specialists will be required in the systems investigation. It may be found that problems exist that are not amenable to solution by conventional methods. Some specialist knowledge is therefore needed and this may not be available within the systems department or the project teams. Typical of such knowledge are operational research techniques, statistical methods and investment appraisal techniques. This specialist knowledge may be provided by persons within the organization or by external agencies if not available internally. It is also possible that one or more of the systems analysts could be trained to cope with the problem.

SAQ 7.4 Which of the following would normally be contained in a feasibility study?

1 Details of the skills required of contracted staff who are to carry out a performance study which would attempt to predict the response time of the system.

2 A full design of the system.

3 Approximate figures about the projected cost of the system.

4 Figures detailing the savings which might be made with the system.

5 The names of the programmers who are to be used for the project.

6 The names of staff who make up the steering committee for the project.

Solution

Items 1, 3 and 4 would normally be included in a feasibility study.

Feasibility survey report

Generally it is best if the results of a feasibility survey are reported to management only briefly. The contents of the report are based on the four purposes described above. Suggestions and reasoned arguments are included so that the steering committee is able to decide whether to sanction further investigation or, alternatively, to reconsider the assignment's aims. One of the difficulties is that management often want an estimate of cost even at this stage. Clearly it is almost impossible to give an accurate figure for developing a system which has not yet been completely analysed. Nevertheless an approximate 'ballpark' estimate may be required with a more accurate estimate for the costs of the next phase of analysing the current system.

7.3 USERS' INFORMATION REQUIREMENTS

As explained in Chapter 1, information is closely connected with the aims and control of the company. And since, in essence, information is the only product of the new system, it is worth giving some thought to the company's need for it. During the systems investigation and later, in the systems design phase, the systems analyst concerns him- or herself with finding the precise information needed by all levels of staff.

In this context information can be taken to include all useful output from the system. That is to say, all the reports, lists, etc., that stem from, and are needed in order to perform, the company's activities. Information varies, for instance, from a financial report for the managing director at a high level to an employee's payslip at a low level.

Since no company is static, it is to be expected that changes in information requirements will be demanded in the course of time. It is part of the systems analyst's job to attempt to predict these changes and make provision for them in the new system. This is obviously a difficult task since even in the medium term, let alone the long term, prediction is hazardous. In most cases all that is definite is that changes are likely to occur and that they will be of a certain type. A straightforward example of an almost certain change is to the income tax rates and brackets. When the new system is designed a simple method of accommodating such changes is built into it.

The following are the main aspects of information requirements.

What information is required and by whom?

Bearing in mind that the information will most probably come from a computer and that this has either to be programmed or software purchased for it, absolute precision in defining the information should be aimed for. Lack of precision means either that changes have to be made later or that the user ultimately receives second-rate information.

There is a clear need for precision in the contents of information, and in this respect it is vital that the systems analyst and the prospective user are talking about exactly the same thing. For instance, the user's understanding of the term 'stock level' could be the quantity of an item in the warehouse unallocated at the end of each day; the systems analyst might see 'stock level' as the current, minute-by-minute, quantity of the item that exists anywhere in the firm. Differences in understanding of this nature may not come to light until the new system comes into operation. In any event, repetition of work, problems and financial cost may result.

Other important aspects of information are its form of presentation, i.e. screen or printed, and its layout (format). It is not usually too difficult to put together a sample print or display for the user's approval. One of the problems in IS is that layouts which seem straightforward to IS specialists sometimes seem complicated to users, and so the layout of information should not be left entirely at the discretion of the analyst. Prototypes are useful in this respect.

The other main factor entering into the preparation of information is its sequence. This applies particularly to lists and tabulations that are to be used for reference purposes. A user may require a list in several alternative sequences dependent on its purposes. Whatever the need, clear agreement regarding the sequence has to be arrived at. Here again a sample is helpful and especially when nested sequences are involved, e.g. order no. within account no. within week no.

Turning to the point of who requires the information, we are concerned with who is really going to make use of it. In other words, to whom should the documents be delivered immediately after printing and/or who should have a VDU on his or her desk? A situation to be avoided is the passage of information via a hierarchy of managers, many of whom may have little or no interest in it, and consequent delay in reaching the person(s) who act upon it.

Every document produced by a computer should have a recipient's name associated with it. Merely addressing a stack of documents to a department often results in no one actually making use of the information thereon.

When is the information required?

The time of delivery or availability of information has to be agreed so that the user knows for certain when it will be available. This time usually depends on source data being received on time by the IS department. This point needs emphasizing especially if the user is responsible for submitting the source data.

A time of delivery or availability may be:

1 The time of day, day of the week, or within a certain time after a variable event, e.g. within 24 hours of a production hold-up.

2 On demand, i.e. when requested by the authorized recipient.

On-demand information is often requested by the user entering a query into a terminal; it is therefore important for the systems analyst and user to agree on the types of query information required.

There are broadly two main types: immediate-access enquiries and longer-access query information. The former is generally a case of the user inputting a small amount of data and getting a small amount of information in return. For instance, entering a sales area number to get the value of sales in the area for the previous week. Longer-access information tends to be larger in volume, e.g. a query calling for a long list of product sales. This might take several minutes even with a mainframe and so the information is better sent later in document form rather than via the terminal.

The important point at this stage is that users are made aware of what is reasonable to demand as immediate-access information and what will take a little time.

3 On exception, i.e. when the information is found to contain figures outside a predecided range. As explained under 'Exception reports' in Section 1.3, the computer is programmed to pick out the exceptions and so these must be clearly agreed and defined during the systems investigation.

SAQ 7.5   The standing orders clerk in a bank is a member of staff who administers the regular payment of monies from the accounts in a bank. A typical standing order might be one which deducts the annual fee of a golf club of which a particular account holder is a member. A bank manager is the member of staff who is concerned with the profitability of the bank that he or she manages. Discuss how the information needs of these members of staff differ in terms of what information they require and when.

Solution

The needs of the standing orders clerk are quite simple: he or she needs to be able to have instant access to the details of the standing order associated with a customer. For example, a customer may telephone and ask when a standing order is due to be paid or when a particular standing order finishes. The standing order clerk requires almost instant access.

The bank manager requires much more tactical information: data about the performance of his or her bank in terms of turnover, the number of new customers taken on, the amount of loans granted over a month, whether there is a pattern of business of which he or she can take advantage. This information is better presented in summary form and is not normally required immediately.

For what purpose is the information required?

Information is sometimes demanded for the recipient to do further work so as to create additional information.

Some users have little understanding of a computer's capabilities and so do not ask for the additional information. This is especially true for situations involving logical decisions. Some people, while accepting that computers can do arithmetic, do not realize that many business decisions are also amenable to computer logic.

Unfortunately, in spite of diligent questioning, users' information requirements are often vaguely stated. This is the 'I don't know what I want but I will know when I get it' problem. Prototyping can often help alleviate this problem by suggesting possibilities to users. If prototyping is not possible, then giving user scenarios or mock-ups of screens and reports does help. There is nothing like a concrete example of how the system will appear for provoking a user into telling you how they don't like it and then, with luck, how it should be.

SAQ 7.6 The management of a university require an information system which helps them to carry out their work. The only sensible reply an analyst can get from the senior member of staff about this system is the statement 'We want to find out how things are going'. Can you detail some of the information needs that this vague statement represents?

Solution

There is a host of possible needs. We have detailed a number of them below:

- How many students have we got?

- How many students are there in each department?

- What students have not paid their fees?

- What is our current salary bill?

- How many students have dropped out in the past year?

- What is the current value of our research grants?

7.4 USAGE AND FLOW OF INFORMATION

The usage of information and its flow within the company are the means through which control and planning are achieved. Information moves around the company mainly in documentary form, and consequently these documents must be carefully analysed during the investigation. The documents may originate internally or come from external sources. After origination, documents are passed from hand to hand before being dispatched, filed or destroyed. By tracing their movements and noting the entries made at each stage, it is possible to obtain a clear picture of the flow and build-up of information in the company.

From each department the following information needs to be obtained:

- What documents are originated and how many copies are prepared?

- What entries are made on each document and what work does this involve?

- Where are the documents sent after origination or processing?

- What documents are received from other departments or external sources?

- What data is extracted from each document and for what purpose?

- What documents are held in files, for what purpose, and for how long?

The answers to these questions are cross-checked between departments and any discrepancies are reconciled by further investigation.

Document usage

In this context documents can be regarded as any form, card, list or sheet used in the company for holding data. A document may also be part of a conventional file, i.e. a permanent document held with a number of similar documents.

Almost all documents consist of two essential parts: headings and entries. Internal documents start life as preheaded but otherwise blank forms, and entries are made during their use in the company. Externally originated documents arrive with entries already inserted and further ones may be added internally. Pitfalls to be watched out for are out-of-date and incorrectly headed documents. These are avoided by inspecting current live documents and asking users to validate their contents. Similarly, unusual symbols or coloured entries on documents need investigation as they might have special importance to the user.

Nowadays many documents used in companies stem from the computer. From the systems investigation aspect these are no different from other, non-computer documents.

Document specification

A convenient and accurate way of noting the meanings of a document's entries is to use a document specification. One such form is filled out for each type of document and a live specimen of the associated document kept with it. Referring to Figures 7.2 and 7.3, the square-outlined reference (T1) is the analyst's

WORKS ORDER

To: *Section No.6*   T1   Date *9/4/94*

Job No.	Part No.	Qty. Req'd	Due start Day/ Week	Due completion Day/Week	Finish	Material code	Material quantity
27159	AM 26318	500	1/18	5/21	Matt black	F475CG	1500 ft
27160	AM 13006	100	1/19	2/22	Dove grey	F4795R	1200 ft
27161	AP 378	1800	3/20	4/23	Matt black	F228	750 ft
27162	BM 2215	50	1/21	5/23	White	D378A	178 ft
27163	BA 57	100	2/20	4/22	None	G36ET	180 ft

Fig. 7.2 Specimen document (works order form T1 in Figure 7.8)

	Document Specification			

Name of document *Works Order Form*			Originating Department *Production Planning*	
Document reference *T1*	Date filled out *5/7/94*	No. of copies *1 + 2*	Systems analyst *H.D.C.*	

Initial distribution

           1. *Supervisor (Yellow Copy)*
           2. *Operator (White Copy) via Supervisor*
           3. *Production Planning (Blue Copy) to W.I.P. file*

Remarks

        Section Supervisor ticks off jobs on his copy on completion

Entry ref.	Item description (heading)	Max items per doc.	Picture	Entered by	Remarks
A	*Section number*	*1*	*9*	*Planner*	
B	*Date originated*	*1*	*99/99/9*	*Job clerk*	*Day/Month/Year*
C	*Job originated*	*5*	*9(5)*	*"*	*Taken from job register*
D	*Part number*	*5*	*AA 99999*	*Planner*	
E	*Quantity required*	*5*	*9999*	*"*	
F	*Due start, day/week*	*5*	*9/99*	*"*	*Usually 5 but sometimes 6 working days per week*
G	*Due completion, "*	*5*	*9/99*	*"*	*do*
H	*Finish*	*5*	*A(12)*	*"*	
J	*Material code*	*5*	*A999AA*	*Asst. Planner*	*Taken from parts spec. file*
K	*Material quantity*	*5*	*9999*	*"*	*Computed from parts spec. file*

Fig. 7.3 Document specification form

document reference and is unique to the particular type of document. The blue circled letters on the specimen document, known as entry references, are entered thereon to cross-reference the entries to the document specification form (Figure 7.3). On some types of document there are several entries of a similar nature, e.g. the jobs on the works order in Figure 7.2. These similar entries are all covered by the one entry reference.

For the most part the document specification (Figure 7.3) is self-explanatory but perhaps 'Picture' calls for some elucidation. A '9' in the picture means that the corresponding position may have any value from 0 to 9. 'A' means A to Z. 'X' means either 0 to 9 or A to Z, and 'B' means a blank position. In order to reduce the size of a picture, it is conventional to put a bracketed number after the picture character to show how many times it is repeated. e.g. A(5) is equivalent to AAAAA. The purpose of noting an entry's picture is in order to plan its storage and usage in a computer system; this is especially pertinent to the creation of a data dictionary (Section 5.4).

SAQ 7.7 The standing orders clerk in a bank receives instructions from a customer to make regular deductions from his or her account. Can you think of some of the item descriptions that would be found on a form used to support this activity?

Solution

A list is shown below:

■ The identity of the account which is to be debited.

■ The frequency of debit: monthly or yearly.

■ The name of the company to which the payment is to be made.

■ The details of the company to be credited.

■ Some unique reference (key) which differentiates this standing order from others.

■ The date on which the standing order should start.

■ The date on which the standing order should finish.

Example 7.2	**Procedure for a works order**

The procedure described below is the basis for the data usage chart in Figure 7.4 and the data flow diagram of Figure 7.8, page 233.

Narrative of works order procedure

1 Three copies (yellow, white and blue) of each Works Order (T1) originate in the Production Planning department. The job numbers are extracted from the Jobs Register (F1) and ticked off therein. The material codes and quantities per part are derived from the Parts Specification file (F2), and the material quantity computed (quantity per part × quantity required).

2 The yellow and white copies of the works order are passed to the section supervisor, who passes on the white copy to the operator and retains the yellow copy. The third, blue, copy is filed in the Work-in-progress file (F3) in job number sequence.

3 On completion of a job, the section supervisor fills out a Job Completed form (T2) with two copies (pink and white). He or she retains the pink copy (for three months), and passes the white copy to the Production Planning department. Here it is checked for scrapped work and, if present, a Scrap Report (T3) is filled out and filed until the end of the month. All the white copies are then inserted into the Jobs Completed file (F5).

4 At the end of the month, the scrap reports are sorted into finish code within part number within section number sequence, and analysed to create the scrap analysis (T4).

▨ Data usage charts (grid charts)

A document entry (data item) may appear on several different documents, probably by having been copied from one document to another. An entry may have been calculated from other entries, created as a result of the company's activities, or been received from external sources. In order to maintain a check on the usage of data items, and to ensure no unnecessary duplication, a 'data usage chart' is created; an example is shown in Figure 7.4 (see Example 7.2).

Each transaction and document is given a reference and wherever a data item appears, a code is inserted in the appropriate position in the data usage chart. This code is cross-referenced to the flowchart as demonstrated by Figures 7.4 and 7.8.

The meanings of these codes are

T = transcribed from the transaction document denoted by the reference, e.g. section number on the job completed form is copied from the works order form (T1).

Name	Ref.	Section no.	Job no.	Part no.	Quantity reqd.	Due start date	Due completion date	Finish	Material code	Material qty.	Qty. completed	Qty. scrapped	Actual completion date	% scrapped	Material per part	Remarks
Job register	F1	T1	I													Job nos. ticked off as used
Parts spec. file	F2	I		I			I	I							I	
Works order	T1	F2	F1	I	C	I	I	I	F2	C						Quantity required is computed from other data
Job completed form	T2	T1	T1	T1							I	I	I			
W.I.P. file	F3	T1	T1	T1	T1	T1	T1	T1	T1	T1						
Scrap report	T3	T2	T2	T2				T1		T2	T2		C			
Scrap analysis	T4	T3		T3				T3			C	C	C			Quantities are summarized totals
Job completed file (pink copies)	F4															As T2
Job completed file (white copies)	F5															As T2
Scrap report file	F6															As T3

Fig. 7.4 Data usage (grid) chart

C = calculated from other entries.

E = received from an external source and first recorded on this document.

I = originated internally and first recorded on this document.

F = transcribed from the file document denoted by the reference, e.g. job number on the works order is copied from the job register (F1).

Dataflow diagrams

The two main methods of illustrating dataflow are dataflow diagrams (DFDs) and ISO/BSI flowcharts. The latter has now largely been superseded by DFDs as these are clearer and neater.

A DFD has four types of symbols, as explained in Figure 7.5. DFDs can be used to model physical processes (e.g. the flow of materials in a manufacturing plant) but physical views should be converted into their logical equivalent, e.g. the flow of data that accompanies the material.

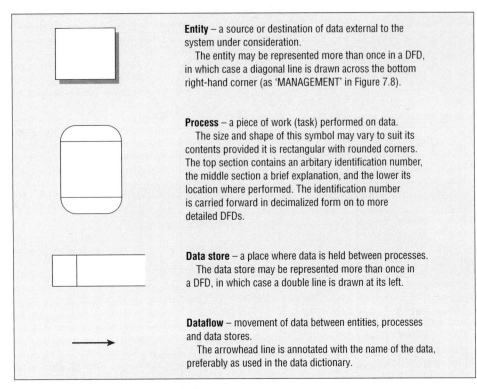

Fig. 7.5 Dataflow diagram symbols

The names shown in the DFD should, in the final version, be identical to those in the data dictionary. The precise nature and contents of each dataflow need to be ascertained. This is done in a similar way to that explained previously for documents even if the dataflow is not in document form, e.g. data keyed into an on-line terminal.

SAQ 7.8 What should the following be regarded as: an entity, a process or a data store?

■ Customer

■ Account details

■ Validate account details

■ Sales data

■ Airport maintenance manager

Solution

Customer is an entity, *Account details* is a data store, *Validate account details* is a process, *Sales data* is a data store and *Airport maintenance manager* is an entity.

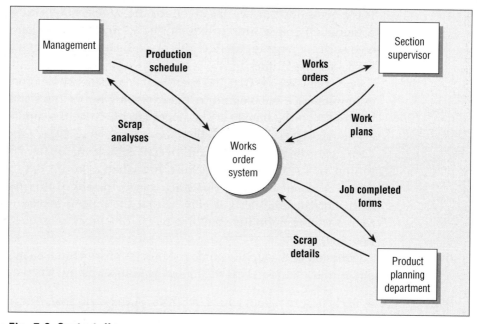

Fig. 7.6 Context diagram

DFDs support a top-down analytic approach. The top-level diagram describes the system as a whole and this is often used at the feasibility stage. The top-level diagram, also called the context diagram, shows the whole system as one process and the input and output connected to the external entities (see Figure 7.6). The context diagram is therefore a useful statement of the scope of the investigation as the external entities will not be analysed. It also shows the system input and output which give useful clues for further decomposition.

The next step is to decompose the system into its main processes. This creates the top-level dataflow diagram (usually called the level – 0 DFD). This shows a small number of quite complex processes (usually about 4–5) and their dataflow connections. The external entities are present as before so the diagram matches the context diagram in terms of its input and output. All dataflows going into the circle (or bubble) at the context diagram level will appear as dataflows going into and out of the level – 0 DFD. This cross-referencing between levels in a DFD hierarchical is called 'balancing'.

There is no absolute mechanism for creating a DFD but some heuristics are as follows:

- Ask the question 'What process needs this input?', which points towards the first processing in a DFD sequence.
- Draw the first process and then ask 'What output does this process produce?', which gives a clue about the identity of the next process in a chain.
- Repeat the first question and draw in the second process. Connect the processes by a dataflow.

These questions can be repeated until the system output to an external entity is found. Of course, things may not be so simple, so further questions can be used such as 'What inputs does this process need?', which may uncover the need for data stores or files.

Once the level – 0 DFD has been completed, the whole process is repeated by examining each bubble in turn and exploding it to a lower level of detail. Again the inputs to the bubble at the upper level become the inputs to the diagrams at the next level down. This produces a level set of DFDs which may be 3 or 4 deep depending on the size and overall complexity of the system. The 'where to stop' point is difficult. Generally, it is when a process has been split down to a level of detail where its procedure can be described in a paragraph of structured English. Another guideline is that the process should not contain more than 3–4 of the primitive building blocks of a program (sequence, selection and iteration).

Cohesion and coupling guide the process of structured analysis so it is important for the analyst to bear the following points in mind when constructing DFDs:

■ Each process should have a single purpose or goal. The process is named with a verb-noun pair, e.g. Check-Order, Calc-Commission. Names should be unique and computer-like terms should be avoided (e.g. process order).

■ Dataflows connect processes. If one process has too many connections (more than 4–5 input and output dataflows), it is probably too highly coupled. This suggests decomposition to a lower level. Dataflows are names with nouns and each name should theoretically be unique. As dataflows proliferate at lower levels this can be a problem.

DFDs are easy to learn and use, although they do have some problems. The technique is best used iteratively, as the first analysis often results in DFDs which are uneven, e.g. some processes are large and complex while others are small and trivial. The diagram may have to be partitioned and the analysis revised. Some analysts like to do analysis bottom up (identify the low-level components first and then compose the higher-level diagrams later).

Structured walkthroughs

One virtue of DFDs is that they can be used to explain the specification to other members of the team and to end users. The diagrams are used in critiquing sessions in which one of the team explains the diagram to other members, one acts as a moderator to ensure critiquing does not get out of hand and another person acts as a recorder. Walkthroughs are good for checking that the data necessary for a process is input to the process and that it produces the necessary outputs. They can also be used to elicit user feedback on the functionality of the system.

Figure 7.7 is the DFD of the works order procedure shown in Example 7.1 above.

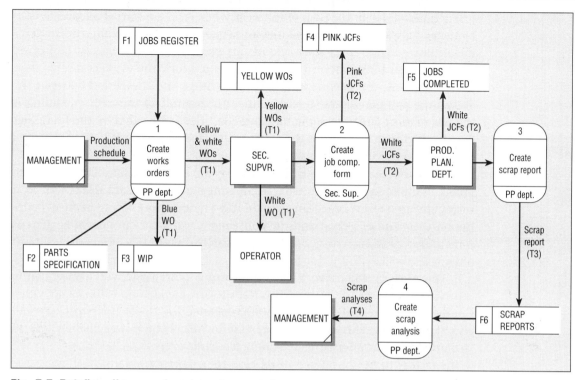

Fig. 7.7 Dataflow diagram of works order procedure

7.5 CURRENT ACTIVITIES

■ Cyclical nature of business activities

Nearly all business activities are performed on a regular cycle. For instance, wages are prepared weekly, sales statements monthly, stock evaluation annually. There are usually good reasons for these cycles in the existing system but they should not be assumed to be immutable. With a new system, advantages might accrue from modifying the activity cycles, and so the true reasons for the existing cycles ought to be ascertained. Prominent among these reasons are:

■ *Agreement* – e.g. with trade unions to pay workers weekly.

■ *Legal* – e.g. financial statements prepared annually.

■ *Phasing* – to interface with other activities, e.g. stock evaluation is needed for the annual balance sheet.

■ *Control* – to provide regular information for decision-making, e.g. monthly factory loading reports to decide extra resources required.

■ *Custom* – it has always been done that way!

233

The times of day or the days of the week when tasks are carried out are decided by the activity's cycle. Some activities are scheduled backward from a finish deadline, others forward from a start point. An example of the former is the payroll routine, which hinges upon the agreed time at which the workers receive their pay packets. All the payroll work has to be fitted into the period between the end of the pay period and this deadline. An example of forward scheduling is the preparation of the periodic sales analysis. This is triggered by the final sales figures arriving at the end of the sales period, and the analysis is then prepared as quickly as possible.

Certain activities proceed without any apparent cycle. Handling customers' orders is often so because orders arrive continuously and are dealt with in a steady stream day after day. The cycle of these orders varies considerably depending on their rate of arrival and the consequent workload but generally there is some target cycle such as goods being dispatched within 24 hours of receipt of order.

A new system will have greater processing power and will consequently achieve results in a shorter time. This leads to an amended timetable of activities. Computers, for instance, are commonly operated for 16–24 hours per day and also at weekends. This enormous increase in processing power enables jobs to be started later or, alternatively, completed earlier.

The factors to be considered in relation to activity cycles are:

■ When are the deadlines, and are they absolute or relative to other activities?

■ When are the start points, and are they decided by internal or external circumstances?

■ Could an activity be cycled either more frequently or less frequently with advantage? Dispatching bills daily instead of monthly, for instance, could result in prompter payments and fewer bad debts.

■ What time allowances are made for uncontrollable factors? Delivery times of goods and letters, for example.

Activity cycles are conveniently recorded on a chart such as the weekly payroll in Figure 7.8. Monthly cycles call for a similar but extended version of this chart.

■ Activity processes

Every business activity consists of several processes (tasks); some are trivial, others are complex, but whichever is the case the analyst must be certain that he or she understands them and that none has been overlooked.

For each process the following information is needed:

■ What numbers of documents, of all types, are handled?

■ What enquiries are made; what information is demanded, how soon, and for what purpose?

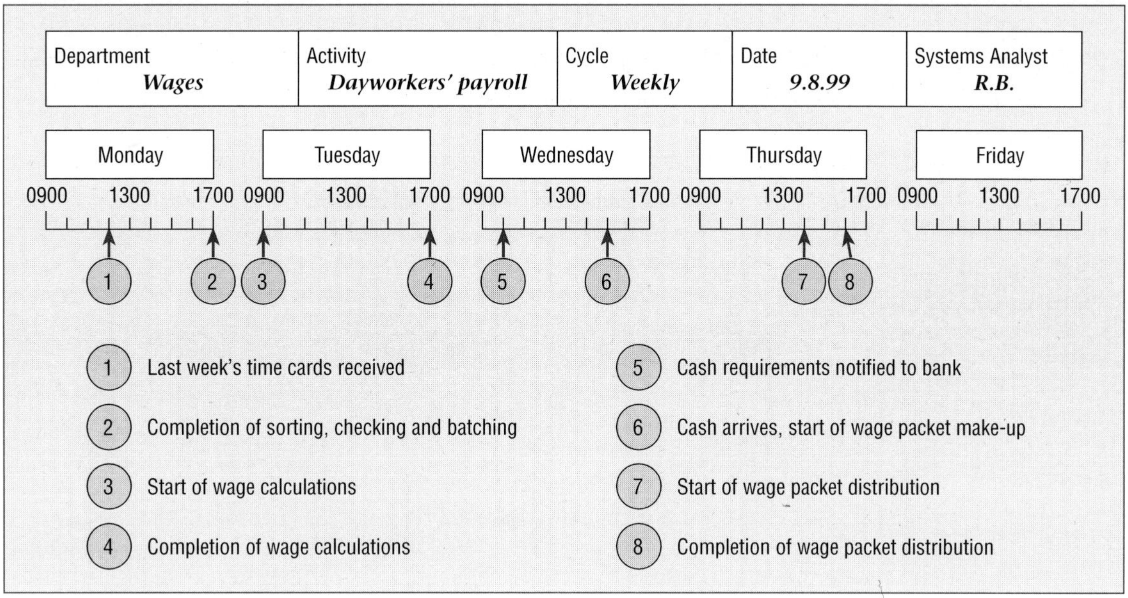

Department	Activity	Cycle	Date	Systems Analyst
Wages	**Dayworkers' payroll**	**Weekly**	**9.8.99**	**R.B.**

Monday	Tuesday	Wednesday	Thursday	Friday
0900 1300 1700	0900 1300 1700	0900 1300 1700	0900 1300 1700	0900 1300 1700

①	Last week's time cards received	⑤	Cash requirements notified to bank
②	Completion of sorting, checking and batching	⑥	Cash arrives, start of wage packet make-up
③	Start of wage calculations	⑦	Start of wage packet distribution
④	Completion of wage calculations	⑧	Completion of wage packet distribution

Fig. 7.8 Activity cycle chart

■ Do variations occur in the above; if so, in what way and for what reasons? Seasonal variations are common, and random variations also occur.

■ What calculations are made in the course of handling documents and dealing with enquiries?

■ What reference is made to file documents (data stores)? This includes the updating of file records and the creation of new ones.

■ Are controls imposed on the process? These could be feasibility checks (Section 8.4) applied to the data handled or the creation of control totals and audit figures.

■ What decisions are made and what follows as a result of each decision?

■ Decision trees and tables

It has already been suggested that certain processes involve decisions; a simple example of this is the scrap report decision above. Where a decision is straightforward and with only a few possible outcomes, there is no difficulty in specifying what is involved. In some cases, however, there are several related decisions leading to a larger number of outcomes.

It is possible to construct a decision tree; in this the decision points are shown as diamond-shaped symbols such as in Figure 7.9. This method soon reaches its limit, however, owing to the sheer size of the tree if there are more than a few related decisions. A more convenient method is to employ a 'decision table'. This

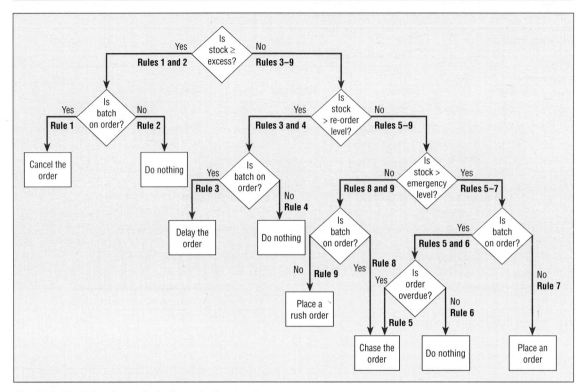

Fig. 7.9 Decision tree of stock control procedure

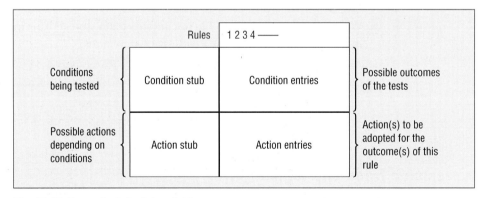

Fig. 7.10 Format of decision table

takes the form of a table on which are shown 'conditions' (decisions) and related 'actions' (outcomes). The four parts of a decision table are shown in Figure 7.10. Each column of entries is known as a rule and is given an arbitrary rule number.

The aims in preparing a decision table are first, to ensure that all conditions and their associated actions are included, and secondly, that all the rules are both logical and clearly understandable. There must obviously be no ambiguity whatsoever as this would negate the whole point of this method (see Example 7.3).

| Example 7.3 | **Decision tree and decision tables** |

A stock control procedure follows the rules outlined below:

■ If the stock in hand is at or above the excess level, any batch on order is cancelled.

■ If the stock in hand is greater than the re-order level but below excess level, any batch on order is delayed.

■ If the stock in hand has dropped to or below the re-order level, an order is placed unless a batch is already on order, in which case this is chased if overdue.

■ When the stock in hand has dropped to the emergency level, any batch on order is chased, otherwise a rush order is placed.

The above procedure is shown as a decision tree in Figure 7.9 and as two different decision tables in Figures 7.11 and 7.12.

As is seen from Figures 7.11 and 7.12, the condition entries and the action entries may be in either 'limited' or 'extended' form. A limited condition entry consists of either a 'yes' or a 'no' or is irrelevant (Y, N and – respectively). Similarly, a limited action entry comprises one or more crosses in each column to indicate the action stub(s) or otherwise a blank. An extended condition entry is either in descriptive or in quantified form.

Rules	1	2	3	4	5	6	7	8	9
Is stock ≥ excess?	Y	Y	N	N	–	–	–	–	–
Is stock > re-order level?	–	–	Y	Y	N	N	N	–	–
Is stock > emergency level?	–	–	–	–	Y	Y	Y	N	N
Is a batch on order?	Y	N	Y	N	Y	Y	N	Y	N
Is order overdue?	–	–	–	–	Y	N	–	–	–
Cancel the order	X								
Delay the order			X						
Do nothing		X		X		X			
Chase the order					X			X	
Place an order							X		
Place a rush order									X

Fig. 7.11 Limited condition/limited action decision table of stock control procedure

Rules	1	2	3	4	5	6	7	8	9
Stock in hand?	≥ Excess	≥ Excess	< Excess and > re-order level	< Excess and > re-order level	≤ Re-order and > emergency level	≤ Re-order and > emergency level	≤ Re-order and > emergency level	≤ Emergency level	≤ Emergency level
State of batch on order?	Placed	Not placed	Placed	Not placed	Overdue	Placed and not overdue	Not placed	Placed	Not placed
Action regarding the order	Cancel order	None	Delay order	None	Chase order	None	Place an order	Chase order	Place a rush order

Fig. 7.12 Extended condition/extended action decision table of stock control procedure

Limited entries inevitably mean that more details are needed in the stubs than with extended entries. They also tend to result in more rows appearing in the action parts of the decision table. It is also possible to have a decision table with limited conditions and extended actions, or vice versa.

There are no hard-and-fast rules as to which arrangement to adopt in a decision table, each situation has to be judged on its own merits. Generally speaking, limited entries are clearer to follow but they become numerous if many decisions are involved.

Structured English

This is an alternative means of specifying the control logic of a process. Each dataflow bubble is described by structured English when the DFD set has been decomposed to its lowest level. There is no standard for structured English which is basically English stripped to its bare essentials. A set of several words are declared for the basic structural components of all computer programs, for example:

Sequence DO
Selection IF . . . ENDIF
Repetition WHILE . . . END–WHILE

Other components are verbs describing actions and nouns describing data items. Other reserved words may be included such as Read and Write to describe the consumption and production of dataflows. Indentation is used to show the span of control for each program component. An example of structured English is shown in Example 7.4.

Example 7.4   **Structured English**

Check-Order
Read Cust-Ref-No From DataStore CustRef
Read Products From DataStore ProductData
WHILE Orders on File
  Read Order DF
  IF Cust-No NOT = Cust-Ref-No
  THEN
    Write Order to DF Reject-Orders
  END-IF
  IF Date NOT = Today-Date
  THEN
    Write Order to DF Reject-Orders
  END-IF
  WHILE Products on Order
  IF Prod-No = Prod-ID
  THEN Add 1 to OKProds
  END-IF
  IF OKProds NOT = Prods-on-Order
  THEN
    Write Order to DF Rej-Order
  END-IF
END–WHILE

Structured English is reasonably close to program code and can be used as a final specification document from which code can be written. Alternatively structured English may be transformed into pseudocode and then into a program language, such as COBOL. Standards for structured English vary, for instance some authors use REPEAT UNTIL and IF THEN ELSE SO as keywords, and some authors may not add detail such as setting counters to zero in structured English.

Structured English, dataflow diagrams and the data dictionary constitute a structured specification. All elements of the specification are cross-referenced for consistency checking, so it is possible to find the dataflows, structured English and data stores used by any DFD process. The structured specification at its lowest level is composed of a series of mini specs. Each mini spec. is composed of a cross-reference set of facts:

- Process identifier, linking it to a DFD diagram.
- Structured English for the process.
- Decision tables/trees if necessary.
- Description of dataflows input and output, cross-reference to the data dictionary.
- Description of any data stores used, again linked to the data dictionary.

Structured specifications can be combined with informal nodes and other diagrams according to the system development method being used. It is common to use DFDs and structured analysis in combination with data analysis and entity relationship modelling.

SAQ 7.9

For the standing orders clerk discussed in the previous SAQ write down a context diagram which describes the fact that they process standing orders from a customer with each standing order being placed in a standing orders database.
In carrying out this process the clerk may encounter errors in the data given; for example, the customer may have specified the wrong account to be debited.

Solution

The context diagram is shown below.

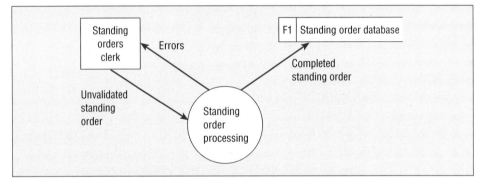

Fig. 7.13 Context diagram

SAQ 7.10

Write down a structured English description of a lecturer receiving a set of essays which the students in his or her class have written, batching them up, marking them and returning them to each student after writing down the marks in a

Solution

WHILE there are more students to hand in their essays
   Receive an essay
END–WHILE
WHILE there are more essays to mark
   Mark an essay
Place the mark in the register
END–WHILE
WHILE there are more essays to hand back
   Hand back an essay to the student who wrote it
END–WHILE

■ Exceptions to the norm

It is the ability to cope with the exceptions that makes or breaks a system. During the systems investigation the analyst is perpetually looking out for exceptions, and indeed, exceptions to the exception. The information he or she acquires about the present system naturally tends to apply to the normal state of affairs rather than the unusual. It is therefore important to appreciate what is meant by exceptions.

Special procedures

These are associated with abnormal situations such as:

- the registration of new employees;
- the paying off and deregistration of leavers;
- the recording of a new product's details;
- the validation of a new customer's creditworthiness;
- chasing potential bad debts;
- *ad hoc* reports demanded by management; and
- sudden and unexpected procedural changes as caused by strikes and other disruptions.

SAQ 7.11   A part of a sales system is meant to process the sales generated by the various departments in a group of stores. Can you think of any exceptions that could occur with this part of the system?

Solution

The main exceptions are a customer returning an item for some reason, perhaps it was malfunctioning. Another exception is where an item has been reduced in price and the computer system has not been formally notified by it; for example, a hand-written label may have been stuck on the item.

Abnormal work loads

These tend to occur at certain times of the year, mostly owing to seasonal fluctuations in sales demand. They also occur randomly due to unforeseen circumstances and chance factors. An example of an unforeseen circumstance is the liquidation of a competitive company causing an increased demand for the company's products. A chance factor might be an unexpected weather pattern creating a sudden demand.

Peak workloads are of particular relevance in the designing of real-time systems because they have to be handled immediately by the computer and without degradation of service.

| SAQ 7.12 | Can you think of an abnormal processing period for the system described in the previous SAQ? |

Solution

The abnormal period is the four weeks leading up to Christmas.

Periodic additional work

This occurs at predecided times owing to annual holidays, stocktaking, change-over of the sales range, and financial year-end work.

Error-correction procedures

These are designed to cope with errors in procedures, missing data, mistakes in calculations and misunderstandings in requirements. By studying the error-correction procedures currently adopted, the systems analyst is better able to design equivalent procedures for the new system.

If the new system's input is on-line, errors are detectable at an earlier stage and are therefore easier to correct. Nevertheless it has to be clearly understood what errors might occur and how best to correct them.

Overtime working

Overtime is caused not only by high workloads but also by bottlenecks delaying work so that insufficient time exists before a deadline. Many businesses have peak workloads intrinsic to their trade at certain times of the year. These demand additional resources not available except through overtime working.

Staff absence and machine breakdowns

Although these contingencies do not usually occur frequently, it is advisable to be aware of the standby arrangements in case they impinge upon the new system.

7.6 CURRENT SYSTEM COSTS

This section should be read in conjunction with Section 9.7 as this discusses the costs and savings of a new system. There are few things that impact upon business managers as the chance of either saving or making money. If management is presented with a statement showing potential savings from a new system, it is more readily accepted. Tangible cost savings, such as are brought about by reductions in staff, equipment and materials, are straightforward to explain.

Intangible cost savings are less easy to quantify, and often no definite monetary value can be placed upon them. Nevertheless intangible costs are important since they may be greater than realized. If at all possible, monetary figures should be put on intangible costs so that they can be judged in comparison with other costs (see Example 7.5).

Example 7.5	**Intangible costs**

- Hold-ups in production caused by the late ordering and/or late delivery of materials.

- Loss of orders due to overpricing of quotations, goods out of stock or overloading of production facilities.

- Overstocking of slow-moving items.

- Obsolete stock clogging the stores or warehouse.

Cash flow

Existing costs can be split into recurring expenditure and capital (fixed asset) costs. Recurring costs are those incurred at regular intervals such as wages, materials, rented equipment and overheads. Capital costs are applicable to the outright purchase of permanent items such as office machines and equipment, computers, furniture and premises. From the financial accounting aspect, capital costs are covered by depreciation incorporated into the financial statements. In other words, the cost of a capital purchase is disposed of on paper by spreading it over a number of years. Nevertheless, in spite of the depreciation concept, money for the purchase still has to be found either directly, or indirectly via a financing institution.

By combining all the relevant costs, a cash flow covering past time periods can be created and these costs are projected into the future. The extent to which future cost predictions are accurate depends largely upon how far the projection stretches. It is not normally too difficult, however, to cover the next few years with reasonable accuracy.

A scientific approach to evaluating and comparing the costs of projects is to discount future cash flows back to their present values. In other words to calculate and total the present value of all future costs and savings. The higher the net present value (NPV), the more desirable the project as an investment venture. Thus when comparing two or more alternative projects, their respective NPVs are computed and, other things being more or less equal, the project with the higher NPV is chosen (see Example 7.6).

A simple case of discounting is to suppose we are to receive £10,000 one year from now and the current borrowing rate is 15 per cent, then the NPV of this amount is £10,000 ÷ 1.15, i.e. £8,696 (to the nearest pound).

Example 7.6

Calculating the NPV of a project

A company intends to purchase a computer system for £450,000 and expects to make savings during the first few years of its use as shown in Table 7.1. The money for purchase has been borrowed at an interest rate of 20 per cent. It is assumed that the system will last for five years.

Table 7.1

Period	Costs	Savings	Discount factor	Present value
Start	£450,000		1.000	−£450,000
1st year	£100,000		0.833	£ 83,300
2nd year		£150,000	0.694	+£104,100
3rd year		£300,000	0.579	+£173,700
4th year		£400,000	0.482	+£192,800
5th year		£400,000	0.402	+£160,800
				NPV = +£98,100

Other methods of investment appraisal are payback period, returns/outlay (savings/costs) ratio and internal rate of return. The former two are simple but crude, the latter is sophisticated but difficult to compute.

SAQ 7.13

Which of the following are recurring costs and which are capital costs?

1 The cost of portable computers used by sales staff in a purchasing system.

2 The cost of renting a large computer.

3 The cost of stationery used to generate reports.

4 The cost of special-purpose communications equipment.

5 The cost of maintenance of a computer.

Solution

1 and 4 are capital costs while 2, 3 and 5 are recurring costs.

■ Sources of cost information

Primary sources of cost figures in large organizations are the costing department and the budgetary control department. These departments are able to provide actual costs, operating budgets and cost variance analysed under cost heads. Unless the system analyst is a cost accountant, he or she is wise to ask for help in interpreting these figures.

Another difficulty entering into the costing of the existing system is that it is unlikely to be supplanted completely by the new system. The miscellany of tasks remaining eventually requires costing along with or as part of the new system so that cost comparisons can be made equitably.

Staff costs

If cost information is either not available or is unsuitable, the analyst must compute the costs him- or herself. There is not a lot of advantage to be gained, however, from a very detailed costing exercise. For the moment, the systems analyst is concerned with approximating the costs of the present activities to be replaced by the new system.

Machine, equipment and other costs

In addition to labour costs, the systems analyst is also interested in existing machine (including computer) costs, material costs and overhead costs. Machines and equipment are either rented, leased, hire-purchased or purchased outright. If one of the first three methods applies, the cost is easy to determine. Nevertheless long-term contracts may prevent immediate reductions in these costs with the introduction of a new system. If a machine has been purchased outright, it is necessary to apportion its cost over its estimated life to obtain an annual cost. The company's depreciation policy gives some guidance here but this may not truly represent the actual annual costs.

Office equipment, such as filing cabinets and furniture, has a long life and therefore its annual cost is low. Generally, office equipment of this nature continues in use with a new system and so, in effect, its cost can be ignored. If it is the intention to purchase new office equipment, the old equipment is written off since it will have long since been paid for and its scrap value is negligible.

Office materials comprise mostly stationery, especially that associated with sales procedures, e.g. order forms, bills, invoices, statements. In a strongly sales oriented company it is likely that these costs are fairly high and will tend to remain so, therefore a careful investigation is desirable.

Overhead costs cover items such as indirect labour, e.g. office cleaners, indirect materials, e.g. cleaning materials, heating, security, and so on. The apportioning of overhead costs between departments is a fairly complicated business and becomes impossibly so if attempted between individual activities. In most cases it is not unreasonable to assume that the overheads of the present and future systems will balance out. Failing this, the systems analyst should enlist the assistance of the cost accountant in estimating, allocating and apportioning overheads.

7.7 ENTITY SETS

In this context an 'entity' is any artefact, person, company or activity with which the company is concerned and about which data is kept. An 'entity set' is a group of entities that have a common characteristic(s) in terms of usage or physical attributes. Entity sets include, for instance, amongst many others:

- the products manufactured by a company;
- the customers of a bank;
- the workers employed in a factory;
- the machines installed in a factory;
- the students in a college.

Entities fall into three types: real, activity and conceptual. Real entities are actual things, e.g. products; activity entities are happenings, e.g. financial transactions; conceptual activities are more abstract, such as cost centres.

The main reason why entity sets are of interest is that they form the data framework for the new system. By discovering their characteristics and understanding their usages, they can then be incorporated into the new system more beneficially.

SAQ 7.14   Can you think of another entity set which would be used in a banking system?

Solution

An obvious one would be accounts. Another, less obvious, one would be standing orders.

■ The characteristics of an entity set

The full understanding of entity sets is of especial importance prior to designing and creating a database and the associated data dictionary (Section 5.4). The following make up the relevant characteristics of each entity set.

How many different entities are at present in the set and how is the number changing over time?

With some sets the number is almost static, whereas others fluctuate violently even within a short period. We are mainly concerned with how many entities in each set might have to be catered for within the foreseeable future. This is relevant to the amount of backing storage needed to hold the database or the files. It is just as significant for a PC holding a few thousand entities on a floppy disk as for a mainframe holding several million entities on hard disks.

How is the set inherently subdivided?

In almost every case there are several ways in which the entities in a set can be separated into groups. Bank customers, for instance, could be split into personal accounts and business accounts, and the former further segregated into deposit accounts and current accounts.

The present size of each group and, as far as possible, its future size should be ascertained. This information relates to the different ways in which these groups will be processed.

What attributes apply to the entities?

The attributes of an entity set give rise to the data items held in the database. Attributes are sometimes extensive in number and, at this stage, the systems analyst may be uncertain as to their relevance to the new system. The answer is to become aware of the existence of all attributes, and to gather information about those that are clearly relevant.

The sort of information needed is the same as explained under data dictionaries in Section 5.4.

What code numbers apply to the entities?

In this context, code numbers are the attribute(s) that in some way identify the entity. Thus, a code number of a bank account is its account number.

By reading 'code number design' (Section 8.2), the reader will become aware of the need for information about code numbers. The main points at this stage are to establish the existence of code numbers pertaining to the entity set and discover what meanings, if any, can be deduced from them. It is also important to establish the precise layouts (pictures) of each set of code numbers.

Bearing in mind that code numbers used in IS must give unique identification, it is important to confirm that this is so for all entities. That is to say, is there a one-for-one relationship between the entities and the code numbers? Any suspicion that this is not the case calls for further investigation and, if necessary, rectification or possibly complete replacement of the set of code numbers.

■ Entity modelling

Entity models, also known as entity relationships and logical data structures, are a method of analysing information acquired in the process of systems investigation. It is a technique that is used in methodologies such as SSADM (Section 8.10), and UML (Section 8.10). The precise approach differs slightly between these methodologies and so the explanation that follows covers the main points only.

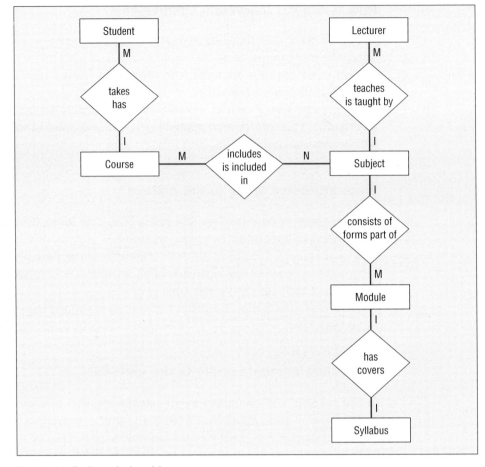

Fig. 7.14 Entity relationships

An entity model shows diagrammatically the relationship between sets of entities. These include concepts, e.g. costs, as well as objects but not unquantifiable things such as 'order processing' since this is a process. The main advantage from constructing entity models is that the analyst is more certain that all relationships have been noted. They reduce the possibility of a system being designed with an important relationship accidentally omitted. Such an omission could seriously delay the implementation of the system, if, for instance, 'subjects' in Figure 7.13 were not known to be composed of 'modules'.

From Figures 7.13 and 7.14 it is seen that entity models may be illustrated in two main ways. One way (Figure 7.13) shows the entity sets as rectangles linked by diamonds that indicate the relationships. The designations M, N and 1 signify whether a relationship is one-to-one (1 and 1), many-to-one (M and 1) or many-to-many (M and N). In this context 'many' can be taken to mean one or more.

A one-to-one relationship is such as module and syllabus, i.e. there is only one syllabus for a module and there is only one module that has a particular syllabus.

An example of a many-to-one relationship is students on a course. A course has many students but a student does only one course at a time. Many-to-many is exemplified by courses and subjects. A course includes several subjects and a subject is often part of several courses.

SAQ 7.15

In a purchasing system a company stocks a number of electrical items such as televisions which can be manufactured by a number of companies. A customer places an order by telephone and the items are delivered by a courier service. The order is accompanied by a delivery note. Categorize the following relationships as either one-to-one, many-to-one or many-to-many.

1 The relationship between electrical goods and manufacturers.

2 The relationship between a customer order and manufactured goods.

3 The relationship between customer orders and delivery notes.

Solution

In 1 each item of goods is manufactured by a number of companies, therefore it is many-to-one. In 2 a customer can order a number of goods, therefore the relationship is many-to-one. Finally, in 3 each order (assuming that they are not split into separate chunks) is associated with only one delivery note, therefore the relationship is one-to-one.

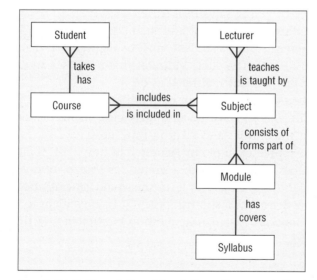

Fig. 7.15 Entity relationships (alternative convention)

Figure 7.15 illustrates the same relationships as Figure 7.12 but is drawn in a different convention, i.e. the IE convention, in which symbols and annotated lines indicate relationships, crow's feet representing a 'many' relationship.

249

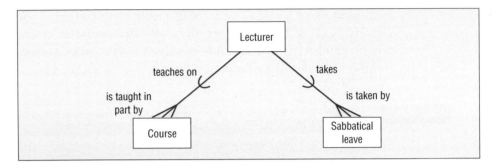

Fig. 7.16 Alternative entity relationships

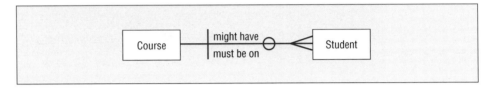

Fig. 7.17 Mandatory and optional entity relationships

Figure 7.16 shows a relationship in which there is an alternative; this is indicated by the short curves across the lines. The diagram demonstrates that at a certain point in time a lecturer may be involved in teaching on several courses or alternatively be taking sabbatical leave.

Figure 7.17 illustrates optional and mandatory relationships. The bar drawn across the line means that a student must be registered as taking a course (even if temporarily absent). The circle indicates that a course can exist (for a time at least) without any student registered as on it.

A point to be remembered about entity modelling is that the diagrams can be misleading. The analyst must be absolutely clear in his or her own mind as to precisely what the entities and relationships mean.

Even in the simple example in Figures 7.12 and 7.14, some degree of ambiguity exists. Does a lecturer teach a subject, as shown, or does he or she really teach a module? In the long term he or she would generally be regarded as teaching a subject, whereas for a shortish period it is convenient to regard him or her as teaching a module. A semantic question is whether a lecturer teaches a subject or students.

■ Normalization

In order to allow for the flexible use of data held in a database and to minimize the effect of application changes on the database structure, a process termed 'normalization' has been developed. It has been established that there are three

Branch no.	Account no.	Name and address	Postcode	Tel. no.	Balance	Transaction			
						No.	Date	Type	Amount
15	36895	Jones, 4 High St., Warwick	WK5 3MS	251–6734	215.93	228651	6.5.94	DR	65.82
						228704	9.5.94	DR	38.50
						228841	11.5.94	CR	15.00
15	36899	Todd, 18 Martin Rd., Stoke	ST2 7GM	158–3541	1265.83	227091	14.4.94	CR	90.00
						228604	28.4.94	DR	78.90
						228935	18.5.94	DR	57.83
17	35206	Carter, 2 Byron Ave., Ryde	IW7 2AH	393–0206	489.21	228711	9.5.94	DR	95.75

Fig. 7.18 Unnormalized records

main types of normalization, i.e. normalized relations. These are known as first normal form (1NF), second normal form (2NF) and third normal form (3NF). There follows a brief explanation of the meanings of these three main forms and also of two lesser-used forms (4NF and 5NF).

Suppose we have several branches, each of which deals with a number of customers. Each customer has an account number and transacts purchases (debits) and payments (credits) from time to time. Each customer also has a name and address, a postcode and a telephone number, and the account has a current balance. In completely unnormalized form this data might be in the form of the records shown in Figure 7.18.

1NF

To convert a record into 1NF the repeating groups must be put into records of their own, and so Figure 7.18 changes into Figure 7.19. It should be noted that each transaction takes with it the account number to which it applies. The process of converting to 1NF in effect turns each variable length record into several fixed length records.

2NF

A record is in 2NF provided it satisfies the conditions for 1NF and also all non-key data items are fully functionally dependent on the primary key. In our example the name and address is dependent only on the account no. not the branch, i.e. the name and address can be found if we know the account no. only. Similarly, the transaction details relate only to the transaction and not to the account, i.e. the amount, etc., is known if we know the transaction no. only.

Accounts details					
Branch no.	Account no.	Name and address	Postcode	Tel. no.	Balance
15	36895	Jones, 4 High St., Warwick	WK5 3MS	251–6734	215.93
15	36899	Todd, 18 Martin Rd., Stoke	ST2 7GM	158–3541	1265.83
17	35206	Carter, 2 Byron Ave., Ryde	IW7 2AH	393–0206	489.21

Transactions details				
Transaction no.	Account no.	Date	Type	Amount
227091	36899	14.4.94	CR	90.00
228604	36899	28.4.94	DR	78.90
228651	36895	6.5.94	DR	65.82
228704	36895	9.5.94	DR	38.50
228711	35206	9.5.94	DR	95.75
228841	35206	11.5.94	CR	15.00
228935	36899	18.5.94	DR	57.83

Fig. 7.19 1NF records

The 2NF is obtained by creating new records for the accounts and transactions as Figure 7.20.

SAQ 7.16   In Figure 7.19 if the *Transactions details* table did not contain a column called 'Amount' would the records in this table still be in 2NF?

Solution

Yes, all the non-key items are still dependent on the primary key.

3NF

A record is in 3NF provided it satisfies the conditions for 2NF and also no non-key data item is functionally dependent on any other non-key data item. In the example the postcode and telephone no. are dependent on the name and address, i.e. if we know the address, the postcode is known, and if we know the name and address, the telephone number is known.

To convert into 3NF these data items are put into another record as Figure 7.21.

It is apparent from the normalization procedures described above that relational databases, although straightforward in principle, can be cumbersome to store and organize. They involve the creation of virtual files, i.e. replicated data

Branch–account relationships	
Branch no.	Account no.
15	36895
15	36899
17	35206

Account–transaction relationships	
Account no.	Transaction no.
35206	228711
36895	228651
36895	228704
36895	228841
36899	227091
36899	228604
36899	228935

Accounts details

Account no.	Name and address	Postcode	Tel. no.	Balance
35206	Carter, 2 Byron Ave., Ryde	IW7 2AH	393–0206	489.21
36895	Jones, 4 High St., Warwick	WK5 3MS	251–6734	215.93
36899	Todd, 18 Martin Rd., Stoke	ST2 7GM	158–3541	1265.83

Transactions details

Transaction no.	Date	Type	Amount
227091	14.4.94	CR	90.00
228604	28.4.94	DR	78.90
228651	6.5.94	DR	65.82
228704	9.5.94	DR	38.50
228711	9.5.94	DR	95.75
228841	11.5.94	CR	15.00
228935	18.5.94	DR	57.83

Fig. 7.20 2NF records

of a temporary nature that is transparent to the user. This replication causes them to occupy a large amount of storage.

4NF and 5NF

It is possible in some cases to continue the process of normalization to 4NF and 5NF. 4NF is relevant to a situation where the key data item has multi-valued dependencies, i.e. several things of the same type are related to it. For instance, a name and address may have more than one telephone number. In order to be in 4NF, the name and address details in Figure 7.21 have to be split into a record for each telephone number.

In doing this the postcode is replicated in these records. 5NF avoids this by putting the postcodes and the telephone numbers into separate records each linked to the name and address.

253

Branch–account relationships

Branch no.	Account no.
15	36895
15	36899
17	35206

Account–transaction relationships

Account no.	Transaction no.
35206	228711
36895	228651
36895	228704
36895	228841
36899	227091
36899	228604
36899	228935

Accounts details

Account no.	Name and address	Balance
35206	Carter, 2 Byron Ave., Ryde	489.21
36895	Jones, 4 High St., Warwick	215.93
36899	Todd, 18 Martin Rd., Stoke	1265.83

Names and address details

Name and address	Postcode	Tel. no.
Carter, 2 Byron Ave., Ryde	IW7 2AH	393–0206
Jones, 4 High St., Warwick	WK5 3MS	251–6734
Todd, 18 Martin Rd., Stoke	ST2 7GM	158–3541

Transactions details

Transaction no.	Date	Type	Amount
227091	14.4.94	CR	90.00
228604	28.4.94	DR	78.90
228651	6.5.94	DR	65.82
228704	9.5.94	DR	38.50
228711	9.5.94	DR	95.75
228841	11.5.94	CR	15.00
228935	18.5.94	DR	57.83

Fig. 7.21 3NF records

7.8 FACT FINDING

Several techniques are available to the system analyst for fact finding. The most commonly used is interviewing.

Interviewing

This is the most common way of finding out about the current system and user's requirements. However, it is important for the analyst to conduct interviews with care. Different questions will be required at various stages and their content depends on who is being interviewed. Some guidelines for interviewing are:

- Obtain permission from management for the interviews.

- Use a quiet room for the interview, although if the interviewee's workplace is suitable, this is often advantageous.

- Prepare for the interview: create a list of topics to be covered, look at notes of previous interviews if they exist.

- Open with a brief preamble of the aims and background of the investigation.

- Ask open-ended questions to elicit user attitudes and opinions.

- Ask more focused questions for details and facts.

- Avoid committing to any opinions or promising solutions.

- Avoid taking sides with staff or departments, be a neutral observer.

- Do not get into exchanges of personal opinions or arguments with users.

- Test understanding with users, tell them what you think the position is.

- At the end of the interview summarize the facts you have gathered.

- Be constructive and encourage the user likewise; make suggestions and elicit suggestions from the user on how to improve things.

Guidelines to interviewing

- The interviewer should brief him- or herself on the position and general duties of the interviewee; and also, of course, on the subject of the interview. Interviews are either structured or unstructured; if the former approach is chosen, a plan for the interview has to be devised.

- Arrange the time, place and subject of the interview well beforehand so that the interviewee is able to make arrangements regarding his or her work, and to collect together documents and information related to the subject of the interview. If there is likelihood of continual interruptions or eavesdropping, the interview should be arranged to take place away from the interviewee's normal workplace.

- Put the interviewee at ease by providing a quiet, interruption-free environment. If a tape recorder is used, explain why and reassure the interviewee that it is not for subsequent 'inquests'.

255

■ Listen more than talk, yet say enough to keep the discussion going along the right lines. Avoid leading questions, aim towards short questions and long answers. Take written notes, as memory is unreliable; if it is felt necessary, prepare beforehand a draft of the questions.

■ Interview only one person at a time, thus eliminating arguments between staff and reducing the diffidence of reserved persons. If a manager calls in his or her staff to answer specific questions, encourage them to depart as soon as this has been done. Similarly, discourage a manager from attending an interview involving one of his or her staff as this tends to inhibit the latter.

■ Control the interview by minimizing digressions, separating opinions from facts, and not allowing generalizations to obscure the true situation.

■ Do not attempt to cover too much ground in one interview. After one hour, at the most, temporarily conclude the discussion and arrange to resume in the near future. Long interviews suggest that either the interviewee is being allowed to digress or that the discussion is becoming bogged down.

■ Conclude with a brief résumé of the ground covered, asking whether any important points have been omitted, and leave an opening for further discussion. Any dubious information is verified by sending a memorandum later; with operational staff this is best done via the department manager.

Note-taking during interviews can be useful to record facts; however, some users may feel uneasy when notes are taken. Permission should always be sought from the interviewee, whether taking notes or tape recording the interview. Note-taking also disrupts the flow of an interview so many analysts prefer to write up interview notes immediately afterwards.

Interviewing technique

In common with many other human activities, the art of interviewing is better learned through practice than from books. Nevertheless it is an indispensable

SAQ 7.17 The following are a set of questions asked by an analyst investigating a possible system for a building society. What criticisms, if any, can you make of them?

1 I intend revolutionizing your systems, where would you like me to start?

2 I found it difficult dealing with your colleague Janet. I hope that you will be able to answer my questions. The first is, on what task do you spend most of your time?

3 Your procedures for handling new accounts seem rather cumbersome, don't you agree?

4 I'm trying to get a good idea of what are the tasks that are carried out manually when an account is closed down. Could you just go through them and perhaps mention where you think the most errors are created?

Solution

1 The analyst should not commit to solutions, even nebulous ones. It is also somewhat lax to expect a member of staff, perhaps unschooled in information technology, to try to answer the second part of the question.

2 The analyst is not being a neutral observer. The second sentence is fine as a question.

3 The analyst is committing him- or herself to an opinion and then forcing it down the throat of the employee who is being interviewed. Perhaps the procedures are cumbersome but this should emerge from careful questioning.

4 This is an excellent question: the analyst explains why it is being asked and then gets the subject to answer.

part of fact finding and, if well conducted, provides valuable information about policies, procedures and situations that might not be apparent from documents.

The need for a flexible approach is paramount. Interviewees range from company directors to junior clerks, and cover a wide spectrum of ages, social backgrounds and personal attitudes. A systems analyst has to some extent to be 'all things to all men'. This does not imply, however, that he or she merely adopts a role-playing posture – this would soon be detected by the interviewees.

Interviewing top management

Top managers are concerned with strategic decisions and this ought to be re-flected in the level of questions put to them. They should be encouraged to explain objectives, major problems, large-scale developments, and the like. Questions about petty details of the company's operations will serve only to irritate them. The discussions at this level cover aspects such as impending mergers, takeovers, new markets, new product ranges, expansion of manufacturing or service facilities, and future organizations; also the present structure of the line management and their responsibilities.

In particular, the systems analyst discovers from top management the informa-tion they need in achieving their aims. At the same time, he or she is in a position to explain the potentialities of IS from their point of view, and particularly if the manager is unfamiliar with computer-based systems.

Interviewing line (middle) management

Here the interviews are more concerned with tactical planning and its related information. Most line managers have ideas for improving the present system if facilities were available. They are also able to elucidate on the organization and staffing of their department, thus enabling the systems analyst to find out who else to interview and what aspects to investigate. This is also a convenient time to request permission to interview the manager's staff.

Interviewing operational staff

Operational staff are best not encouraged to put forward policy suggestions even though, in some cases, they are only too keen to do so. Questions are restricted to details of the interviewees' duties and, if necessary, they are asked to demonstrate their work rather than describe it. It is important at this level to allay the feelings of apprehension that arise when changes are on the horizon.

Another worthwhile point to remember is that operational staff are more likely to be politically sensitive towards management and it is therefore advisable for the systems analyst to maintain a strictly neutral attitude in this respect. By friendliness, avoidance of condescension and showing an interest in their work, the co-operation of operational staff is generally secured.

SAQ 7.18
An analyst interviews the staff of a department store chain with a view to investigating a new system for handling sales. Who do you think the following questions are aimed at: top management, line management or operational staff?

1 What happens when a customer returns an item after buying it?

2 How do you know when an item is out of stock?

3 What information do you need when hiring staff for a rush period such as Christmas?

4 What information about the profitability of each store in your group do you need?

5 How do you go about ordering items for your department within the store?

Solution

The first question is to operational staff: those on the shop floor selling goods. The second question is aimed at an individual department manager so is really aimed at middle management. The third question again is aimed at whoever is dealing with personnel issues in each store, hence it is aimed at middle management. The fourth question is aimed at top management and the final question is aimed at middle management.

■ Questionnaire

Sometimes the users are remote from the analyst. Indeed the development of products for an anticipated market, the users may not be accessible at all. In these circumstances questionnaires can be used to gather facts, attitudes and some suggestions about the system. Questionnaires are inevitably limited in the information detail that can be collected when compared with interviews;

however, they are valuable in acquiring information from a large sample of users.

The adoption of questionnaires as a fact-finding tool is not as straightforward as the inexperienced systems analyst might assume. It is wise to give careful consideration to their employment and design. The main difficulty lies in the avoidance of misunderstandings because the respondent cannot easily query vague or confusing questions. Similarly, the systems analyst cannot always pursue answers that might lead to further information. Questionnaires should be used only when no other method of fact finding is practicable: they supplement rather than replace other methods. If time and circumstances permit, a trial run should be carried out using a small number of representative persons prior to the full survey.

Situations suited to questionnaires

■ Where the systems analyst is located at a considerable distance from the staff to be questioned, e.g. managers of widespread branches.

■ Where there is a large number of respondents such that interviewing is prohibited by the time available, e.g. a large sales force.

■ As a means of verifying information found by other methods. The questions in this case must be framed so as to avoid leading the respondent.

■ When the questions are simple and call for direct answers, preferably when they are to be selected from a list of answers shown on the questionnaire, i.e. multiple choice.

■ When a full set of replies is not necessary in order to determine the facts. People tend to give low priority to filling out questionnaires, and therefore the response is rarely 100 per cent. The sample returned must, however, be large enough to provide a reasonably accurate picture of the situation.

Guidelines to questionnaire design

■ Give a brief explanation of the questionnaire's purpose if this is not self-evident from its contents.

■ Bear in mind the level of intellect and likely interest of the respondents.

■ Keep the questions short, unambiguous and unbiased, and, whenever possible, give multiple choice answers from which one is to be selected. In any case do not pose questions that necessitate a long answer.

■ Avoid overmuch branching and skipping in the structure of the questionnaire, e.g. 'if the answer is "yes" go to Question 8; if "no" omit the next two questions'.

- ▪ If the survey is extensive, give consideration to the use of scanning documents in order to reduce the manual work in analysing the answers.

- ▪ Impose a deadline by which the questionnaire has to be returned. A prepaid addressed envelope is beneficial if the respondent is without secretarial help.

- ▪ Prepare the ground and identify the target population.

- ▪ Motivate the users to return the questionnaire, either by offering a bribe or by making it clear how they will benefit from the results.

- ▪ Draw up the objectives of the questionnaire, what information is required, what is the size of the user population sample, and what return rate is acceptable.

- ▪ Design the questions (see below).

- ▪ Group questions by topics and structure the questionnaire so that the questions and groups follow a logical order as far as possible.

- ▪ Give instructions for answering questions, and if the reply set is known, then make this explicit and ask the user to underline or circle their choice, e.g. please select one of the following: novice/experienced/expert.

- ▪ Avoid biasing the answer in a question.

- ▪ At the end thank the user for their co-operation.

Design of questions

The questions, too, must be designed. These may be of three types as follows:

1 Quantitative replies in which the user circles a number from 1 to 7 to represent a scale from 1 = very negative to 7 = very positive. It is important to ask the question so that it can be answered in a scale and to ensure that only one question is asked at once. For example:

   'How important is it for you to have spell checking in a word processor?'

1	2	3	4	5	6	7
not important				very important		

   These types of question can acquire information on users' attitudes, opinions and priorities. The advantage is that replies can be quantified to give an average and range for a population of users.

2 List type questions to elicit user requirements, with replies which may be ranked. For instance:

   'Please list in decreasing order of importance the features you would like to see in a word processor (a feature is something like spell checking, multi-fonts, integration with graphics, etc.). Please number your list 1, 2, 3, etc., to show the importance of each feature.'

3 Open-ended questions which acquire description of problems and user requirements. For example:

'Please describe any problems you have encountered with the current system. Please number each problem separately and give any suggestions you have to explain why the problem exists and how it may be cured.'

Observing

While interviews are useful in gaining information users do not automatically tell the analyst everything. Much knowledge is tacit, i.e. it is never verbalized because the user assumes the analyst already knows commonplace facts. Unfortunately, facts assumed by the user may well be vital for the analyst's understanding. One way to tackle the tacit knowledge problem is to use 'observing'. The analyst observes the user in their workplace, while trying to be as inconspicuous as possible.

Observing is time consuming but it can uncover much information on working practices. Activity analysis is helped by observing as people rarely give complete descriptions of how they do things. It also gives details of exceptional conditions, informal communication, and shows when people adopt ways around the system to deal with exceptional problems. Unfortunately, exceptions may be rare, hence a considerable amount of observing may be necessary for this technique to pay off. A further problem is the Hawthorne effect – people are usually aware of the observer and modify their behaviour accordingly. The observed activity may therefore be better than normal and many informal practices may be hidden. Nevertheless observing is a useful backup to interviews, especially when group interaction needs to be analysed.

Observing entails watching the departmental staff carrying out their various tasks. It is a time-consuming activity and therefore not to be indulged in without a definite purpose. As a rule, people do not take kindly to being observed at their work, so this is one of the systems analyst's more delicate tasks. Any attempt to quantify the staff's activities by timing their actions is likely to result in a distorted measure of their work and so a work-study approach is inappropriate in this context.

These remarks must not be taken as implying that observing is always a waste of time and effort. When done discreetly, this method yields information unobtainable through other methods and much depends upon the nature of the work in the department. Observing the operational staff carrying out a routine clerical function yields little new information. On the other hand, watching people at work in a diversified department often brings to light numerous tasks and problems not discovered previously.

It must be borne in mind, however, that even an extensive period of observation may not expose all the problems. At best it is possible to get only snapshots of a continuously changing scene, and these may lead to a distorted impression of the true picture if not supported by other methods of fact finding.

The aspects of a department's work revealed by observing are:

- **Interruptions to the normal flow of work**: these are caused by callers from other departments, telephone calls received and made, and visitors from outside the company.

- **Informal communication of information**: this is between members of the department, callers and visitors, and over the telephone. Since no paperwork is involved, this type of information flow might go undetected if not observed.

- **The usage of files and documents**: included here is the non-routine reference to file documents in order to handle queries, often received by telephone. In an already computerized system, it is possible to observe how the staff make use of computer printouts and VDU displays.

- **The balance of the work load**: this applies to the different times of the day or week, and between the various members of the department.

- **Operational inefficiencies**: the observable factors here are bad working conditions, machines and equipment in poor condition, absence of authority or leadership, and insufficient understanding of the procedures.

■ Reading

The problem with reading as a fact-finding tool is in knowing what to read and when to stop. Companies have virtually unlimited amounts of documents and literature, and it is therefore important that the systems analyst is not swamped by them. It is best to be guided by the department staff who appreciate the problem and are willing to select the relevant material.

Literature worth pursuing comprises:

- *Reports of previous surveys and investigations*: the more recent of these are likely to contain valid information and conclusions, but even the older reports may contain facts and ideas bearing upon modern systems. The older the information, however, the more stringently must it be verified before being acted upon.

- *Company instructions (or the equivalent)*: these documents provide useful information regarding the company's organization, administrative procedures, policies and future developments. Company instructions are likely to exist only in the larger, highly structured companies and in government departments.

- *Sales literature and company information booklets*: these enable the analyst to acquire an overall and broad view of the company's activities and organization.

- *Job descriptions*: these should confirm the positions, duties and responsibilities of the staff as noted during interviewing.

- *Existing IS documentation*: this includes the documents as described in Section 8.9.

- *Management reports*: these encompass a wide range of management information and form a good starting point for discussions re future requirements.

- *Procedure manuals*: it is not likely that procedure manuals are in great profusion as most companies do not have them. Where they do exist, they should form a helpful background to present procedures.

- *Forms and documents*: the need to investigate these has been explained in Section 7.4.

■ Measuring

When facts are unobtainable through other methods or when their accuracy is suspect, measuring or estimating is employed. It is not a method recommended for general use since it absorbs considerable time and demands great care. In most cases measuring yields an approximate figure but nevertheless this is acceptable for its purpose.

Example 7.7   **Measurements and counts**

Quantities (numbers):
of stock record cards held; of staff employed in the wages department.

Times:
to prepare a set of sales statistics; to deal with a query from management.

Intervals:
between the month-end and dispatching the last sales statement; between the issue and return of job tickets.

Rates:
of arrival of customers' orders per day; of issue of parts from stores per hour.

Each of the factors shown in Example 7.7 has certain inherent characteristics, i.e. maximum, minimum, average (mean, mode or median), spread (e.g. standard deviation) and distribution pattern, and so the analyst needs to be clear as to precisely what he or she is attempting to measure. For instance, merely to find that an average of 200 customers' orders arrive per day is not necessarily of great importance. More consequential is the fact that this figure rises on occasions to 500 per day, and that all orders have to be dealt with on the day of receipt.

■ Sampling

A systems investigation often necessitates determining a figure that can be found only by measuring a sample drawn from the entire group. This is usually because the entire group (known as the 'population') is too large to be measured completely and so a sample is measured – in the hope that this is representative of the population (see Example 7.8).

Example 7.8	**Sampling**

The average number of items on sales invoices is found by inspecting a sample of, say, fifty invoices and counting the number of items on each of these. If it is found that the numbers of items are closely grouped, it is safe to use simply the average of the sample. A wide spread of the numbers of items in the sample leaves considerable doubt as to the average being sufficiently accurate to represent the population.

Sampling theory enables us to calculate the size of the sample needed to attain a given probability of being within an acceptable tolerance of the true (population) average. For instance, in counting the invoice items, we might be content if we are 90 per cent certain that the sample average lies within plus or minus one of the population average. Alternatively, sampling theory tells us the probability of our sample's average lying within a predecided acceptable tolerance.

■ Verifying facts

Verification of the facts acquired from the systems investigation covers two aspects. One is to verify that the accuracy of a quantifiable fact is within an acceptable limit. It must be realized that it is almost impossible and not really productive to aim for absolute accuracy in most figures. They are constantly changing and so a completely accurate measure applies only to one moment in time. The main point of verifying this type of fact is to eliminate large errors, e.g. a quantity accidentally recorded as ten times too big.

The other aspect of verifying is to ensure that activities are performed and situations exist as reported. There are obviously slight discrepancies between one person's account of events and the next person's, nevertheless all accounts should be substantially in agreement and any major differences reconciled.

The answers a systems analyst receives in response to his or her questions may be erroneous, and mainly for the following reasons.

Ignorance of the subject matter

If the respondent does not fully understand the matter under discussion, he or she may make guesses at the answers. An isolated guess can go undetected but a string of guesses should be obvious. There is also the occasional erroneous reply even though the respondent knows the subject matter quite well: he or she simply makes a genuine mistake.

Misunderstanding the question

In many cases, an experienced analyst detects a misunderstood question and so rephrases it. The worst state of affairs is a discussion diverging from the true subject owing to a succession of erroneous answers and consequent inappropriate questions. This situation should not occur for long with an experienced systems analyst.

Deliberate misrepresentation

Fortunately this is an uncommon reason for incorrect answers. It is possible, however, that the respondent has something either to hide or to gain by distorting the truth. The former reason is usually associated with incompetence, idleness or even fraud, the latter with 'empire building', promotional prospects or job security.

Mistakes are less likely in documented information, but even here the systems analyst should not assume that 'everything in print is correct'. The greatest dangers are obsolete documents and out-of-date information on documents.

Methods of verifying

The prime method of verification is by comparing the factual information obtained from two or more sources. The snag is that the sources might stem from the same origin and so are not really independent.

Another means of verification is to analyse the information acquired. This is done by ascertaining that the constituent figures in the information combine together to give the supposed result.

A third means of verification is common sense. This is a combination of business experience and intelligent supposition.

▣ Analysing facts

After collecting the facts and figures pertaining to the existing situation, the systems analyst studies these so as to formulate ideas for the new system. Apart from the qualitative aspects of the facts, he or she can use two main statistical techniques to detect certain characteristics in the figures: trend analysis and correlation analysis. A systems analyst needs a general understanding of these, even if only to appreciate what is involved and to call for specialist assistance if need be. Brief explanations of these techniques follow; fuller details are available from most books on statistics.

Trend analysis

Trend analysis is based upon time series analysis, and comprises a number of techniques helpful to the systems analyst in predicting future figures from past figures. Prediction is important, for instance, where there is a changing situation such that there could be a considerable difference between the present work load and that of a few years hence. It cannot be emphasized too strongly that prediction is no substitute for good intelligence about future events. It is more useful to acquire information than to make the most sophisticated forecasts.

The main aim is to remove seasonal, cyclical and random variations from the past figures in order that any steady trend is revealed. There are three main methods of revealing a linear trend in a set of figures: graphing, semi-averages, and least squares. The method to adopt depends largely upon the computing power available and the amount of data to be analysed.

Graphing

This is simply a matter of plotting past figures on a graph, and drawing by eye the best line through them. It is difficult to define what is meant by the 'best' in this context except to say that it is the line which, on the whole, is nearest to all the points on the graph. The more recent points should take preference if the points lie in a markedly curved area. This is tantamount to saying that, if necessary, the best curve should be drawn through the points.

Whatever line is drawn, it can be projected for a few time periods into the future with, hopefully, a reasonable degree of accuracy. In any event, it is likely to be more accurate than merely assuming that the recent figures will continue to hold.

Semi-averaging

This is a crude but simple method of assessing the linear trend in a set of figures. The principle is to split the past figures into two equal or nearly equal groups and to compute the average value of each group. The difference between these averages is then divided by their distance apart, i.e. the number of time periods between them. The result is a rate of change of the figures that can be projected into future time periods.

The least-squares method (regression analysis)

This is more scientific in that it determines mathematically the best straight line running through a set of figures. Put another way, it finds the equation of the straight line that minimizes the average distance between itself and the figures as plotted on a graph. The equation can then be used to compute the figures applicable to future periods.

Seasonal variations

Superimposed on the linear trend of a set of past figures may be a seasonal trend, occurring to a greater or lesser extent each year. This is often masked by random fluctuations and may not be immediately apparent from a visual inspection of the figures or a graph.

By computing the linear trend, as already described, and subtracting the trend figures from the actual figures, we are left with the deviations. In order to detect a seasonal trend the figures must be available for monthly or at least quarterly periods. The average deviation for a given month or quarter taken over a number of years is a measure of seasonal variation. In practice it is advisable to use the figures for five years or more before coming to any conclusions about the significance of seasonal variations.

Correlation analysis

Correlation is a measure of the strength of the relationship between two variable quantities. That is to say, if one quantity changes, does the other quantity also change, either in the same or the opposite way? The strength of this relationship is measured by the correlation coefficient (r). If $r = 1$, there is a perfect positive relationship, i.e. the two variable quantities are linearly related, one increasing at a proportional rate to the other. If $r = -1$, there is a perfect negative relationship, i.e. the rate of decrease of the one is proportional to the rate of increase of the other. If $r = 0$, there is no linear relationship, i.e. the quantities are random relative to each other or have an irrelevant relationship. As a rule of thumb, if r is above 0.7 or below -0.7, there is a relationship worth investigating but, of course, common sense needs to be applied.

It must be emphasized that merely because two variables are shown by correlation analysis to be statistically related, they are not necessarily related in any other way. That is to say, they might not be causally or logically connected in any way whatsoever. A favourite pastime of anti-statisticians is to demonstrate the existence of significant correlation coefficients between obviously unrelated variables.

Within the sphere of systems investigation, however, there is a reasonable chance that two statistically related variables are either causally or logically connected; for instance, the amounts due on bills receivable and the times taken before payment. If a statistical relationship is established, then a causal connection is worth looking for and investigating further.

Multiple correlation analysis

This is a more sophisticated technique enabling a measure of correlation to be obtained between several different sets of variables, e.g. vehicles in use, road deaths and the maximum speed limit.

It is unlikely that a systems analyst will need to employ multiple correlation analysis but if he or she does, the assistance of a statistician may well be necessary.

| SAQ 7.19 | Can you think how correlation analysis might be used when investigating the way that items are ordered for a department store? |

Solution

There are a number of relationships that can be investigated:

■ The relationship between the volume of overall sales and the time of year.

■ The relationship between the sales of individual items and the time of year.

■ The relationship between two items where there might be a high degree of correlation between the items being bought together.

■ The relationship between the decreasing sales of a particular item and time.

7.9 PROTOTYPING

The conventional approach to systems development is through the systems development life cycle (SDLC), after the style of Figures 7.1 and 8.1. These diagrams show several iterative paths that are, of course, necessary in order to obtain further information or make changes to previous design features. The need for changes is often caused by misunderstandings of the prospective user's requirements. It is also caused by the user not being clear as to what it is possible for him or her to expect. Although for simple applications the amount of iteration is generally minimal, for a large and complex application it can be everlasting – or so it seems. Thus any method that minimizes replication of effort on the part of both the systems analyst and the user is obviously desirable.

Prototyping is a collection of techniques for reducing misunderstandings and clarifying requirements. It is usable during both the systems investigation and systems design stages – and the earlier in the SDLC it is used, the greater the advantage.

Prototypes are amended and presented several times during the SDLC in a form dependent on the stage reached. By so doing there is the potential for early amendments to weaknesses in the designed system and, in an extreme case, for its abandonment. These possibilities exist because prototyping involves the close participation of prospective users, and consequently their reactions are readily perceived.

Prototyping is particularly beneficial in situations where the application(s) is not clearly defined. On the other hand, for well-understood, fully definable applications, prototyping is probably not worth while. An example of the former situation could be a firm of estate agents setting up a system for maintaining and interrogating a database appertaining to properties on their books. Although the estate agents have previously done this by means of a card filing system, they find it difficult to imagine a computerized system. They would be greatly reassured by the sight of their own records on a screen and by the opportunity of calling it up themselves through a hands-on method. If the prototype makes them believe that the computer is totally under their control, they become more confident and so more inclined to suggest relevant information and suitable screen formats.

An instance of a well-defined application could well be sales invoicing and accounting. The work is well understood as it has been carried out for a long period and there is nothing to be gained from changes. Thus the new system is to be a replica of the previous system from the input/output aspects, and so prototyping is unnecessary.

SAQ 7.20 Would a system for processing accounts in a bank be a candidate for prototyping? What about a system for providing interactive shopping by a supermarket via the Internet?

Solution

Banking systems are well defined and would not normally require prototyping. Interactive shopping is less well defined and could require a degree of prototyping.

Tools for prototyping

In order for the systems analyst to be able to construct prototypes promptly, a speedy and straightforward method has to be available. In this respect fourth-generation technology (4GT), and, in particular, fourth-generation languages (4GLs) (Section 6.2), forms the cornerstone of prototyping.

There is a host of techniques available, though, strictly speaking, there is no absolute definition of what constitutes 4GT. In broad terms, however, 4GT could be said to include any technique or language that furnishes sophisticated results with a minimum of human effort. Types of 4GT are given below.

Screen generators (screen painters)

A screen generator is a tool for creating displays on VDU screens quickly and easily. The principle is that the analyst 'draws' or 'paints' the required layout on the screen by means of a mouse, a pointer, a palette and a keyboard. This enables lines, icons and coloured areas to be represented on the screen and then headed and annotated with text and data from the keyboard. By this means it is feasible to create any document or other layout on the screen, and to modify the layout and its contents according to the wishes of the user.

Facilities are available for inserting items from a data dictionary such as titles, headings and annotations. Also available are preformed skeletons of menus and forms for the subsequent entry of data, and icons for helping with the choice of requirements.

Graphics

In business terms graphics means the representation of data in the form of graphs, histograms, pie charts, bar charts, pictograms, etc. (see also Section 9.5). These representations are generated by the 4GT without recourse to programming on the part of the analyst. Thus it is possible, for instance, to show a histogram based on real data so that a manager can confirm that this prototype meets his or her needs.

Report generators (report writers)

A report generator is software for quickly creating a report from data in the database. The content and format of the report are specified through the use of a non-procedural language, i.e. a 4GL, or alternatively by filling in screen forms. The report generator retrieves, sorts and summarizes the appropriate records; it is also capable of performing rudimentary processing, e.g. calculating percentages. The report's format, headings, annotations and totals are controlled through the 4GL or alternatively are entered automatically by default. In this way a user is encouraged to set up his or her own reports during prototyping.

Fourth-generation languages

4GLs are explained in Section 6.2, and from there it is clear that they are invaluable for prototyping owing to their flexibility and ease of use.

■ Categories of prototypes

There is the danger in categorizing that the things in question become regarded as completely dissociate. With prototyping this is a significant point because in practice some or all of the categories described below and shown in Figure 7.22 might be incorporated into the prototyping of the one system. The various applications and procedures may be prototyped in different ways or perhaps not at all. The reader should therefore accept the following methods as being approaches rather than clear-cut categories.

Non-working prototypes

With this approach the prototype is a dummy and is usable only for the purpose of demonstrating input procedures and/or input formats. The prototype is incapable of actually processing data but merely reproduces results that have been predetermined and then incorporated into the programs. Accordingly the results are unreal, i.e. false, and so a non-working prototype is unsuitable for realistic user interfacing.

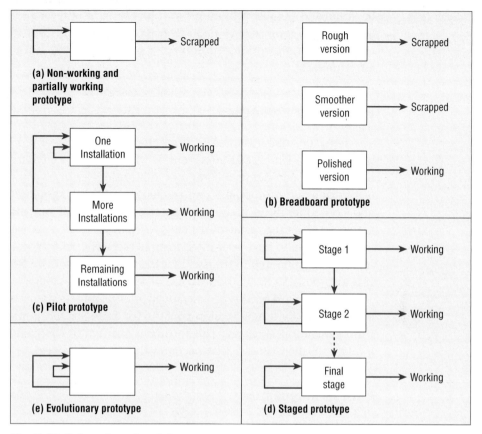

Fig. 7.22 Categories of prototypes

Although this is deceitful, it is nevertheless a valid approach so long as the users are aware of the truth of the matter. Its main purpose is to illustrate the layout of screens, documents and reports irrespective of their contents. As a matter of interest this method has been used for thirty years or more in sales environments to show computer printouts (see Section 7.3).

A non-working prototype is scrapped after the end of its usefulness.

Partially working prototypes

This method could also be termed an 'interfacing' prototype because it is intended to allow users to operate it as a system that gives responses to certain input. Any input to which the prototype cannot give a correct response is rejected so that there is no chance of users being misled.

The prototype is arranged to check input, giving appropriate responses and moving through a succession of dialogue messages. When an input message calls for a substantial piece of output, e.g. a longish report or list, this may have to be simulated because the necessary processing is not yet programmed. This is acceptable so long as it is made clear that the output is not real.

As with a non-working prototype, a partially working prototype is scrapped after fulfilling its purpose.

Breadboard prototypes

The term 'breadboard' is taken from electronic engineering, it implies a rough-and-ready but working arrangement. It was originally a circuit held on a wooden board but in our case is a computer application.

A breadboard prototype is constructed mainly with speed of creation in mind, and consequently is inefficient in terms of speed of operation and usage of storage. This is the result of it being written in a 4GL.

It is possible to have user interfacing because the prototype actually works but it must be emphasized to the user that the final system will be much more efficient. It is also possible for it to make use of an existing database.

The breadboard prototype is scrapped after noting its strengths and weaknesses and replaced by a new version constructed from scratch, probably written in a third-generation language. This in turn is eventually scrapped to be replaced by a final, polished version that becomes the working system.

Pilot prototypes

A pilot or first full-scale prototype is applicable to situations involving a number of installations doing the same work, e.g. a point-of-sale system in a chain of supermarkets. The principle is for the prototype to be introduced into one or a few of the installations so as to enable weaknesses to be detected. These are removed before the prototype is installed in further locations. Generally an application that is suitable for pilot prototyping is fairly straightforward technically; the problems tend to arise from the human aspect.

A pilot prototype is likely to encompass the full range of activities of the application. This means that most of the systems design and programming will need to have been done before initiating the prototype. It is also likely that the complete database is needed, e.g. with a point-of-sale system all the prices, descriptions, etc., must be available from the outset.

Staged prototypes

Staged or incremental prototyping implies that a start is made with only certain features of the full system. Further features are added stage by stage, and the prototype is checked at each stage. A stock control application, for instance, could start simply with the updating of the stocks in hand, then move on with stock evaluation, usage analysis, demand forecasting, automatic re-ordering, and so on.

At each stage of the prototyping, user interaction is important in order that weaknesses and misunderstandings are not compounded in the next stage.

Once a stage has been shown to be fully correct and complete, using real data, it can be rewritten in a 3GL so as to speed up the whole process of prototyping. The contents of the files may have to be enhanced at each stage but this need not be a problem if allowed for from the start.

With staged prototyping it is preferable that iteration occurs within each individual stage rather than between several stages. The final stage, when fully accepted, becomes the working system.

Evolutionary prototypes

An evolutionary prototype is in some ways similar to a staged approach. But whereas staged prototyping entails adding a succession of separate but closely associated stages, evolutionary prototyping allows the one integral application to evolve through a succession of increasingly refined phases.

It is obviously not always practicable to employ evolutionary prototyping because many applications do not lend themselves to this approach.

As an example, suppose we have a hotel room reservation system for which a file exists holding details of the rooms and their current bookings. The first phase is for the user merely to make enquiries regarding room availability. When this procedure has been agreed, the next phase could be to insert the procedure for making bookings. Thereafter the prototype is refined by the addition of procedures such as cancellations, changes, special requirements, and so on.

SAQ 7.21 What categories of prototype are the following:

1 A prototype made up of paper versions of the computer screen that a user would see.

2 A program which implements the stock re-ordering part of a purchasing and sales system.

3 A system for an airline which consists of three parts delivered over a period of three years.

Solution

The first is a non-working prototype, the second is a partial prototype and the third is an evolutionary prototype.

EXERCISES

Exercise 7.1   Decision tables

Prepare the following:

(a) a decision tree;

(b) a limited condition/extended action decision table; and

(c) an extended condition/limited action decision table;

to specify the circumstances described below:

A company's customers are either home or foreign, a company customer or an individual person, known or unknown to the company's credit controller, and either able or not able to provide an acceptable reference.

Credit is allowed to some customers depending on the above factors as follows.

Home companies known to the credit controller are allowed credit of up to £1,000, if unknown £400. Foreign companies with an acceptable reference get £700, otherwise £300. Individuals who are home customers and provide an acceptable reference get £100, otherwise no credit. Foreign individuals are allowed £50 on providing an acceptable reference, otherwise nothing.

Exercise 7.2   Data usage chart

Prepare a data usage chart to incorporate all the documents, files and data items involved in the following situation:

Orders are received by telephone and an order form (T1) filled out with the details of the order, i.e. commodity code, quantity ordered, order date, also customer's account number looked up in the name and address file (F1). When the goods are subsequently dispatched a dispatch note (T2) is filled out with the same data as on the order form plus dispatch data, quantity dispatched, and customer's name and address (from F1). Later an invoice (T3) is prepared; this is copied from the dispatch note but quantity ordered is omitted. Additionally the commodity price and VAT rate are looked up from a price file (F) and entered; also the commodity value, VAT amount, and amount payable are all calculated and entered.

Exercise 7.3   Dataflow diagram

Draw a dataflow diagram for the following procedure:

▪ When a goods received note (GRN) is received from the goods inwards section, the corresponding order copy (OC) is extracted from the goods ordered file.

▪ The data on these two documents are compared, and if not entirely alike, a discrepancy form (DF) is filled out and this together with the GRN and OC are sent to the enquiries section.

  If the data are alike, the OC is stamped 'goods received' and attached to the GRN; they are then filed pending receipt of the corresponding purchase invoice (PI).

▪ When a PI arrives, it is checked for various errors, and if any are found the PI is photocopied and this is filed for future reference. The erroneous PI is returned to the supplier with a covering letter.

If the PI is error-free it is stamped 'checked OK' and the corresponding GRN and OC are extracted from the file.

- The data on the PI is checked against that on the GRN and OC. If there is any discrepancy, all three documents are attached together and sent to the enquiries section. If the data are alike, the PI is stamped 'passed for payment' and filed for subsequent payment. The GRN and OC are filed together in another file.

Exercise 7.4   Entity relationships

Construct an entity relationship diagram in the IE convention for the following situation:

A company consists of several departments, each having a number of employees. Each department has a manager who must be on the monthly payroll. Other employees are on either the monthly payroll or the weekly payroll, and are members of the sports club if they so wish.

Exercise 7.5   Dataflow diagram

Create a dataflow diagram to illustrate the following procedure:

Mowell Ltd is a company manufacturing lawn mowers and allied equipment. These products are assembled from sub-assemblies (made-in) and parts and raw materials (bought-out). Sub-assemblies are made from parts and raw materials.

At quarterly intervals management issues the forward production programme (PP); this gives the quantity of each product to be manufactured in the programme.

The PP is first passed to the production planning department (PPD) where it is broken down into its consistent sub-assemblies, parts and raw material gross requirements. The sub-assembly gross requirements are passed to the stock control department (SCD) where they are netted against unallocated stocks.

The sub-assembly net requirements are passed back to the PPD for breaking down to give the parts and raw materials gross requirements.

All the parts and raw materials gross requirements (both from sub-assemblies and from products) are passed to the SCD for netting against unallocated stock. The net requirements are then passed to the purchasing department (PD) for forming into purchase orders to be sent to the appropriate suppliers.

Exercise 7.6   Systems investigation/prototyping

Luxury Furs Ltd has a chain of twelve retail shops selling ladies' fur coats. It operates a low volume, high price business and each garment is allocated a separate stock number in the manual stock records system operated at head office. All sales are for cash which is banked intact each day. Each shop renders, by post to head office, a daily sales report which is supported by copy sales dockets and a bank paying-in slip. The following are the head office routines:

- The daily sales sheets and supporting information are received by the sales/stock audit department who check the sheets, stamp them to show that they have been checked and update their manual stock records. They then pass them in daily batches to the cashier who enters the amount banked into his or her analysis cash book, the daily total of which is transferred to the appropriate column in the main cash book.

- Each week the cashier calculates the VAT element of sales cash, transfers it to a separate column in his or her analysis cash book and prepares a sales report for the business showing the sales for each shop for the week and for the year to date compared with similar information for the previous year. This is submitted to the sales manager.

- Each week the cashier receives a bank statement which he or she checks against the cash book and prepares a bank reconciliation.

- Each week the cashier prepares a cash book summary which is passed to the accounts department to be posted into the nominal ledger. In addition to updating the nominal sales and VAT accounts, the accounts department apply the fixed cost of sales percentage to the sales figure to provide the cost of sales figure which is used to update the respective shops' stock and cost of sales accounts.

- All purchases are delivered to the head office warehouse where they are checked, ticketed and assembled for dispatch weekly to the branches under cover of priced dispatch notes. Copy dispatch notes are passed to sales/stock audit department for use as a stock record and to the accounts department to update the nominal ledger.

- Suppliers' delivery notes are passed by the warehouse under control of pre-numbered goods received notes to the sales/stock audit department to update the head office stock records before passing the delivery notes to the accounts department to hold pending receipt of the respective invoices.

- At the end of each four-week period, each shop takes stock and submits stock sheets to the sales/stock audit department who value them at cost, having checked the cut-off and goods in transit. The accounts department sends a copy of the nominal ledger stock accounts for each shop for the period to the sales/stock audit department who agree the closing balances of these with the closing stock values shown in their manual stock records. When these have been agreed, the totals are compared with evaluated totals of the stock sheets submitted by the shops and any discrepancies between the two are reported, on a standard form, to the sales manager.

Requirements:

(a) Prepare an overview manual procedures flowchart showing, in outline, the involvement of the various departments concerned in the above routines.

(b) Design a monthly branch stock discrepancies report in a suitable form for the sales manager.

(ICAEW PE1, Audit sys. & DP, May 1987)

Exercise 7.7   Normalizing

A company manufactures products by assembling components together. The components are either bought-out, i.e. fully completed, or made-in from raw materials. The latter take various forms such as sheets, liquids, bars, etc. Each form has a unit-of-measure (UOM) code, i.e. 1 = metres, 2 = square metres, 3 = kilograms, 4 = litres. Components have a designation to indicate whether bought-in (= 1) or made-in (= 2).

The company's production planning file holds the data items shown below. Transcribe these into 3NF and list the figures as per Figure 7.21. They are deliberately unordered.

- Product no., description, assembly time
- Component no., description, quantity per product, supplier no., supplier name, BO/MI designation } repeated for each component in product
- Raw material no., description, quantity per component, supplier no., supplier name, UOM code } repeated for each made-in raw material in component

Product no. 325 is a trolley and contains components B1378 (six off), M496 (one off) and B2284 (four off), and takes a time of 0.35 to assemble. B1378 is supplied by AGD Ltd, B2284 by White & Co., M496 is made-in from three raw materials, i.e. 0.6 square metres of sheet steel (no. 43750), 1.5 metres of steel bar (no. 40605) and 1.2 litres of black paint (no. 32809).

All steel materials are supplied by VG Steel Ltd (no. 4381), all paint by Premier Paint Co. (no. 2659).

B2284 is a 6-inch wheel, White & Co.'s no. is 1126, M496 is a chassis, B1378 is a bracket, AGD's no. is 1759.

Exercise 7.8   Fourth-generation languages

Many software developers are now using 4GLs to build their systems. How does the introduction of 4GL affect the following:

(a) The role of the programmer?

(b) The role of the system analyst?

(BCS, GPI, April 1992)

Exercise 7.9   Recruitment procedures

(a) Using as many of the following files as you consider necessary, draw a dataflow diagram to show the likely recruitment procedures which would be used by the personnel department of a large company.

- a job specification file, holding a job description of every job in the company;
- an establishment file, holding, for each department, the maximum number of staff permitted for each job type;
- a personnel file, holding one record for each existing employee;
- a vacancy file holding details of unfilled vacancies;
- an applicants' file holding details of all job applications;
- an interview file holding a diary of all forthcoming interviews.

(b) At least two of the files listed in part (a) above, will contain personnel data. Briefly describe the scope of UK legislation intended to protect individual privacy and prevent misuse of confidential information.

(ACCA, Man. Inf. Sys., Dec. 1992)

Exercise 7.10   Decision logic

(a) Express the logic of the following procedure using ONE of the following three tools:

   (i) a decision table;

   (ii) a decision tree;

   (iii) structured English.

Policy for course admission to a college department dealing with students who may possess Advanced levels, a part-professional qualification, or both, is as follows:

■ applicants with two Advanced levels are eligible only for a degree or HND place;

■ applicants with one Advanced level are eligible only for an HND place;

■ applicants with a part-professional qualification only are eligible only for a professional course;

■ applicants with both Advanced levels and a part-professional qualification are eligible for a professional course or the course their Advanced levels qualify them for.

Course rejection letters are sent to candidates failing to satisfy the admission requirements.

(b) Assess, using example situations, the relative strengths and weaknesses of the three tools referred to in (a) above.

(ACCA, Man. Inf. Sys., Dec. 1992)

Outline solutions to exercises

Figures 7.23–7.37 can be found on pages 279–288.

Solution 7.1

(a) Refer to Figure 7.23.

(b) Refer to Figure 7.24.

(c) Refer to Figure 7.25.

Solution 7.2

Refer to Figure 7.26.

Solution 7.3

Refer to Figure 7.27.

Abbreviations used:

GRN   Goods received note
OC   Order copy
PI   Purchase invoice
DF   Discrepancy form

Processing notes:

P1 OC corresponding to GRN is extracted from file (F1).
Data compared:

■ if alike, OC stamped, attached to GRN and both filed (F2);

■ if not alike, DF filled in and sent to enquiries section with OC and GRN.

P2 PI checked:

■ if correct, PI stamped and corresponding GRN and OC are extracted from file (F2);

■ if erroneous, PI photocopy made and filed (F5), original PI returned to supplier with covering letter.

P3 PI checked against corresponding OC and GRN:

■ if no discrepancies, PI stamped and filed (F4), OCs and GRNs are filed together (F3);

■ if discrepancies, all three documents are attached together and sent to enquiries section.

Solution 7.4   Refer to Figure 7.28.

Solution 7.5   Refer to Figure 7.29.

Solution 7.6   (a) Refer to Figure 7.30.
                     (b) Refer to Figure 7.31.

Solution 7.7   Refer to Figure 7.32.

Solution 7.8   Refer to Sections 6.2 and 7.1.

Solution 7.9   (a) Refer to Figure 7.33.
                     (b) Refer to Section 9.9.

Solution 7.10   (a) See Figures 7.34 to 7.36.
                     (b) See Figure 7.37.

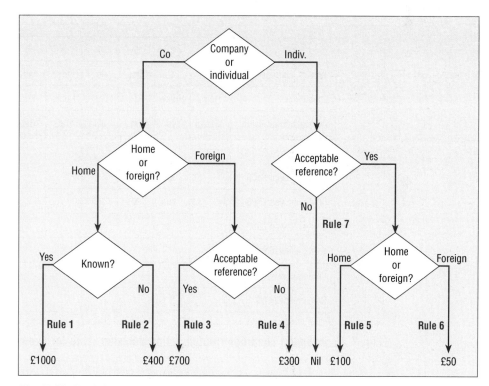

Fig. 7.23 Decision tree for Exercise 7.1

Rules	1	2	3	4	5	6	7
Company customer?	Y	Y	Y	Y	N	N	N
Home customer?	Y	Y	N	N	Y	N	–
Known?	Y	N	–	–	–	–	–
Acceptable reference?	–	–	Y	N	Y	Y	N
Credit allowed £	1000	400	700	300	100	50	Nil

Fig. 7.24 Limited condition/extended action decision table for Exercise 7.1

Rules	1	2	3	4	5	6	7
Type of customer	Co.	Co.	Co.	Co.	Indiv.	Indiv.	Indiv.
Location	Home	Home	For.	For.	Home	For.	–
Information available	Known	Nil	Ref.	Nil	Ref.	Ref.	Nil
Credit allowed £1000	X						
Credit allowed £400		X					
Credit allowed £700			X				
Credit allowed £300				X			
Credit allowed £100					X		
Credit allowed £50						X	
Credit allowed Nil							X

Fig. 7.25 Extended condition/limited action decision table for Exercise 7.1

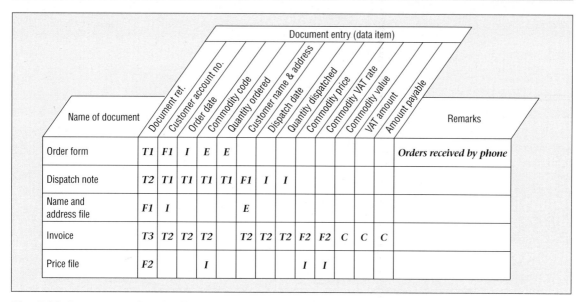

Name of document	Document ref.	Customer account no.	Order date	Commodity code	Quantity ordered	Customer name & address	Dispatch date	Quantity dispatched	Commodity price	Commodity VAT rate	Commodity value	VAT amount	Amount payable	Remarks
Order form	*T1*	*F1*	*I*	*E*	*E*									*Orders received by phone*
Dispatch note	*T2*	*T1*	*T1*	*T1*	*T1*	*F1*	*I*	*I*						
Name and address file	*F1*	*I*				*E*								
Invoice	*T3*	*T2*	*T2*	*T2*		*T2*	*T2*	*T2*	*F2*	*F2*	*C*	*C*	*C*	
Price file	*F2*			*I*					*I*	*I*				

Fig. 7.26 Data usage chart for Exercise 7.2 (meanings of codes as in text)

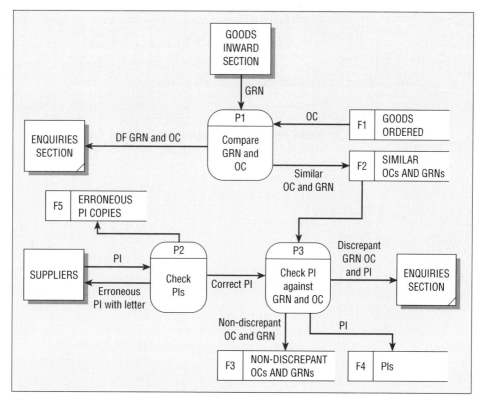

Fig. 7.27 Dataflow diagram for Exercise 7.3

281

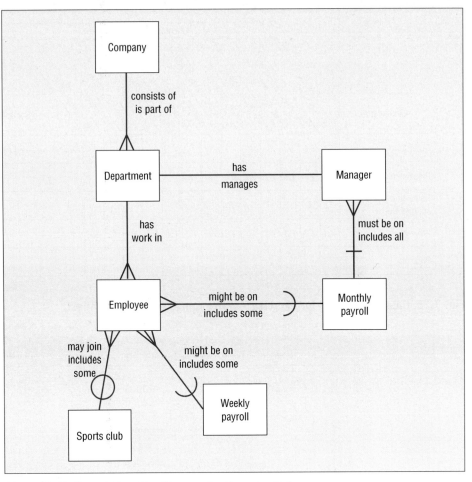

Fig. 7.28 Entity relationship diagram for Exercise 7.4

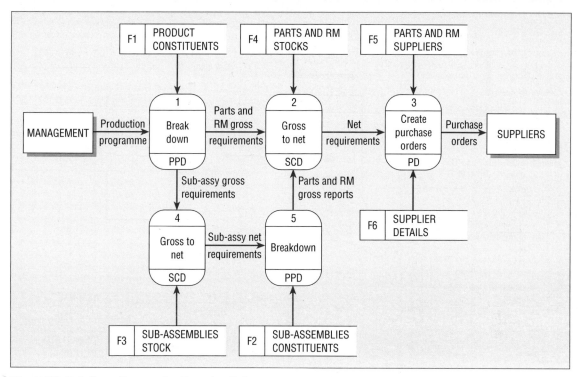

Fig. 7.29 Dataflow diagram for Exercise 7.5

Branch stock discrepancy report as at 8.10.99 (period 15)				
Branch No. Name	Branch stock value at cost	HO stock value at cost	Discrepancy Dr	Cr
1 Bedford	8050.60	8595.75	545.75	
2 Reading	7948.50	7593.00		355.50
3 Winchester	6853.00	6853.00		
12 Bristol	5262.00	5675.00	413.00	
Totals	95311.75	95586.95	2433.50	458.30
Total net discrepancy 1975.20				

Fig. 7.30 Branch stock discrepancy report for Exercise 7.6

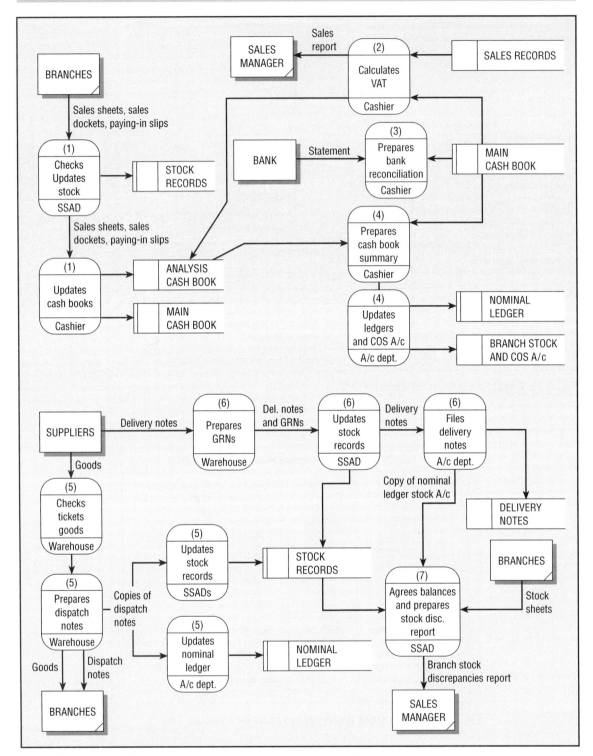

Fig. 7.31 Dataflow diagram of manual procedures in Exercise 7.6

Product details

Product no.	Description	Assy. time
325	Trolley	0.35

Product–component relationships

Product no.	Component no.	Quantity per product
325	B1375	6
325	B2284	4
325	M496	1

Components details

Component no.	Description	Supplier no.	BO/MI des.
B1375	Bracket	1759	1
B2284	6 inch wheel	1126	1
M496	Chassis	–	2

Suppliers details

Supplier no.	Supplier name
1126	White & Co.
1759	AGD Ltd.
2659	VG Steel Ltd.
4381	Premier Paint Co.

Component–raw material relationships

Component no.	Raw material no.	Quantity per component
M496	32809	1.2
M496	40605	1.5
M496	43750	0.6

Raw materials details

Raw materials no.	Description	Supplier no.	UOM code
22809	Black paint	2659	4
40605	Steel bar	4381	1
43750	Steel sheet	4381	2

Fig. 7.32 3NF records of Exercise 7.7

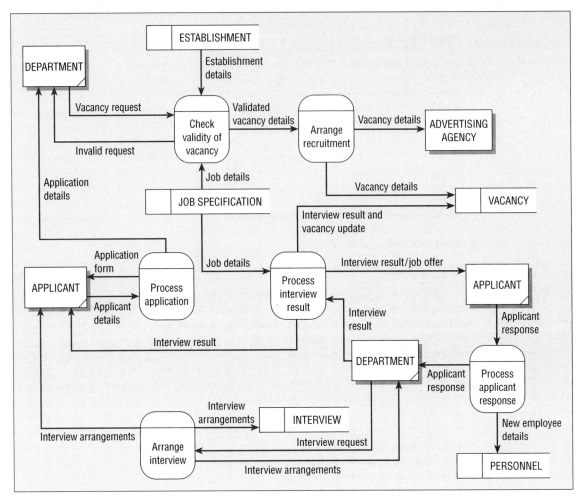

Fig. 7.33 DFD for Exercise 7.9

Rule #	1	2	3	4	5	6
2 A-levels?	Y	N	N	Y	N	N
1 A-level?	–	Y	N	–	Y	N
Part-professional?	N	N	Y	Y	Y	N
Degree	X			X		
HND	X	X		X	X	
Professional			X	X	X	
Rejection						X

Rule #	1	2	3	4	5	6
Number of A-levels	2	1	0	2	1	0
Part-professional?	N	N	Y	Y	Y	N
Course(s) offered	Degree and HND	HND	Professional	All	HND and Professional	None

Fig. 7.34 Decision tables for Exercise 7.10(a)

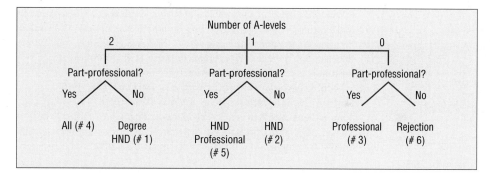

Fig. 7.35 Decision tree for Exercise 7.10(a)

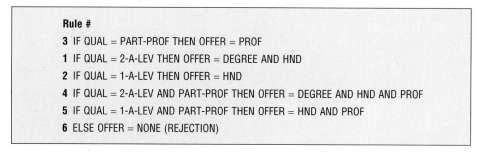

Fig. 7.36 Structured English for Exercise 7.10(a)

	Decision table	Decision tree	Structured English
Clarity	Good	Very good	Poor
Machine readability	Very good	Poor	Moderate
For complex procedures	Good	Poor	Poor
For simple procedures	Moderate	Good	Good
Ease of construction	Moderate	Good	Moderate
Ease of checking	Very good	Good	Poor

Fig. 7.37 Solution to Exercise 7.10(b)

References and further reading

7.1 Britton, C. and Doake, J. *Software System Development: a Gentle Introduction* (McGraw-Hill, 1996).

7.2 Currs, G. and Cutts, G. *Structured Systems Analysis and Design Methodology* (McGraw-Hill, 1997).

7.3 Flynn, D. *Information System Requirements* (McGraw-Hill, 1997).

7.4 Griffiths, G. *Essence of Structured Analysis Techniques* (Prentice Hall, 1998).

7.5 Hoffer, J.A., George, J.F. and Valacich, J.S. *Modern Systems Analysis and Design* (Addison Wesley, 1998).

7.6 Jackson, M. *Software Requirements and Specifications* (ACM Press, 1995).

7.7 Kaujalgi, V.B. *Structured Systems Analysis and Design* (Sangam Books, 1994).

7.8 Kovitz, B.L. *Practical Software Requirements: A Manual of Content and Style* (Manning, 1998).

7.9 McDermid, D. *Software Engineering for Information Systems* (McGraw-Hill, 1990).

7.10 Perrone, G. *Structured Analysis with CASE Tools* (Prentice Hall, 1994).

7.11 Pfleeger, S.L. *Software Engineering* (Prentice Hall, 1998).

7.12 Pressman, R.S. *Software Engineering* (McGraw-Hill, 1998).

7.13 Schach, L. *Software Engineering with Java* (McGraw-Hill, 1997).

7.14 Skidmore, S. *Introducing Systems Analysis* (MacMillan, 1993).

7.15 Somerville, I. and Sawyer, P. *Requirements Engineering* (John Wiley, 1997).

7.16 Thayer, R.H. and Dorfman, M. *Software Requirements Engineering* (IEEE Press, 1997).

7.17 Yeates, D., Shields, M. and Helmy, D. *Systems Analysis and Design* (Financial Times, 1994).

8 Systems design

AIMS

After reading this chapter you should:

- be able to enumerate the objectives of system design;
- be familiar with the processes that make up input design;
- be familiar with the processes that make up output design;
- understand the nature of human–computer interface design;
- be familiar with the main interfaces to computer programs;
- be familiar with the process of designing the processes that make up an information system;
- know the main notations used for documenting processes;
- be familiar with the main techniques used in securing a business information system;
- perceive the importance of system documentation;
- understand the nature of system development methodologies;
- be familiar with the main categories of CASE tools.

8.1 DESIGN PHILOSOPHY

In Chapter 1 the iterative nature of systems analysis and, in particular, problem definition are discussed against the background of business organization and the need for information. Chapter 7 describes the various aspects of systems investigation: the facts found in this stage form the basis for the design of a new system. As is shown in Figure 8.1, systems design and implementation is also an iterative procedure. Except for the most straightforward of systems, it is not feasible to design a system by working through a series of precise steps. Each decision taken during systems design tends to result in a rethink of the previous steps, and this continues until eventually the whole system takes shape.

This philosophy is epitomized by the procedures involved in prototyping. As explained in Section 7.9, prototyping inevitably causes rethinking and, to some

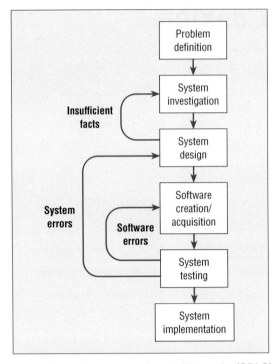

Fig. 8.1 Systems development life cycle (SDLC)

extent, redesigning of the system. As is seen from Figure 8.2, there is a variety of loopbacks connected with the different categories of prototyping. It is quite likely that several of these alternatives are employed in the one system's development life cycle (SDLC).

Figure 8.2 depicts in broad outline the relationships between the conventional SDLC and the five categories of prototyping. Following an examination of the results from a prototype, the SDLC either moves forward to the succeeding activity or returns to a previous activity. The latter is either to correct or complete work done previously or to start the next cycle of the SDLC.

Designing anything, be it physical, artistic or social, is a creative process absorbing something of the designer's personal stamp. With a data system, the design process is logical and yet calls for lateral thought. A logical approach implies systematic moves towards the end-product, each step being the result of decisions based upon the previous steps. At the same time consideration of the capabilities of people and equipment enters into each decision.

Lateral thought means the encompassing of ideas in the broadest sense, perhaps outside the usual IS functions and equipment. The systems designer should endeavour to free his or her mind from preconceived notions, and look for each problem on its own merits. A danger with systems design is that the systems analyst overplays his or her experience by immediately adopting previous solutions for new situations.

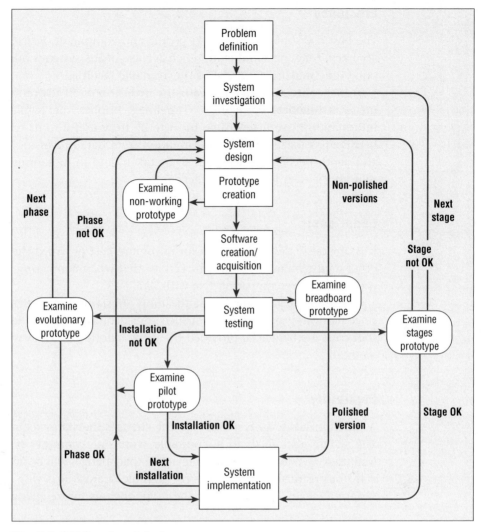

Fig. 8.2 SDLC and prototyping

Objectives of systems design

Practicality

The system must be capable of being used over a long period by competent but average-intelligence persons. It is no use designing a system that needs geniuses to operate and maintain it.

In this respect it has to be borne in mind that the eventual users of the system may have had no involvement in its design and prototyping. To them it is completely new and not a little daunting; thus practicality and, if possible, simplicity are of great importance.

Efficiency

The best use should be made of people and equipment. Efficiency involves the accuracy, timeliness and comprehensiveness of the system's output whilst at the same time making good use of the available facilities.

In this context, accuracy means the inclusion of all relevant data rather than mere arithmetical correctness. Timeliness implies the regular production of information in sufficient time for it to be truly useful to its recipient. Comprehensiveness means that the information from data processing is of the breadth and depth needed for decisions to be taken; and in more mundane procedures that all the data is taken into account.

Least cost

It is obviously desirable to aim for minimum cost provided the system fulfils its other requirements. We must be certain that when comparing different systems costwise, we are comparing like with like.

It is probable nowadays that the most substantial ingredient in cost is staff since hardware costs have diminished considerably. Thus any factors that reduce staff costs are beneficial provided the work is done at least as well by automated methods.

Flexibility

A system needs to be responsive to the changes inevitably requested by its users. The system also needs to be 'portable' from one computer system to another. No business is static and so it is inevitable that changes will be necessary. Although it is not practical to cater for every possible change, it is wise to build into the system a means of introducing the more obvious future changes.

Security

Several aspects are included here: hardware reliability and the associated fallback and standby procedures; physical security of data; and the detection and prevention of fraud and abuse (Section 8.7).

■ Planning and control of systems design

As with systems investigation, it is pragmatic to employ a project team approach to systems design. A project team may be responsible for the design of one or several applications. Within a team it is usual to find different skills, some members being business oriented, others technically minded. This mixture tends to induce a balanced approach and consequently a practical design evolves.

If the overall system is extensive, several project teams may be involved, in which case it is imperative that they maintain close contact. This eliminates misunderstandings, omissions from the designed system, and duplicated effort.

As seen from Figures 8.1 and 8.2, systems development as a whole follows through a life cycle. Associated with the SDLC are the categories of prototyping adopted according to the circumstances. A complex situation may well arise in which the activities in the SDLC and the various aspects of prototyping are closely interconnected.

An IS system can be regarded as consisting of three components: output, logical files and input, surrounded by its environment. The actual processing aspects of business IS, although important, are usually the least difficult to design. Provided the above three components are correctly designed, processing falls into place automatically.

A logical approach is to start with the output since the end requirement defines what is needed in the input and logical files. This is still true even when a database already exists since by working backwards from the output requirements it will soon become evident if data is missing from the database. It is sometimes argued that to start with the output is not a good idea because future output requirements cannot always be foreseen at the time a system is being designed. This is manifestly true, but the alternative approach of analysing all the data that might possibly be needed at some time in the future is far more problematical. In some cases there is a vast amount of irrelevant data which if incorporated into the database would impose a drag on its efficiency and thus on the system as a whole.

As with most things, it is a matter of pragmatism and compromise. Whereas a large organization has an army of systems analysts and programmers to build up a comprehensive database, a small firm has neither the time nor the resources to so do.

The methodologies described in Section 8.10 have been designed to make systems development more accurate, logical and complete. This is laudable but it remains to be seen to what extent these methodologies are adopted, especially by smaller organizations.

SAQ 8.1 Which of the following items do you think are tasks that a designer should carry out?

1 Specifying the individual software modules of a system.

2 Managing the project which develops the designed software.

3 Talking to customers about their needs.

Solution

Only the first item would really be within the scope of the designer. The only exception might be in *very* small projects where some other task such as 2 might be carried out by the designer.

8.2 CODE NUMBERS

In a computer-based business system it is inevitable that code numbers (which can include alphabetic characters) are needed; this is in order to identify uniquely every entity. The greater the number of entities within sets, the more significant are the code numbers, since there is an increased chance of misidentification. For practical purposes it is impossible to identify, uniquely and unerringly, even as few as a hundred different entities if only their descriptions are available for this purpose.

It is probable that sets of existing code numbers will have been found during the systems investigation, and that these are satisfactory for the present system. Unfortunately they are not always suitable when moving from a manual system. Manual codes do not need to be as precise in their layouts as those for computer use. A computer utilizes a code number as the sole means of both identifying and locating a data record and there are no half measures in this respect.

■ Required characteristics of code numbers

Uniqueness

Each entity type should have one unique code number so as to eliminate the possibility of misidentification. If an entity type needs to have two (or more) different sets of code numbers as alternative means of identification, e.g. a part number and drawing number, then they must all be unique and the two or more sets need to be differentiable, i.e. a part number should not be thought to be a drawing number.

Brevity

The layout of a set of code numbers should not be longer than is necessary for its purpose and structure. Since seven digits is the most that people can hold in their short-term memory, it is therefore advantageous if the code number is no longer than this.

Consistent pictures

Within a given set all the code numbers should have the same picture (Section 7.4). That is to say, the numeric digits and the alphabetic letters lie in the same relative positions in all the numbers. By having this consistency it is simpler to check a code number's accuracy and completeness. In this respect leading zeros are best avoided. And it is generally the custom to use only upper-case, i.e. capital, letters.

Distinct pictures

Where various sets of code numbers are used in the one system, it is preferable that each set has its own unique picture. This allows for clear recognition of the set and so prevents misidentification due to two identical numbers being in two different sets, e.g. a product number 2751 and a part number 2751.

Symbols and spaces

Code numbers of over six characters are easier to use manually if broken up into groups of characters by hyphens or obliques. Other symbols should not be used.

Visual and audio recognizability

If a set of code numbers is to be used extensively in manual procedures, it is worth considering choosing characters that are as distinct as possible from each other.

Avoidance of letters such as I, O, Q, S and G is helpful because these letters are similar in appearance to other letters and digits, e.g. 1, 0, 8, 6 and others. Similarly the letters rhyming with 'e' are best avoided if code numbers need to be spoken either by persons or by computers.

Expansibility

The structure of a set of code numbers must allow for additions if there is any chance of expansion in the number of entity types. If it is the intention to re-use the code numbers of obsolete items, great care has to be taken that the items are gone forever.

▆ Assigning code numbers

The designing of code number systems and the assignment of code numbers to entities must be done centrally. The systems department or organization and methods (O & M) department is responsible for these procedures so that there is no chance of confusion arising. If departments are allowed to create their own code number systems, duplicated codes and double usages soon appear. A central record should be kept of all code numbers assigned, these being clearly linked to the corresponding entities.

The means of linking codes to their entities may be through drawings, written specifications, suppliers' or manufacturers' descriptions, photographs or microfilms.

In many instances the biggest problem is to ensure that an ostensibly new entity has not been assigned a code number previously. This applies particularly in the engineering industry where many similar components and materials are used.

Before adopting any code number system for an entity set, it is advisable to examine the set for any inherent classifications. That is to say, do the entities have any characteristics that are usefully incorporated into their code numbers? Books are a good case in point; these are classified according to the well-known Universal Decimal (Dewey-decimal) Code (UDC) system. This is obviously advantageous in libraries for indexing and locating purposes.

Classifying entities and then assigning classification codes does not necessarily provide unique identification. This is apparent in a library, where many different books have the same UDC number. Thus, since unique identification is necessary with business entities, further coding has to be appended to the classification code.

Described below are the better-known classification and coding systems. These can generally be combined in order to obtain the most suitable system.

> **SAQ 8.2**   In the part of a banking system used for account processing what code number is used for the accounts entity?

Solution

The account number is used.

▪ Classification schemes

Hierarchical classification

The UDC system, mentioned above, is a good example of hierarchical classification. The concept behind this method is that every entity falls into a sub-group of a larger group, which in turn forms part of an even larger group.

In a properly designed hierarchical classification system (see Example 8.1 and Table 8.1), each entity has one place only where it fits into the system. It is convenient if each level of classification can be catered for by one digit position in the code, i.e. not more than ten groups in each level (coded 0–9).

The assignment of digits is entirely arbitrary and their meanings are discernible only in relation to the classification at higher level.

Example 8.1   ## Hierarchical classification

Suppose we wish to classify the various means of passenger transport.
Table 8.1 is an example of transport classified into three levels. Thus the
classification codes are, for instance, airliners 111, cars 322, sailing yachts 212.

Table 8.1

Higher level (first digit)
1 = air transport
2 = sea transport
3 = land transport

Middle level (second digit)
air   ⌈1 = winged aircraft
      ⌊2 = helicopters

sea   ⌈1 = displacement vessels
      ⌊2 = surface vessels

land ⌈1 = railway trains
     ⌊2 = road vehicles

Lowest level (third digit)
air, winged   ⌈1 = airliners
                ⌊2 = light aircraft

air, helicopters   ⌈1 = large
                     ⌊2 = small

sea, displacement ⌈1 = powered
                        ⌊2 = sailing

sea, surface   ⌈1 = hydrofoils
                ⌊2 = hovercraft

land, rail   ⌈1 = electric
                | 2 = diesel
                ⌊3 = steam

land, road   ⌈1 = coaches and buses
                | 2 = motor cars
                ⌊3 = motor cycles

Faceted classification

In this system each position in the classification code has its own independent
meaning. In contrast to hierarchical classification, a position does not have to
be associated with a higher level in order to ascertain its meaning. This enables
faceted classification to be more easily interpretable both by human beings and
by computers (see Example 8.2 and Table 8.2).

Example 8.2

Faceted classification

A range of machine screws is to be classified according to four characteristics: material, diameter, head shape and finish (Table 8.2). Thus 5-mm chromium-plated brass round heads are classified 2312.

Table 8.2

Material (first digit)
1 = stainless steel
2 = brass
3 = steel

Diameter (second digit)
1 = 3 mm
2 = 4 mm
3 = 5 mm

Head shape (third digit)
1 = round head
2 = countersunk
3 = pan head

Finish (fourth digit)
1 = no finish
2 = chromium plated
3 = zinc plated
4 = painted

▪ Code number schemes

Serial code numbers

Serial code numbers are assigned to entities in an entirely arbitrary way with no meaningful information conveyed by the code number itself. This is quite a usual method and serial code numbers are also appended to classification codes in order to provide full identification of every entity. Serial coding has the supreme advantage of simplicity and low redundancy, i.e. a high proportion of the available numbers can be utilized (see Example 8.3).

Example 8.3

Serial code numbers

All cars have code numbers starting with 322, followed by a serial code of three digits to give absolute identification of each particular model, e.g. a Ford Escort might be coded 322001.

Non-transposable code numbers

This is a variation of serial coding and is intended to minimize errors in copying and keying. The concept is that no pair of adjacent digit positions have values that are interchangeable with another code number. For example, if code 135 has been assigned, 153 will not be. This method considerably reduces the number of available code numbers.

Sequential code numbers

The code numbers are assigned to the entities incrementally after the latter have been arranged into some useful sequence. Typically, customer account numbers are assigned in alphabetical order of customers' surnames. This method is not viable if new entities are to be introduced later unless gaps are left between the code numbers in the set.

Block code numbers

A block coding system splits up a set of serial or sequential code numbers into blocks as determined by some general characteristic of the entities (see Example 8.4). The blocks of code numbers when assigned should allow for further additions to each block.

Example 8.4

Block code number

Parts used in a factory might, for instance, be block coded as follows:

General bought-out parts	1000 to 3999
Special bought-out parts	4000 to 4599
Made-in parts	4600 to 9999

Interpretative (significant digit) code numbers

In an interpretative code number, all or some of the digits are equal to one or more of the actual quantitative characteristics of the entity (see Example 8.5). This method of coding, although obviously useful, calls for careful thought in design. A clear understanding of the characteristics of the entities is needed – both present and future – otherwise new characteristics can appear that are not compatible with the system as devised.

Example 8.5

Interpretative code numbers

Interpretative coding applied to machine screws could indicate their diameter and length. Thus a 6-mm diameter screw 35 mm long would be coded 635. Additional digits could be used to indicate its colour or other characteristic(s) needed for unique identification.

Mnemonic codes

A mnemonic code, as the name suggests, is intended to act as an *aide-mémoire*. Generally all or part of the code number is derived from the entity's description or name, thereby bringing it to the user's mind. There are no particular rules for creating mnemonic codes except that the most obvious parts of the description are used in the code number (see Example 8.6).

Example 8.6

Mnemonic codes

Mnemonics for cities might include LDN for London, NYK for New York, TKO for Tokyo, etc. Mnemonic codes are suitable only for small sets of entities otherwise they defeat their own purpose.

SAQ 8.3

Airlines identify their flights by codes such as BA101 for flight 101 from British Airways and LH777 for flight 777 from Lufthansa. Of what sort of code is this an example?

Solution

It is an example of a mnemonic code. It gives airport staff a good idea of what is associated with a particular flight.

8.3 DESIGN OF OUTPUT

All output from an information system, be it a complex management report or a simple error message, is derived from one or more of three sources: input data, stored data and computation. We are therefore interested in the sources that lead to each item of output. We are also concerned with the time lapse for output from receipt of source data, the updating of logical files and the demands of the system's environment.

■ Output analysis chart

Figure 8.3 shows the paths followed in deriving the origins of each output data item. By working backwards from the output, the derivation of data items becomes apparent. This procedure is followed for each set of outputs and ensures that no items are overlooked. From this procedure it is possible to create an 'output analysis chart' as shown in Figure 8.5. This document is filled out for each group of associated inputs, stored data and outputs. The stored data is,

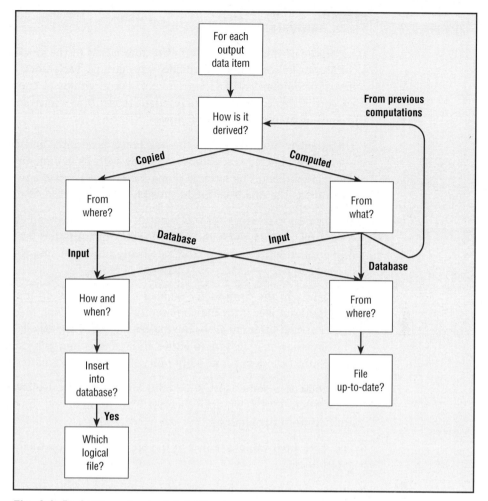

Fig. 8.3 Derivation of output data items

of course, also used for other outputs but needs to be entered on each output analysis chart.

These charts are a fundamental part of the data dictionary or encyclopaedia, either as they stand or after conversion to a standardized form.

Figure 8.5 relates to the stock analysis routine in Example 8.7. In Figure 8.5, a tick means that the data item is in the source at the head of the column. Letters and numbers indicate source references. C means computed or deduced from data item(s) with references beneath the C. Parentheses indicate that the data item may be erroneous or incomplete. The letters and digits are used again in Section 8.6 to illustrate the design of a computer-based stock analysis system.

Example 8.7 Stock analysis

The purpose of this routine is to create an analysis of the stock in hand according to its cost value sub-divided into groups. Each stock item has a commodity number and falls into one of several groups. A grand total of the stock cost is also computed. The resultant report is as shown in Figure 8.4. The steps in the routine are as follows:

1 The quantity of each item in the warehouse is counted and entered on a serially numbered stock form (S1) along with its commodity number and rack number. The latter is noted merely to facilitate any subsequent checking as the one item might possibly be held in several separate racks.

2 The data on each stock form is input to the computer and checked for correct pictures (layouts) and that the stock quantity has been entered. Any errors result in error message P1 being displayed, and the input data is rejected.

3 For each item the commodity number is used to look up the cost price and the group number from the commodity master file (D1), held on disk. If the master record is absent from the master file, error message P3 is printed. This omission may be due to either an invalid commodity number on the stock form or a new commodity not yet added to the master file.

4 The stock item value (cost price × stock quantity) is calculated; if this is greater than £9,999.99, error message P2 is printed, and the commodity is omitted from the analysis.

5 The stock item value is added to the appropriate group total and to the grand total.

6 At the end of the analysis, the group totals along with their group numbers and the grand totals are printed to form the stock analysis report (P4), as shown in Figure 8.4.

Phasing of output

Each output document or report needs to be scheduled in relation to the times of:

▪ the receipt of the associated inputs;

▪ the updating/amendment of the master files; and

▪ the completion of preceding computer runs.

The receipt of source data is usually connected with activities taking place in the system's environment, e.g. jobs completed in a factory, goods received into the stores.

STOCK ANALYSIS	
Group no.	**Stock value**
1	4963.82
2	5013.09
3	687.52
4	1739.76
Grand total	63580.64

Fig. 8.4 Stock analysis report

Data item ref.	Data item	Picture	S1	D1	P1	P2	P3	P4	Source refs
			Stock form	Commodity master	Error message	Error message	Error message	Stock analysis report	
1	Serial no.	9 (6)	✓		SI	SI	SI		
2	Commodity no.	A999 A	✓	✓	(SI)	SI	SI		
3	Rack no.	A99	✓		(SI)	SI	SI		
4	Stock qty.	999	✓		(SI)	SI	SI		
5	Cost price	99.99		✓		DI			
6	Group no.	9		✓		DI		DI	
7	Stock value	9999.99				$\frac{c}{4.5}$			
8	Group total	99999.99						$\frac{C}{7}$	
9	Grand total	999999.99						$\frac{C}{7}$	

Fig. 8.5 Output analysis chart

Computer output may rely on data derived from another system, and so cannot proceed until this has been completed. Job costing, for instance, often depends upon the apportionment of factory overheads derived from previous computations.

Sequence of output

The sequence (order) of the items in an output report or the documents in a set, is an important factor in designing the IS system. In general it is faster and more straightforward to hold file records in the same sequence as the predominant output. This, in turn, tends to imply that the input is also best if in the same sequence. These simplistic rules are less valid if the output is required in several different sequences.

The sequence(s) of an output report should be shown on the report itself as this may not be apparent from its contents.

■ Design of documents and displays

It is important that computer-printed documents and VDU displays (screens) are well designed because, in many cases, they are the only contact between end-users and the IS. The layout and contents must be absolutely clear, and the presentation of a good standard. Several guidelines can be followed when deciding what information to place in a report and how it should be presented.

The first step is to decide on the report or display contents. These should be related to the user need. This will have been captured during requirements analysis and may be information necessary for decision-making, data to help accomplish a task, or information that has a less focused requirement, e.g. monitoring data and exception reports. The report should be linked to some purpose, otherwise its need is questionable. The grouping of data into reports can therefore follow the principles of cohesion. All data belonging to one user goal is placed together. Once the information requirements have been decided, the next step is to group and order data. This will subset the information to make it easier to use. The following two basic principles should guide design of reports and displays:

■ *Structure*: data should be sub-divided, placed in blocks, sorted, and formatted to show the structure. For instance, sales figures should be grouped to reflect the organization of sales teams into areas, districts and sales patches.

■ *Compatibility*: data should conform to the user's expectations. If users expect to have data in a certain order then it should be presented in a compatible way.

To achieve good layouts which conform to these principles the following guidelines can be used:

- Group data according to entity sets, with all the attributes belonging to one entity.

- Place data in order of importance or priority as set by the user.

- Place data in order of frequency of usage.

- Sort data to make the structure clear.

- Summarize data where possible by adding totals and sub-totals.

- Summarize using graphics to show trends, associations and differences in numeric data.

- Use the above layout guidelines according to compatibility with the user's needs.

Design features of printed documents

- All information must be unambiguous, thus necessitating clear headings and annotations. These are either preprinted, as with payslips and invoices; or computer-printed, as with most reports and analyses.

- Each page of output is numbered so as to detect the loss of a page; the final page is annotated 'last page'.

- The document's width is dictated by the printer's print bank width and by external requirements, e.g. enveloping. The width of the printing cannot be greater than that of the print bank and so the information must be laid out accordingly.

- The number of copies required and any differences between them need to be considered. Most printers can manage up to four or five copies but more than this demands additional copying facilities. The methods of making printed copies are interleaved carbon sheets, carbon-backed stationery and chemically coated stationery.

  Differences between the contents of copies can be catered for by either blacked-out areas on the copy stationery or blank areas in the carbon backings.

- The quality of the printing relates to the usage of the documents. Those going to external recipients, such as customers, need high-quality printing, whereas internal documents of a lower standard are usually acceptable.

- The vertical spacing of the lines of print. Single spacing (six lines per inch) enables about sixty lines to be printed on a sheet of standard continuous stationery. Double spacing (three lines per inch) reduces the lines to thirty per sheet but enhances readability.

Design features of visual displays

The display of what is presented to the user calls for careful thought since it is more transient than printing. The design guidelines for this are the same as those for printed reports. There are a number of principles that should be borne in mind when designing windows-based data entry and output; some of these are described below:

- Make sure that there are variable speeds of scrolling on windows. The user should be able to scroll slowly through a window when entering data sequentially or quickly when browsing through data. Slow scrolling can be achieved by scrolling bars while fast scrolling can be achieved via fast-forward buttons.

- When developing a system having multiple windows do not have too many windows cluttering up the screen and ensure that, ideally, the windows are separated by space rather than overlapping.

- For large quantities of data such as text from a book ensure that the user can scroll through this data by using scrollable windows.

- Use different colours and styles of presentation to highlight important data that is to be entered. For example, if a window requires text to be entered into a field before other fields are filled the field can be highlighted while the other fields can be greyed out or disabled.

- Use different icons for indicating different functions. For example, if the user is prevented from dragging data from one part of a window to another an icon which has something akin to a no-entry sign can be displayed where the mouse pointer would normally be.

- Use menus in preference to other entities such as buttons when exercising new functions such as opening a new window. This guideline is a specific example of a more general one which is that since the vast majority of users will be used to using modern operating systems such as Windows 98, the systems that they employ for tasks such as data entry should not be widely different in interface from them.

- When an untoward event occurs make sure that this is displayed prominently. For example, the fact that incorrect data has been typed in should be indicated by the appearance of a warning window and the emission of a sound (most data entry staff tend to look down most of the time until the record they are entering has been completed).

- Use restful colours such as light blue for a background.

- Avoid using red for text or any detail; red is poorly processed by human operators and does not stand out if the background lighting is high.

Handling printed stationery

The output sub-system also includes the handling and distribution of the continuous stationery coming from the printer. Various processes are involved such as those listed below. Machines are used if justified by the amount of output.

Decollating: This is the removal of the interleaved one-time carbons from multi-part sets, and the separation of the continuous sets.

Bursting: The pulling apart and stacking of the perforated sheets.

Recollating: Recombining the copies after removal of the one-time carbons.

Guillotining: Cutting instead of bursting the continuous stationery gives a better finish as the rough edges left by the perforations are removed.

Trimming: The stationery is cut vertically to remove the sprocket holes; it may also be cut into several widths or reduced in width.

Folding: The separate forms are folded, usually to fit a standard envelope.

Enveloping: Insertion of forms into envelopes and sealing the envelopes.

Mailing: The automatic franking of envelopes, recording the cost, and handover to the delivery firm.

Copying: The reproduction of multiple copies of output documents off-line, usually by photocopying.

Distribution: Arrangements are made for expeditious delivery of the output documents to the recipients. This is facilitated by a distribution list printed by the computer at the head of each document or batch of documents.

8.4 DESIGN OF INPUT

The design of the input sub-system starts with the origination of the source data. This may be internally such as in offices and factories, or possibly from management. Alternatively, source data may originate externally from customers, suppliers, the general public or government departments.

■ Data acquisition

Acquisition of source data may be via the mail, telephone, telex, orally, automatically (source data automation) or from terminals. Whichever is the case, the systems designer must ensure that the source document or on-line input is well structured and fully understood by its users. This is particularly important when the source document is filled out by hand for subsequent keying and, in some cases, this necessitates redesigning the document.

The problem of illegible entries on a document is alleviated by better form design, typing instead of handwriting, scanning or OCR. The entries on externally originated documents are more difficult to control but may be improved by originating these documents internally before dispatching to the external agency, i.e. turnaround documents. This allows the documents to be preprinted with most of the pertinent data. A customer order form, for instance, contains a complete list of items alongside which the customer merely enters the order quantities. Also preprinted are the customer's account number, name, address, etc.

Batching

Source documents are made up into batches of a suitable size by the originating department. For each batch control counts and totals are created. Care should be taken that this does not become too onerous.

Typical controls are:

■ the number of batches;

and for each batch:

■ the numbers of documents;
■ the batch serial number and date of origination;
■ the total number of items on the documents;
■ totals of certain data item values; and
■ totals of suitable code numbers (hash totals).

Batches sometimes occur naturally, such as the orders arriving in one mail delivery. In other cases a batch is created either at a certain point of time or whenever sufficient documents have accumulated.

Code number entering

Wherever possible the originators of source data should be spared the onus of entering code numbers. This aim is achieved by having the code numbers pre-entered, as with turnaround documents. In many cases this is impracticable, and so the originator must be assisted in getting the code number correct. The points to remember are:

■ Keep the code numbers as simple as possible (Section 8.2).
■ Provide boxes on the document to encourage neatness.
■ Ensure that the user has a clear and up-to-date list from which to transcribe the code numbers.
■ Consider the use of check digits (later this section).
■ Build feasibility checks into the data acceptance run (see below).
■ Incorporate a proper procedure for dealing with erroneous or missing code numbers.
■ With on-line data, provide a displayed list from which to select the relevant code number or, alternatively, select the item from its description without the code number being involved, e.g. using a touchscreen (Section 4.7).

■ Data validation

Ideally all data presented to a system is fully correct and complete. Unfortunately, in practical situations, this aim cannot always be achieved, so we must do what we can to detect errors at an early stage. Data received from external sources, such as customers' orders, is generally less controllable than internally originated data, and so this demands even greater checking. Computer-based systems lend themselves to extensive checking procedures, and this is particularly so with interactive systems, e.g. dialogues (Section 8.5).

The main types of error that might occur are source recording, data preparation (although data verification eliminates most of these), incorrect batches of input data, missing data, duplicated data and incorrect file records (e.g. out-of-date records). The three main types of data checks are input validation, feasibility checking and check digits.

Input validation

This is more of an absolute proof than is feasibility checking because the computer-based system checks the input data against known values. Generally this entails the computer looking up a record from its files in order to ensure that the input relates to an existent entity. Before processing a customer's order, for instance, the computer makes an initial check to ensure that the customer number and the catalogue numbers thereon do actually exist. By discarding orders containing non-existent customer numbers and catalogue numbers at an early stage, difficulties are minimized later. Similarly there may be only one date that is acceptable to a particular processing run and any others are rejected as out of date.

Validity checks are well worthwhile since they greatly reduce confusion and backtracking, especially in complex and sophisticated systems (see Example 8.8).

Example 8.8 | **Input validation**

Certain on-line input systems, such as those employed by mail-order companies, impose immediate checks for accuracy and validity on the customers' orders.

The actual procedure is that orders received from sales agents are keyed into a PC and vetted by the computer for acceptable agent number, correct item (by comparison of keyed and stored price of the item) and valid unit of quantity. Where applicable, a list of options is then displayed from which the order clerk selects the option, e.g. the colour of an item of clothing. This method reduces errors by imposing a limit on option selection and allowing the order clerk to use common sense when the order is unclear.

Flashback (echo) checks

With this method of input validation the computer responds to an input by returning data that in some way confirms or at least strongly implies the accuracy of the input. For instance, the input of an account number could result in the account holder's name being flashed back. In most cases this would provide confirmation of the correctness of the input as the operator already knows the account name.

SAQ 8.4 A clerk in a bank processes cheques which have been received by the bank and which should be credited to their customers' accounts. What sort of data validation would you carry out on such cheques?

Solution

You would check that the bank issuing the cheque was correct, and the account number correct.

Feasibility checking

Feasibility checks look for the likelihood of error as well as for definite errors. From the explanations below of the various checks, the student will understand their purposes and practicality.

Picture checks (layout checks)

The data item's picture (Section 7.4) is checked against the acceptable picture(s) and any difference results in rejection of the data item. Ideally there would be only one picture for the data item but in some cases several are applicable, e.g. UK vehicle registration marks.

Picture checking is mostly applied to code numbers as these often have an intricate layout. It is also possible that a particular position (character(s) or digit(s)) within a data item can be checked as having one or a limited number of values.

Limit checks (range checks)

Every data item has a minimum and a maximum value whether it be input, output or at an intermediate stage of processing. These limits may be as wide as the data item field allows, e.g. 0000–9999 in a four-digit field, but often this is not the case and so a check against the acceptable limits is advantageous (see Examples 8.9 and 8.10).

Example 8.9   **Input limits**

The heights of army recruits are measured and subsequently this data item checked as being within the limits 1.5–2.0 metres. Although it is possible that heights outside this range are encountered, this is unlikely and, in any case, they can be validated before being accepted into the system.

Example 8.10   **Output limits**

The amounts of domestic electricity bills are checked as being within the range £1–£300 before being dispatched to consumers. Those of less than £1 are then annotated with a message telling the consumer to withhold payment until the next bill. Those of over £300 are scrutinized for errors and possibly withheld pending rereading of the meter.

SAQ 8.5   Can you think of a limit check which the clerk described in the previous SAQ would encounter?

Solution

The only limit check would be on the amount to be debited. Rarely are cheques issued over some large amount, say £50,000, and so some check would be made on this to enable the clerk to examine whether the amount to be debited was, in fact, large.

Fragmented limit checks (subrange checks)

These are basically similar to limit checks except that there are several smaller ranges within which the data item should lie. A block coding system lends itself to this type of check if there are definite gaps between the blocks of code numbers (Section 8.2).

One part of a code number may indicate the range within which another part must lie, e.g. if the code number starts with any of the letters D, M or W, the four digits that follow must lie within the range 2,500–6,999.

Restricted values checks

In this case, the data item can legitimately have only certain predetermined values, any others being invalid (see Example 8.11).

Example 8.11   **A restricted value check**

A commodity is sold in round dozens only, an order quantity is therefore restricted to either 12, 24, 36 or multiples of 12 up to, say, 1200.

Combination checks

In this context, 'combination' means the joining together of two or more data items, such as by adding or multiplying their values. After combination, the result is checked by one of the above methods, usually a limit check. The point of combination check is that the data items may pass their own limit checks but become infeasible when combined together (see Example 8.12).

Example 8.12	**A combination check**

In costing stores issues, the quantity is checked with limits 1–50, the cost price with limits £1–£80. The cost value (quantity × cost price) has a combination check imposed with limits £1–£100. Thus, an issue of twenty units whose cost price is £10 each, although passing both limit checks, would fail the combination check.

Compatibility checks

The concept here is that two, or possibly more, data items are checked for mutual consistency. The parameters of the check imposed on one data item are determined by the value of the other data item (see Example 8.13).

Example 8.13	**A compatibility check**

A customer's order for goods worth £50, although normally acceptable, is checked for compatibility with the customer's existing debt. This is found to be £300 and of long standing. Since these amounts are incompatible within the company's trading terms, the sale of the goods is stopped.

Compatibility checking often necessitates the setting up of a table to be stored within the computer. This contains sets of compatible figures and is referred to when carrying out a compatibility check (see Example 8.14).

Example 8.14	**A compatibility check against tabled data**

Data resulting from medical inspections of children includes their ages, heights and weights. By compatibility checking this data against a stored table, ridiculous errors are detected, e.g. a four-year-old weighing 150 lb.

Probability checks (reasonableness checks)

These can be any of the above checks, the outcomes of which are judged against a table of probability of the data being erroneous. The purpose of probability checking is to minimize the investigatory and corrective work following the detection of a possible error. Data only just outside its limits would merely be reported and allowed to proceed. A greater variance would result in the data being held in a suspense file pending investigation, and a large variance in rejection of the data.

Dependency checks

This check means that if one data item is present then so must be another. It could be extended to cover groups of data items dependent on other groups (see Example 8.15).

Example 8.15 **A dependency check**

A series of discount prices each of which must have the appropriate quantity accompanying it.

SAQ 8.6 What sort of checks do the following represent?

1 A check that a customer has not ordered a very large amount of an item which is normally ordered in small amounts.

2 A check that the number of digits in a phone number is correct.

3 A check that the age of an employee in a personnel system is between the statutory limits for hiring.

4 A check that the order number of a delivery note matches that in an order.

5 A check that the item number in a purchasing system represents an item stocked by the company.

Solution

1 is a limit check, 2 is a picture check, 3 is a limit check, 4 is a compatibility check and 5 is another compatibility check.

SAQ 8.7 In a banking system would you use check digits for account numbers? Typically such a number will be at least six digits long.

Solution

Yes, this is a good use of such digits: one of the easiest things that you can do is to mistype a digit or transpose a digit when typing in a number more than three digits long.

8.5 HUMAN–COMPUTER INTERACTION (HCI)

Human–computer interaction (HCI) concerns the design of the parts of a computer system that people use. HCI adds design principles for computer screens and the human–computer dialogue, i.e. the way the communication between users and the computer is structured. As well as design guidelines, HCI also provides advice about wider issues of system design such as how the task should be partitioned between people and computers, and organization of the system environment in terms of work activity, communication and procedures. These issues have already been dealt with in Section 8.1. First we need to examine the relative merits of computers and people before partitioning activities in a system.

Compared with computers, humans excel at problem-solving, using heuristics (rules of thumb) and associative reasoning, but are poor when processing high volumes of data and repetitive tasks. People deal with complexity in the environment by imposing order on it. Classification, structuring of information, and skills are consequences of this propensity to organize and automate. The great advantage people have over machines is a vastly more complex knowledge even for things which we consider to be simple common sense and everyday knowledge.

The human system is an associative reasoning machine. It deals with vast quantities of data from the environment by filtering it and abstracting interesting qualities from basic data. HCI aims to create human tasks that are neither too demanding, nor too simple, which may lead to the operator becoming bored. Variety is desirable in any task. Task complexity also has to be matched to personnel ability, hence the capabilities of the user have to be considered when designing tasks. It is no use giving someone a stimulating yet over-demanding task in relation to their abilities. A compromise has to be reached which ideally should give people tasks that stretch their abilities, thereby encouraging them to develop new skills and widen their experience, while not going beyond their abilities.

Within each task, actions are allocated to either the computer, or the users, or to both. Generally users should receive tasks which require initiative, judgement and heuristic reasoning. On the other hand computers should get repetitive checking, calculations and data handling tasks. Data entry, data retrieval, and decision support are examples of mixed tasks in which human and computer interact to achieve an objective. Mixed tasks require further refinement to the design.

■ Basic principles

From knowledge of human psychology it is possible to draw up basic principles to guide design of human–computer interfaces. Unfortunately psychology does not lend itself to such a venture as many explanations of human behaviour are still models and hypotheses, and in some areas little definite proof exists. However, some principles can be derived in spite of this limitation, as follows.

Consistency

This is similarity of patterns of operating a computer system, in presentation of information and other facets of an interface design. Consistency reduces the human learning load and increases recognition by presenting a familiar pattern. The more consistent patterns are, the less we have to learn, and the easier an interface will be to use.

Compatibility

Compatibility in interface design attempts to match the interface to the user's expectation of operation and appearance. New designs should be compatible with, and therefore based upon, the user's previous experience or mental model of the task. Once again recognition is enhanced, learning is reduced and the interface should be easier to use.

Adaptability

Interfaces should adapt to the user in several ways. The computer should adapt to the user's speed of work, to individual user characteristics, skill levels, etc., and even to individual ways of working. Adaptability, however, must not be overdone otherwise the consistency of the interface is reduced.

Economy

Computer dialogues should be economic in the sense that they achieve an operation in the minimum number of steps necessary to support the user and save users work whenever possible.

Guidance not control

The computer should guide users through a task with prompts and feedback information. The system should function at the user's pace according to the user's command and should not attempt to control the user. This principle has two sub-components: predictability, users should be able to forecast what to do next; and reversibility, users should be able to backtrack at will when mistakes are made.

Structure

Interface designs should be structured to reduce complexity, because people process information by classifying and structuring it within a framework of understanding. Structuring also implies simplicity and relevance; information should be organized so that only relevant information is presented to the user in a simple manner.

315

Principles are intended for overall guidance during design and as a set of criteria against which interfaces may be evaluated. To apply principles in the design process, they have to be translated into guidelines which are applied in a particular context. Unfortunately systems and people are complex and to issue a simple set of guidelines for all situations may be appealing but in reality would only be misleading.

SAQ 8.8 Often the interface to a computer system is designed as a series of hierarchical menus where each menu refers to another menu which again refers to others. For example, in a banking system a menu might contain the entries Current account processing, Overdraft processing and Standing order processing. What basic principle of HCI design is addressed here?

Solution

The principle is that of structure where the interface is partitioned into a framework which matches the functions of the system.

SAQ 8.9 In a banking system each screen presented to the user contains the same bank logo in the top right-hand corner, the text describing the various fields which need data entered have the same font and font size and navigation between each screen is carried out in the same way. What HCI basic principle is addressed here?

Solution

The principle is that of consistency.

■ Interface design

Design of human–computer interfaces raises the question of assessment. The quality of interface designs is frequently measured with terms such as usability, utility and efficiency. There are three basic concerns about the quality of an interface design:

1 How well does it fulfil the users' objectives?

2 How easy is it to learn and use?

3 How acceptable is it for users?

A design should aim to provide users with what they require in order to fulfil their objectives. This concept is common to systems analysis and interface design, i.e. the matching of user requirements to the facilities provided in the system.

This is called 'task fit', providing the appropriate tool to carry out a required task. A system may be easy to use and learn but if it does not do what the user wants it will be useless. Utility is difficult to measure. Attitude data from questionnaires can give some feel for task fit, but more comprehensive analysis necessitates the observation of persons using the interface and then questioning them about where they encountered problems and why the system did not do what they wanted. Utility is often hard to separate from usability which is more concerned with ease of use rather than what the system does.

There is no ideal measure of a good interface but some important qualities have been described under the heading of 'usability' as follows:

Effectiveness

This is a measure of how well the interface, and hence system, performs in achieving what the user wants to do. This can be measured in terms of:

■ error rates lower than a target level;

■ task completion time within a set target time; and

■ usage of system facilities above a minimum target frequency.

Learnability

This measures how easy to learn a system is and how well it is remembered after a period of disuse. Learnability can be quantified with measures of:

■ decreased error rates over time from the start of system usage;

■ decrease in task completion time from the start of system usage;

■ correct recall of system facilities, operational procedures and comman names; and

■ increase in user knowledge about system facilities over time.

Flexibility

This is a measure of how well the system accommodates the user's way of working. This is hard to measure as it can encompass how flexible the system is when used by either novices or experts, and how flexible the system is when being used for different tasks, as well as how flexible it is in style of interaction. Flexibility may be related to the overall use of the system. If some facilities are never used by any users there may be design problems. Coverage is measured as facility usage by x per cent users within a set period.

Attitude

Attitude is the subjective part of usability which quantifies user satisfaction with the system. This is usually measured from a questionnaire or during an interview as:

317

■ user satisfaction exceeds a target rating;

■ user-perceived problems are kept below a set level; and

■ user motivation to use the system exceeds a set baseline level.

Generally it may be thought that there is a trade off between ease of use and ease of learning, but evidence points the other way; interfaces should be easy to learn and easy to use. Efficiency is a consequence of the economy, consistency and compatibility principles.

■ Data entry dialogues

The data to be entered into computer systems may come directly from the source, which may be a person, or a measuring device, or another computer system. Alternatively, data may have already been captured on a paper document, i.e. a form. As result form design tends to be an integral part of data entry design for many computer systems. Data entry screens should be designed to model the input forms as closely as possible. If no input form exists or the old input form is poorly designed and difficult to use, a new screen layout has to be designed. Data entry can be an error-prone process because people may mistake instructions, skip fields, give information in the wrong format, make transposition errors, or write illegibly. Good design of data entry dialogues can reduce these problems.

Data entry, whether onto a form or into a computer, is preceded by data capture. When designing data capture procedures the following guidelines should be considered:

■ Data should be collected at source as far as possible.

■ Data should be entered onto the data capture document (a form) by the originator of the data.

■ Avoid transcription of data from one form to another. Transcription is an error-prone process which should be avoided if possible.

Form design

Forms should be designed for ease of data collection rather than extraneous factors such as fitting into envelopes, and saving printing costs. Data collection is an expensive running cost, so if savings can be achieved by quicker, and more accurate data collection, these will far outweigh capital costs incurred by good forms and data entry design. Forms should have a consistent design as far as possible within a system and organization. The more consistent designs are, the more uniform users' expectations become, and consequently their learning burden is reduced.

Forms have to be designed to capture optional as well as essential information. The all-purpose form suffers from errors due to people filling in irrelevant information and completing the wrong sections. Tailored forms, on the other hand, suffer from people having difficulty getting the right form for their needs and accidentally filling in the wrong one. How many individual data sources to target on one form is a trade-off decision. Generally one form should have one purpose and the number of alternative form types should be kept to a minimum. User analysis should be carried out for form designs as with other interfaces.

Forms consist of three main components:

1 Data entry areas.

2 Supporting information, and instructions.

3 Titles and headings.

Form layout

Layout of data entry forms and entry screens follow the same guidelines although more data can be put on a printed form than on a VDU screen. Data should be grouped together either according to their frequency of use, or their importance, or sequence of entry. The choice of which grouping criterion to use should be made in consultation with the user. Most data entry tasks involve a transaction. Transactions in most information systems are described by a paper document; this is a distinct document which is created and then processed by the system (e.g. customer order, hospital admittance record, export shipping document). When entering transactions, data grouping is usually by sequence of entry. Data entry is a tedious task hence every effort should be made to save the user unnecessary work.

Placing groups of data into boxes is an effective technique for structuring layouts, and background colour can also be used to differentiate data. Fill-in areas should be lightly coloured, while more stimulating colours should be reserved for titles and instructions.

Form-filling (entry forms)

In this dialogue data item names are displayed, each having an adjacent blank box. The values of these are entered from the keyboard and displayed immediately, thus allowing the user to check for correctness and completeness. Alternatively only one data item is displayed at a time. Form-filling is a suitable method for repeatedly entering a large amount of data such as customers' remittances.

An example of a form for such a procedure is shown in Figure 8.6. As each data item is keyed, the form is filled in and, at the same time, feasibility checks are applied. By having boxes, as shown, each data item is restricted in length to the size of its box. Other checks are as described earlier in this section. On detecting an error, the screen is overlaid by a window giving details, thus enabling the

319

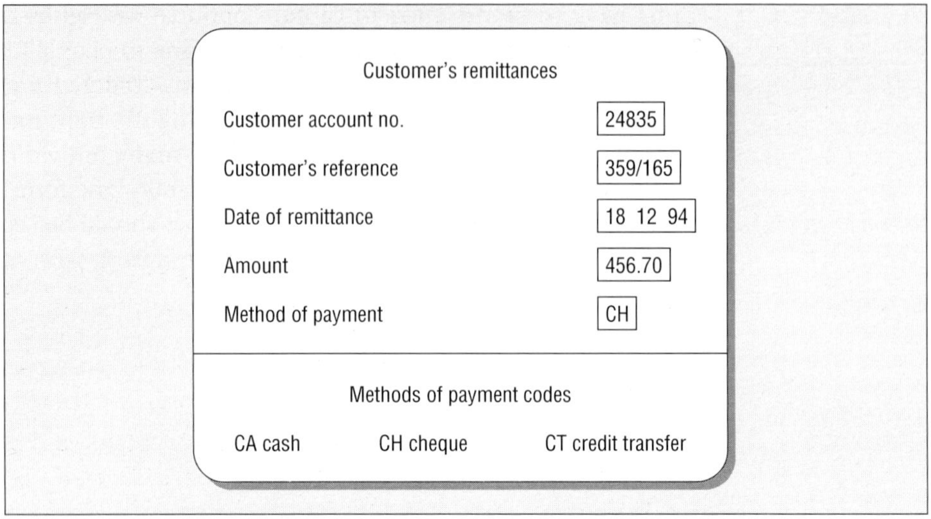

Fig. 8.6 Form filling

user to make corrections. Windows can also be used to obtain help or further instructions in case of difficulty. The lower portion of a form-filling screen often displays a list of possible entries, such as the methods of payment in this example. When designing a more detailed layout it is important to bear in mind the following points:

- Captions and prompts should either precede the fill-in area or be left justified above the entry area.
- Data entry fields should be aligned left justified, and if possible with a right justified margin.
- The filling-in area depends on whether handwriting or typewriter completion is anticipated. For handwriting allow $\frac{1}{4}$" width per character with extra space for separation and a height of $\frac{1}{3}$". For typewriters this can be reduced to $\frac{1}{10}$" by $\frac{1}{6}$". Separating the filling-in area into character boxes can help legibility.
- When type of reply is known, the data entry area should be sub-divided to format the reply field, e.g. enter date in DD/MM/YY format.
- If units of measure are being requested the unit should be specified, e.g. enter product weight in kg.
- Highlighting should be used for titles, mandatory fields, important prompts and instructions for filling in.
- Three rules apply to prompts, titles and instructions: keep it simple; be explicit; and exclude anything not directly relevant.
- Titles must describe concisely the purpose of the form or screen and should be centred at the top.

■ Completion instructions must be clear, brief and use easily understood words, especially in forms for public use. Brief instructions may be located before the entry field to which they pertain, while more complex instructions should be placed at the top or bottom of the form.

Guidelines for data entry dialogues

General guidelines for data entry dialogues are as follows, but the influence of the design context should be considered when putting these into practice:

■ *Explicit enter*: validation and entry occur only when the user presses the enter key; this allows checking within the entry for errors.

■ *Explicit movement*: autoskip/autotab between fields is not usually advisable, as unskilled users find the unexpected movement distracting. Use TAB or CR to move between fields.

■ *Explicit cancel*: if the user interrupts an entry sequence, the data already entered, even in the current field, should not be deleted. This allows reconsideration of a cancel action which may have been a mistake.

■ *Explicit delete*: make deletion an obvious action which is not easy to take without an extra confirming step – Delete order. Are you sure? (Y/N)

■ *Provide feedback*: users should be able to see what they have entered. If several entries can be placed on one screen, the previous transactions should still be displayed. Feedback messages should be given to users to inform them of the next action which is expected.

■ *Allow editing*: editing, ideally, should be allowed both within a transaction and after it has been completed; hence users should be able to edit the current field and to go back and change fields entered previously. A consistent method of editing should be adopted.

■ *Provide undo*: allow users to backtrack to the previous 'before' state. This is often useful in edit and command sequences to correct mistaken courses of action.

■ *Auto format*: users should not have to enter redundant digits and characters such as leading zeros, e.g. 79 not 0079 to fit a picture 9(4). Similarly the entry should not be space sensitive.

■ *Show valid entry responses or values in prompt*: either the range of set or valid replies if a limited data set, e.g. enter discount value in the range of 1 to 10.

■ *Entry at user's pace*: users should be able to control the speed of data entry because forced work schedules will be resented.

■ *Setting defaults* for commonly entered items.

■ *Using codes and abbreviations*.

- *Automatically filling in previously entered items*, e.g. customer name and address from file.
- *Using pointing responses and selection from a list* if entry is from a limited set of choices.

These general guidelines are applied in specific data entry dialogues. The most common type is form-filling in which data is initially captured on a paper document. Systems analysts and interface designers often have to design paper-based interaction for data entry as well as computer dialogues, as elaborated in the next section.

■ Command and control dialogues

The end-user converses directly with the computer – either a PC or a mainframe and intelligent terminal. This is termed an interactive or human–computer dialogue. Ideally the dialogue would be in natural English or another natural language. There are, however, insurmountable problems with this, mainly because natural language relies heavily on colloquialisms and prior understanding of the subject matter of the dialogue. Even though there have been many attempts to utilize natural language, grammatical and syntactic variations and, of course, metaphors completely confuse a computer.

It is therefore necessary to adopt and adapt a dialogue that is the most suitable for the application and situation. The choice in this respect depends on two main criteria: the nature of the work, and the attributes of the dialogue user. The former covers a wide range from occasional interrogations of file records to continuous stream of multivarious transactions demanding rapid service. Users vary from casual users who need complete guidance in achieving their needs, to highly trained dedicated users who are capable of utilizing a computer with great expertise. It is obviously imperative that the dialogue provides the user with his or her precise requirements but also that it does so quickly, efficiently and with minimal effort on his or her part.

When designing a dialogue the following points must be borne in mind.

Feedback: Always provide users with messages to inform them what is going on, especially if there is going to be any significant delay in response. Failure to do so leaves users wondering if they or the machine is at fault and often causes them to press keys to find out what has happened.

Active guidance: The user must never be left high and dry, i.e. they should always be aware of what to do next and when to do it. In this respect it should be made abundantly clear as to what responses are acceptable to the computer so that the user is not tempted to make up his or her own syntax.

Natural conversation: The dialogue between human and computer should be continuous, like human communication, so a message from one party is followed by a reply from the other. Gaps in conversation lead to uncertainty and attention being diverted from the task.

Clear identity: Commands should always be linked to a single function. Multiple commands for one function only serve to confuse the user and are redundant anyway. As far as possible commands should be unique and have a good direct link to the function they evoke.

Status: Provide a message informing users which part of the system they are in. In large systems users may forget which facility they are using resulting in them issuing the right command in the wrong context. This can have unfortunate consequences.

Clear messages: All computer messages should be completely clear, i.e. without computer jargon. In this respect the terminology employed should be consistent. It must also be apparent as to whether a message is a statement, a comment, a command or a question. Messages must remain on a VDU display for as long as they are needed. It is no use instructing the user to make a choice or carry out a procedure if the necessary data has disappeared. Similarly, messages must be clearly seen, i.e. not hidden amongst other data. User messages or instructions to the computer should be acknowledged.

Errors prevention: Errors on the user's part must be detected as far as is practicably possible; the error message should be clear, describe what the problem is and suggest corrective action. Users should not be allowed to proceed until the error is corrected.

Escape: Allow users a method of terminating an operation and escaping from options. Many operations are selected accidentally and one of the most frustrating features of a bad interface design is being locked into an option you never wanted.

Minimal work: Try to save users' effort when operating the interface. This can be effected by using the minimal number of dialogue steps necessary (e.g. do not use two question-and-answer steps where one will do) and by reducing the amount of typing for users with abbreviations and codes. Long-winded dialogues may be supportive at first but users quickly learn dialogue steps and slow multi-step dialogues soon become frustrating.

Default: Set default replies where there is a predictable answer; this again saves the user work.

Help: Provide on-line help whenever possible. Help has two functions: first, as a learning aid for users who are too lazy to read manuals; and second, as an *aide-mémoire* for experienced users who need confirmation of some detailed aspect of an operation. Help should be layered or nested so the information pertains directly to the option or facility which the users want to know about.

Undo: Mistakes will be made and users will want to backtrack in a dialogue sequence and start again. The interface should provide the ability to go back and recover a previous state, e.g. in word processing the previous version of the paragraph being edited.

Consistency: The format and execution of commands should be consistent throughout the interface. For instance, the escape command should use the same code (E to exit) at all levels and should have the same effect, e.g. terminate the operation and return up one level in the interface hierarchy. Consistency reduces the amount users have to learn about an interface.

User's model: Dialogues should be modelled on the user's tasks as far as possible. Although the human–computer dialogue did not exist in the previous manual system, it may be implicit in the way users performed their tasks. They will expect to see computer operations in groupings and sequences with which they are familiar.

Guidelines, however, are only useful if they are applied, but their application will often require compromises between two or more conflicting factors, e.g. should feedback and acknowledgement be given at every step of the dialogue or will over-attentive messaging merely annoy the user? Design decisions will remain human value judgements involving trade-off between contradictory demands of a design.

Initiative in control should be given as much as possible to users. However, the more initiative provided, the more sophisticated and potentially more difficult a dialogue will be to use. Initiative, therefore, has to be constrained for unskilled users with simple dialogues which present only a few choices. Menus and simple question-and-answer dialogues are therefore more suitable for novice users, whereas command languages are more suitable for skilled users. A general rule is that computer systems should not seize the initiative and force users to perform actions according to the computer's command. In practical terms this means that a dialogue should proceed only when the user wants it to; users should not be locked into options without an escape route, and users should rarely be forced to give replies within pre-set time constraints.

Computer control should be by explicit action on behalf of the user and computer. Implicit actions and 'built in' short-cuts in a dialogue may appear to save time, but such implicit changes are unlikely to match users' expectations and hence cause confusion.

SAQ 8.10 This section has described a number of points that should be borne in mind when designing a dialogue. Which points do the following address?

1 A section of a screen containing messages which tell the user what data should be entered next and its format.

2 A section of the screen which tells the user whether a particular action has been successful.

3 A section of the screen which tells the user what major function is being performed, for example in a banking system the message STANDING ORDERS PROCESSING might appear.

4 The designation of a particular key which enables the current set of data entries to be aborted and a new function started.

5 The provision of a particular area of a screen which when clicked will restart the data entry.

Solution

1 is concerned with active guidance, 2 is concerned with feedback, 3 is concerned with status, 4 is concerned with escape and 5 is concerned with undo.

■ Dialogue styles

Simple control dialogues

The most simple dialogues are the question-and-answer type in which the computer asks whether a particular option is required or not and the user simply gives a YES/NO reply. Slightly more complex examples can move towards a menu-based system. These dialogues, although easy to use, are tedious for experienced users and slow to operate. Because each step has to be answered each time, users can quickly become frustrated with repetitions which they know are just wasted effort. Consequently these dialogues should only be used with naïve users or novices.

Menu interfaces

A menu selection dialogue requires the user to select one, or sometimes two, items from each of a succession of displayed lists (known as pages or frames). This type of dialogue is typified by the information systems available to the general public via domestic television, such as CEEFAX and ORACLE. The user indicates his or her choice of item by keying the number (reply code) alongside, by pointer and mouse, by lightpen, or by touchscreen, and this results in a new list appearing. This procedure continues until the page is displayed that contains

325

the user's required information. The successive pages can be either hierarchical or, less frequently, independent.

Hierarchical menu

Hierarchical means that each page after the first stems from its predecessor, e.g. a list of countries followed by the counties or states within the selected country, followed by towns within the selected county. Hierarchical menu selection may well necessitate the storage of a large number of pages if several levels are involved.

Independent menu

With independent menu selection each displayed list covers a set of items that are unrelated to those in the previous list. Consequently it is a good idea to display the lists concurrently, and if windows are available they can be scrolled independently.

A useful variant of menus is the pull-down type. With these the procedure is to position the pointer over one of the several options along the top of the screen. This causes a menu to appear beneath the option, from which an item is selected by means of the pointer and mouse. When the pointer is positioned on a menu item, this switches into inverse video and is then selected by pressing the mouse button.

Menu selection is straightforward but is tedious for a dedicated user if the same lists have to be inspected repeatedly. This is alleviated by incorporating a short-cut such as direct page mode. Here the user is permitted to enter the number of the page he or she wishes to see, and the computer retrieves and displays this page directly. Direct page mode and conventional menu selection may be combined in the one dialogue, i.e the user moves between the two methods as convenient.

Since menu selection is generally intended for casual users, it is important that the dialogue is 'user friendly', i.e. provides good guidance and control in leading the user towards his or her needs. In this respect a number of facilities can be included such as 'return to start' and 'return to previous page'.

Menus are the ubiquitous computer interface, yet sufficient attention is rarely given to their design. Menus work by users associating a reply code with an option displayed on a screen. Reply codes may be either numeric or characters. Character codes can be mnemonic and suggest the meaning of an option: however, this method has the problem of running out of letters to represent options, e.g. the E for edit and E for exit. The solution to duplicates is to use a longer code but this hinders the advantage of giving a response in a single keystroke. Numeric codes, although they contain no meaning, are not a hindrance to efficient menu operation.

An alternative to using a reply code from the keyboard is to use a pointing response with a mouse with pull-down or pop-up menus.

In most systems there are more options than can be easily placed on one menu. This enforces hierarchical organization of menus. It is important that the grouping of options and functions conforms to the user's model. Navigation in menu hierarchies presents two problems for users:

1 Keeping track within the hierarchy, the 'where am I?'.

2 Tracing a path through the hierarchy, the 'where have I been?'.

To help users navigate, status information about the hierarchical level and the sub-system being used should be displayed on the top of the screen. To improve pathway tracing, a backtrack facility is helpful so that users can page back to the last menu with a single keystoke. A further extension of user control in the hierarchy is to give users 'escape to the top' commands.

How many options to display on a menu has been the topic of considerable research. There is a trade-off between depth and breadth in a menu hierarchy. Making the hierarchy broad by placing many options on one menu means that users have to spend longer searching through the list of options. If the hierarchy is deep with fewer options per menu then the search time per menu is shorter although the menu level descent time is increased. Intuition suggests there must be an optimal compromise and some studies indicated that this is with menus containing 7–9 options.

However, the efficiency of broad menus can be increased by structuring the options into groups as exemplified by the Microsoft Word menu. A problem with all hierarchical menu structures is that users soon learn part of the tree and wish to traverse from one option to another without going up and down the hierarchy. To accommodate this desire a menu bypass facility can, as mentioned earlier, give direct access to options. If numeric reply codes have been used, options can be addressed using a page number principle, with the numbers being derived from the menu response at successive levels, e.g. option X has an address of 134 and used to be accessed by typing 1 at the top menu, 3 at the second level menu, etc.

Pull-down menus allow menus in different levels of an access hierarchy to be called directly from a single command line. The options are only present when they are being used and do not obscure the work area. After the choice has been made the menu is removed.

Guidelines for menu design

■ Group logically related options together either as menu blocks or in separate menu screens.

■ Order menu options by the usual criteria, e.g. operational sequence, frequency of use, importance, etc.

■ Indicate the reply expected and associate it with the option.

■ Title and menu according to its function.

■ Give the user feedback about menu levels, errors, etc.

327

■ Provide escape routes and by-passes.

■ Bullet proof the replies, e.g. if 1 to 7 are options and 0 is escape, make sure any other keystroke is followed by an error message and not a program failure.

Function keys

Function keys are a hardware equivalent of menus with options allocated to special keys on the keyboard to save screen space and alleviate the reply coding problem. Function keys can either be hard-coded or soft-coded. Hard-coded function keys have an operation permanently allocated to a particular key. This approach is common for single applications on dedicated hardware, such as word processors, when functions are not going to change. For most systems function keys are soft-coded.

With soft-coded keys the command call is allocated to the function key by the application program. One or more commands can be allocated to each key; but as more commands are linked to a single key, users can become confused. To help users keep track of the function-key mapping, a partial menu can show the allocation of options to keys, mapping the keyboard layout onto the screen.

Most computer hardware suppliers provide 10–12 function keys. Important keys should have a constant function in any context (e.g. F1 is always help, F2 is always escape).

Icons

Icons are an essential component of graphical user interfaces (GUIs). The great advantage of icons is that they are realistic, so we do not have to learn what they represent, and instead can immediately make an informed judgement about their meaning. On the other hand an icon may be ambiguous or can be misinterpreted even though it appears to be a realistic image. As a precaution against ambiguity most icons have some text explanation associated with them.

Icons were pioneered by Xerox in the Star system and later by Apple in the Macintosh interfaces. A key idea in these designs was that pictures of objects in the system could be used to create a visual metaphor of the user's everyday experience. In the well-known desktop metaphor icons represent objects in the office, e.g. in-trays, filing cabinets, folders, calculators and waste paper baskets. The user picks icons and moves them by pointing with a mouse. For example, to delete a file you move a folder into the waste paper bin, following the metaphor of everyday life of throwing waste paper into a bin.

Icons, however, present some problems when functional operations need to be displayed; for instance, cut-and-paste or global find-and-replace operations in a word processor. Some advice to follow on the design of good icons is as follows:

- Test the representation of the icons with users.
- Make icons as realistic as possible.
- Give the icon a clear outline to help visual discrimination.
- When showing commands give a concrete representation of the object being operated upon.
- Avoid symbols unless their meaning is already known.

The size of icons is a matter of compromise. As icons are not a particularly space efficient means of representation they run into similar problems as menus with hierarchies. Size is integrally related to complexity of the image. Simple symbols can be effective in dimensions of 0.5 cm square; more complex images need to be 1 cm.

Direct manipulation (DM)

This term was coined to refer to interfaces which include icons, pointing and features which have now become associated with WIMP (Windows, Icons, Mouse, Pop-up menu) interfaces such as the Apple Macintosh. The central idea of such interfaces is that the user sees and directly manipulates representations of objects in the system, rather than addressing the objects through an intervening code as in command languages or menus.

In direct manipulation a mouse is used to select and move objects around the screen using a dragging operation. In this way new associations between objects can be formed, e.g. a file can be placed in a folder (a sub-directory in non-DM interfaces). The advantage of direct manipulation is that the computer system models everyday operations; the more directly an interface models reality, the easier it is to learn. The essential features of direct manipulation interfaces can be summarized in the following set of principles:

- **Explicit action**: the user points at and manipulates objects on the screen.
- **Immediate feedback**: the results of the user's actions are immediately visible, e.g. when an icon is selected it is highlighted.
- **Incremental effect**: user actions have an analogue/sequential dimension, e.g. as an icon is dragged across the screen display it moves continuously following the user's movement of the mouse rather than suddenly jumping to a new position.
- **Intuitive interaction**: interaction matches the user's conceptual model of how the system should operate and the display shows pictures of familiar objects.
- **Learning by onion peeling**: the complexity of the system is gradually revealed in layers as the user explores system facilities.
- **Reversible actions**: all actions can be undone by reversing the sequence of manipulations.

■ **Prevalidation**: only valid interactions have an effect, so if the user points at an object and this makes no sense in terms of the current task, nothing happens on the display.

The interface portrays a realistic 'virtual world' on the screen. The DM idea has given rise to another acronym, WYSIWYG (What You See Is What You Get), which was initially applied to word processors. This refers primarily to the output in which the results of the user's actions are immediately apparent in the display. WYSIWYG editors use direct manipulation to format text delivering the exact image of what the user sees, and hence eliminating the necessity to remember format control commands.

Windows

Another facet of direct manipulation interfaces is the ability to have several different views on a single object supported by windows. Windows come in two basic types:

1 *Tiled*: the screen is divided up in a regular manner into sub-screens with no overlap.
2 *Overlapping*: windows can be nested on top of each other to give an illusion of depth.

Windows have many uses. Screen areas can be separated for error messages, control menus, working area and help. Two or more processes can be run at once in different windows. In this manner windows allow multi-task processing in a suspend-and-resume manner. The status of background or suspended tasks can be held in a window so the user can periodically monitor what is going on.

Although windows are very useful they have some disadvantages. If too many windows are created the screen becomes cluttered. Increased window clutter incurs the penalty of an unstructured display, and search times increase with complexity. Windows no longer directly related to the current task should be deleted to avoid clutter.

Direct manipulation interfaces and windows require advanced interface software to control the screen display and a high-resolution VDU. Such software, often called a Graphical User Interface (GUI), acts as interpreter between the application software and the user, managing all the interaction and communication.

■ Command languages

Command languages are potentially the most powerful command interface, but more power brings with it the penalty of difficulty of learning. The main advantages of command languages are the economy of screen space, and the flexibility of systems functions which a combination of commands can provide.

All command languages have a word set, called a lexicon, and rules which state how words may be combined together. Command languages need words to identify objects and operations. Objects will be devices, files, etc., which the commands of the language operate on. Word sets should be as meaningful as possible as one objective of command languages is brevity of input, hence coding is necessary. The basic choice when shortening a word is to truncate or abbreviate.

Truncation

Truncation removes the latter part of a word leaving a few characters at the front, e.g. DIRectory, CATalogue, DELete, DISplay, DEVice. This is an effective technique if the front of a word communicates its meaning. The problem with truncation comes from duplicates between words sharing common leading characters, e.g. DELete/DELay, DISplay/DISconnect. When this happens further characters may have to be added to remove the ambiguity. Unfortunately this becomes inconsistent as users have to type in either 3 or 4 characters depending on the word. Ideally, users should be able to invoke commands with a single economical keystroke; however, single letter commands are most likely to create aliases as the command word set grows. Most operating systems use three-letter commands to prevent ambiguity.

Compression

The alternative to truncation is compression. Characters are removed at various points in the word leaving sufficient letters to convey the meaning. The resulting compressed words should, as with all codes, be the same length. Simple elimination of vowels or consonants rarely produce good code words; instead, mnemonic techniques of front-middle-back compression using syllabic emphasis give the best results.

Command languages tend to favour truncation coding. Compression codes are more advantageous in large systems with extensive word sets in which truncation is no longer a viable option.

Command language syntax

The rules which govern how command words may be combined together vary from simple association rules to complex grammars. Command keywords may be used in simple combinations, such as the command/object construct of a verb/noun pair, e.g. TYPE FILENAME, PRINT FILE. No complex grammatic rules are present, consequently the word combinations are limited. Keywords may be qualified by added parameters to enhance the behaviour of the basic command, e.g. DIR/SIZE, DIR/OWNER, DIR/PROT. This gives more flexibility too, as one command can now be used to do several different things. Rules are introduced to govern the set of permissible parameters per command and how they are combined.

Keyword and parameter command languages still have relatively simple rules for combinations of words. In a full command language a set of rules is introduced to formulate the phrases which may be derived by combinations of command words.

Many command language grammars mimic natural language grammars to help learning although the types of sentences are simple when compared with natural language. The types of grammatic constructs required can be grouped in functional categories:

Assignment: This associates two objects, or an object with an attribute or value. For instance, to give a device some property, or setting read-only protection on a file. Command phrases of this type are constructed in the form Verb-Object to Object; for example:

SET LINE-PRINTER TO CHANNEL-2

Assign Disc A = USERfred

Imperative: This command invokes system operations and may be qualified by objects for the destination of results using the Verb-Qualifier-Object paradigm; for example:

nroff -Tlp -ms myfile | Lpr -Plp&
(the UNIX command for formatting a file and then printing it in batch mode)

p ws 4

Locate: Commands which search or find a data item within a list or file. Locate commands are common in data retrieval command languages and take the form Verb(find) Object with Qualifying conditions; for example:

FIND CUSTOMER WHERE ACSTATUS = NEW
(find new customer records)

Accept input: Commands which get input from the user and use that input in an operation; for example:

CREATE BACKUP[FILENAME = ******]
(create a backup file with a 6-character name entered by the user)

Command languages become much more complex if simple phrases can be built up into more complex expressions. In the case of English the grammar dictates the way in which phrases are composed into sentences. Command languages with a hierarchical grammatic syntax approach the complexity of programming languages and are indistinguishable from them. Most operating systems provide facilities allowing users to extend the system's functions by producing new combinations of commands in programs. Examples are COM, EXEC files and shell programs within UNIX. Full syntactic command languages are powerful and flexible, but they impose a considerable learning burden on the user.

Generally, full syntactic command languages should be reserved for sophisticated users. Command language design involves design of an input parser, error-message interpreter and run-time system. These are systems and compiler design issues which will not be treated further here. In summary, design guidelines for command languages are:

- Command word codes should be consistent. If E(XIT) has been used for the escape command, do not use Q(UIT) in another part of the system.

- Punctuation and use of delimiters should be minimized.

- Entry should be flexible and forgiving. Double spaces between words should be ignored and misspelling corrected if possible.

- Command language words and syntax should be economical. Use of the smallest combination of words for a function should be traded-off with word clarity for ease of learning and remembering.

- Command words and syntactic sequences should be natural and familiar; e.g. use COPY from fileA to fileB, and not PIP destinationfile = sourcefile.

SAQ 8.11   Would you use a command language interface for staff entering seat booking details for an airline?

Solution

No, such an interface is normally targeted at sophisticated users. It is probably best to use either a menu interface or the sort of direct manipulation interface described later in the chapter.

Query languages

Query languages are intended as a rapid means of interrogating databases, i.e. logical files, by the use of pseudo-English statements. The computer is enabled to pick out certain keywords from the query and to use these to answer it.

Example 8.16   **Keywords**

'DISPLAY products IN NUMERICAL ORDER OF product code THAT HAVE stock value GREATER THAN £10000'

In Example 8.16, the words in capitals are commands telling the computer what to do; the lower-case words refer to data items in the database. This is a simple example, needing access to one or perhaps two files only. More complex queries would involve several files.

Query languages are non-procedural, i.e. the user specifies what is required rather than how to obtain it. The computer merely detects the keywords in order to achieve what is needed, and so in many cases it would not matter if the keywords were not in exact order.

Query languages are generally specific to the particular type of database they interrogate, in this respect SQL has emerged as a standard for relational databases.

SAQ 8.12   A queries clerk in a bank usually receives phone calls from customers asking about basic information such as the current balance of the account and whether a particular cheque has been cleared. Would such a member of staff normally use a query language?

Solution

No, query languages are quite complex and the job of the query clerk could be provided by a simple screen which provides summary data on an account and the ability to scroll through account details sequentially.

Mnemonic command languages

A dialogue of this type employs a series of messages consisting of abbreviations, and/or other easily remembered words or codes, i.e. mnemonics (see Example 8.17). The purpose behind this method is to make each message brief and informative but without being too difficult to compose or understand. A mnemonic dialogue demands that the user is trained to be conversant with the structure of and the mnemonics used in the messages. In practice, this means that the user is dedicated, i.e. using the dialogue frequently and as a main part of his or her occupation.

Example 8.17   **A mnemonic message**

A typical message from an airline reservation clerk to a computer is:

A 8SEP LON NYC 1500

In this the 'A' is an action code meaning that the message is a request for the availability of seats on flights on 8 September from London (LON) to New York (NYC) departing at about 3 p.m. (1500). The computer responds in similar vein, and so by an interchange of such messages, the customer's requirements are fulfilled.

The main point about mnemonic commands is that the user must not be expected to remember too many words. If a large number is unavoidable, there should be some straightforward method of recalling and interpreting them, e.g. by means of a list that can be displayed when needed.

Directory searching

When there is a large number of items from which one is to be selected, the search can be narrowed down to the required item by displaying a series of lists in a similar manner to menu selection. Each list is in some logical arrangement, e.g. alphabetical order, so that the required item or the one logically preceding it can be seen.

An instance of directory searching is where it is necessary to find the precise name and details of one of a large number of customers, many of whom have similar names. The search commences by entering the first few letters of the customer's name, whereupon a list is displayed of all names that start with these letters.

This allows the user to scrutinize the names and, hopefully, select the one for which details are required. If on seeing the customer's full name and address, the choice is perceived to have been incorrect, others can easily be tried.

8.6 DESIGN OF PROCESSING

The design of processing is largely a matter of interconnecting the sub-systems of output, input and logical files so as to create a system that provides the requirements of each application. In most cases this also entails the design and creation of a database to hold the application's master data. In other cases the database already exists and the new system is designed to interface with it. It is also possible that the system needs to interface with software purchased from external suppliers; this applies particularly to the use of a database management system.

In most businesses the work falls into a number of applications. By and large, these are separate areas of responsibility and effort, generally by reason of the inherent departmentalization of most companies. There is no harm in designing an overall system in the form of applications provided the data in the database is not thereby replicated. The applications interface with one another at certain points in their logic and at certain points of time; this arises naturally through their use of the logical master files in the database.

A business application splits naturally into several routines (procedures).

Routines and jobs

A routine is a piece of IS work that produces some usable output, and is carried out as a series of related processing jobs on the computer. A routine is generally time-bounded by the arrival of source data and the production of the resultant information, especially documents. Routines vary in complexity from one organization to another but typically they relate to the applications as in Example 8.18.

Example 8.18

Routines or procedures

Application	Routines (procedures)
Sales accounting	Order acceptance (daily)
	Credit control (daily)
	Invoicing (weekly)
	Statement preparation (monthly)
Payroll	Gross wage computations (weekly)
	Payslip printing (weekly)
	Payroll analysis (quarterly)
Production control	Job scheduling (daily)
	Labour planning (weekly)
	Materials requirements (monthly)
	Machine loading (monthly)

A routine usually comprises several separate jobs each of which is a continuous computer process involving one program, one main lot of input, and one main lot of output. This definition is a little dogmatic but is typical if not absolutely true. Routines are split into jobs simply because computers are often incapable of carrying out the complete routine in one go. The main limiting factors in this respect are:

■ the number of program instructions that the main store can hold at the one time;

■ the availability of peripheral devices; and

■ the time relationships between jobs – these are normally under the control of external circumstances.

Routines and jobs both entail the carrying forward of data from one processing cycle to the next. This data is left stored on magnetic media so that the time taken to write and subsequently read it is minimal. With routines, disks or tapes are removed from the peripherals and held in a secure place between processing cycles. With jobs, data is sometimes left in main store while a new program is loaded, e.g. order analysis figures accumulated during order acceptance are left stored for printing by a subsequent job.

Types of processes

The processes, jobs or runs used in conjunction with each other to form routines are mainly of the following types:

336

- **Data acceptance**: this job validates, edits and totals the source data, and prints error messages and control tasks. It also transfers the edited source data onto disk or tape so as to form a transaction file.

- **Sorting**: if the transaction records are to be processed sequentially, the second job sorts them into the appropriate sequence.

- **Referencing**: a further job is used to look up data from a master file(s), e.g. prices, descriptions, balances, and to apply these in some way to the transaction records. This may result in the creation of a working file, e.g. holding priced transactions, but this is often omitted nowadays.

- **Updating**: the master file records are updated by data from the transaction records, e.g. stock issues and receipts applied to stock balances.

- **Amending**: the master file is brought into a correct and up-to-date condition by the insertion of new records, deletion of obsolete records, and changes made to existing records. For instance, a record is inserted for each new product, a record is removed for each dead product, and alterations made to the prices of products.

- **File printing**: updated records are copied from the master file or are summarized or analysed before being printed, e.g. a list of the stock levels of all the materials.

- **Document printing**: master files and transaction files are used to print documents such as invoices, payslips, statements, etc. The output is frequently spooled (written) to an output file prior to being printed.

Diagramming the system

Flowcharting

System (routine) flowcharts

The highest level represents the way in which the whole routine holds together, as in Figure 8.7. This is often referred to as either a system or a routine flowchart, and this example relates to the stock analysis routine described in Section 8.3.

It is probably better not to incorporate too many different types of symbols into a system flowchart but to stick to the four basic symbols.

The individual jobs or runs shown on a system flowchart need further definition and this brings us to the next level of flowchart – a job or run flowchart.

Process, job or run flowcharts

Figure 8.8 is an example of a process flowchart and applies to the stock analysis. This flowchart shows the processing in a single job or run, and the only symbols really needed are the processing symbol and the decision symbol. The main point is that each symbol must be explained in detail elsewhere since it is not possible to do this properly within the confines of the symbol outlines.

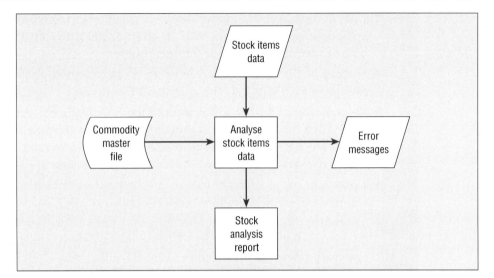

Fig. 8.7 System flowchart of stock analysis routine

The philosophy of flowcharting is that all the necessary steps are incorporated but that the flowchart does not dictate the programming methodology.

Dataflow diagrams (DFDs)

DFDs are explained in Section 7.4 as applied to manual systems; they are also used to show the fundamental processes in a computer-based system. Bear in mind that a DFD is based on the flow of data between processes and data stores, and that three main symbols are involved (Figure 7.5, page 230).

Figure 8.9 on page 340 shows the DFD for the stock analysis example and, as is evident from this diagram, the annotations need to be amplified before being passed for programming; in this respect DFDs are similar to flowcharts.

Structure diagrams

The concept of a structure diagram (SD) is that it is composed of a number of modules each of which reports to a higher-level module, thus tying in with the principles of structured programming (Section 6.1). Each module should be as self-contained as possible, i.e. have the maximum binding and minimum coupling. As explained previously, binding means cohesiveness, that is to say, the module performs one clear-cut function, such as those depicted in Figure 8.10. Coupling means that interplay between modules, and by minimizing coupling, modules are independent from the program aspect.

A structure diagram is created as a hierarchical tree of modules and Figure 8.10 shows the stock analysis example split into modules in this way. The modules are executed from left to right and are shown in increasingly greater detail from

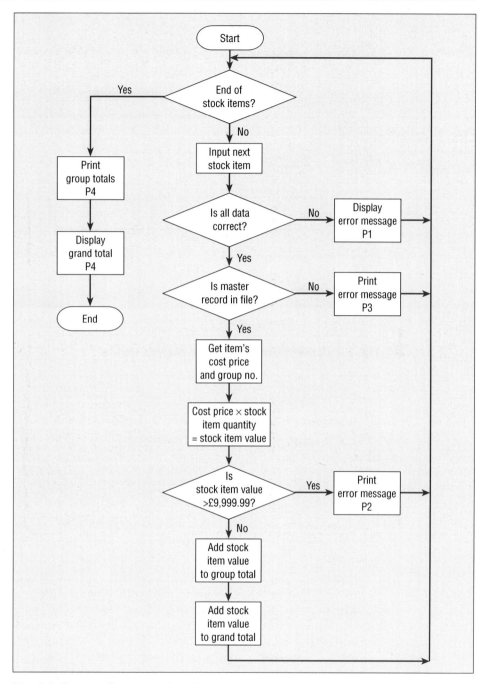

Fig. 8.8 Process flowchart of stock analysis routine

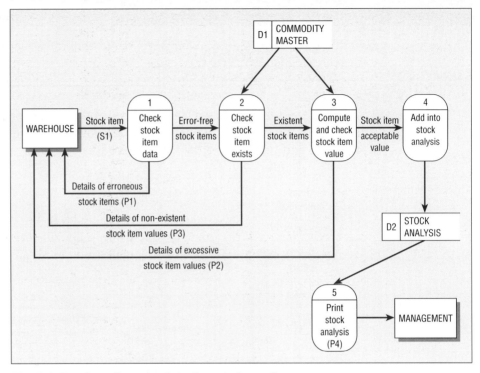

Fig. 8.9 Dataflow diagram of stock analysis routine

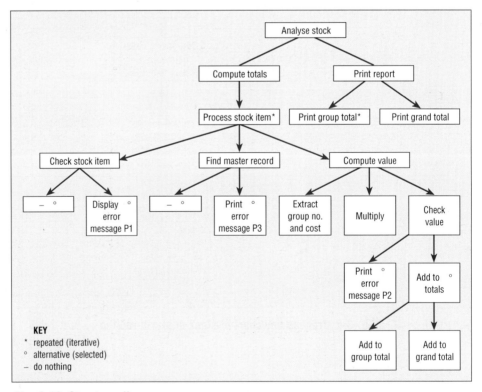

KEY
* repeated (iterative)
° alternative (selected)
− do nothing

Fig. 8.10 Structure diagram

top to bottom. Thus after completing a module the process normally goes back to the next higher level and then takes the next path to the right. The exceptions are the alternative modules (shown containing circles); only one of these is executed during the one processing cycle. When a module contains an asterisk, it is repeated as often as necessary before proceeding to the next module. This is the case for 'process stock item' in Figure 8.10 because of the large number of stock items, and similarly for the 'print group total' module.

Structure charts

Figures 8.11–8.13 depict, in increasing detail, another type of diagram known as a structure chart (SC). The principle is the same as described above but the final SC has more details and is, in effect, an amalgam of an SD and a DFD.

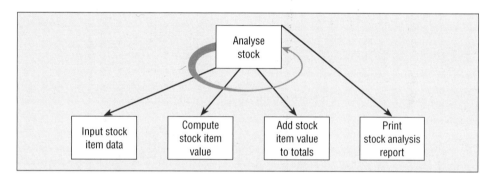

Fig. 8.11 Structure chart: first stage

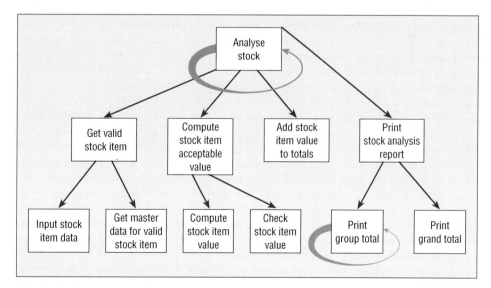

Fig. 8.12 Structure chart: second stage

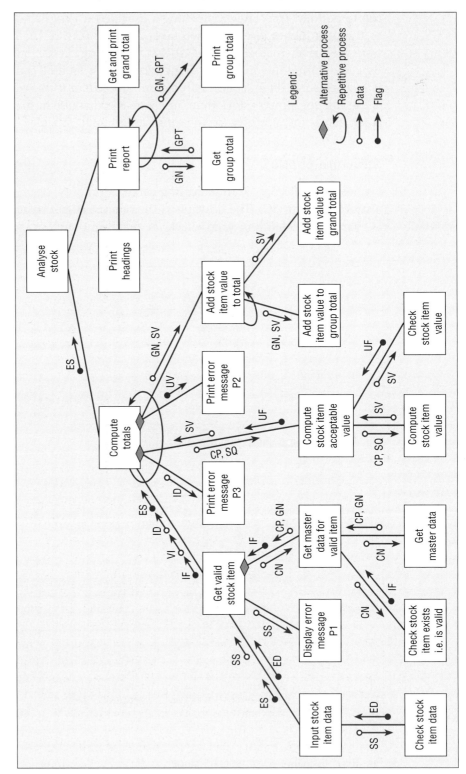

Fig. 8.13 Structure chart: third and final stage

Abbreviation	Name	Description
CN	Commodity number	Identification of stock item's commodity
CP	Cost price	Cost price per unit of the commodity
ED	Erroneous stock item data flag	A designation indicating that the stock item is erroneous or incomplete
ES	End of stock items flag	A designation to indicate the end of the stock items
GN	Group number	The group into which the commodity falls
GPT	Group total	The total cost value of stock within a group
GT	Grand total	The total cost value of all the stock
IF	Invalid stock item flag	A designation indicating that the stock item is not in the master file
ID	Invalid stock item data	SN, CN, RN, SQ
RN	Rack number	The location of the stock item in the warehouse
SN	Serial number	The serial number pre-printed on the stock form
SQ	Stock item quantity	The quantity of the stock item in the warehouse
SS	Stock item source	SN, CN, RN, SQ
SV	Stock item value	The cost value of the stock item in the warehouse
UF	Unacceptable value flag	A designation indicating that the stock item's value is over the top limit
UV	Unacceptable value data	SN, CN, RN, SQ, CP, GN, SV
VI	Valid stock item	SN, CN, RN, SQ, CP, GN

Fig. 8.14 Abbreviations used in Figure 8.13

Referring to Figure 8.13, the arrows show movements of data and flags between the modules. Conceptually each module receives data and instructions from its higher-level module and passes results back to it. The higher-level module may pass the data either to its higher-level module or to another lower-level module for further processing. The modules are operated from left to right and those with connecting lines lying within the curved arrowed line are repeated as often as necessary; 'compute totals', for instance, is performed once for each stock item. At the end of the stock items, denoted by the flag ES, control passes back to the highest module (analyse stock) which then initiates the 'print report' module.

The diamond shapes denote alternative processes so that, for instance, either the error message P3 is printed or the stock item value is computed.

A convenient way of constructing an SD is to start with a broad outline showing just the main functions (Figure 8.11) and therefore to amplify the modules into more modules working downwards (Figures 8.12 and 8.13). This may have to be repeated several times until it is not worth while or indeed possible to amplify the modules further. It is convenient and advantageous to abbreviate the data items and to create a list showing their definitive names and descriptions (Figure 8.14); this then forms the basis of a data dictionary.

SAQ 8.13 Which of the following statements are true and which are false? If a statement is false then outline why it is so.

1 Structure charts are normally specified in great detail and then more abstract versions of the structure chart are prepared.

2 A structure chart is normally shown by the analyst to a customer.

3 A structure chart will show the movements of data and flags in a system.

4 A structure diagram will describe the software modules in a system.

Solution

False, the reverse happens. False, a structure chart is a design document and the vast majority of customers will not understand this form of documentation. The final two statements are true.

Pseudo-code

A number of similar methods have been devised to specify processes, these enjoy various names such as pseudo-code, action diagrams and structured English. An example is shown in Figure 8.15 from which it is apparent that there are strong similarities to structured programming. The names of the data items (in italics) and of the processes should conform to those in the data dictionary, thus eliminating any doubts as to their precise identification.

SAQ 8.14 In Figure 8.15 what does the chunk of text starting with the identifier PROCESS-ITEM do?

Solution

It obtains a record containing details of a stock item, it then gets a record from a master file, if it finds the record then it carries out the process described by COMPUTE-VALUE and if it doesn't then an error is flagged.

▪ Program specifications

The various methods described previously, such as DFDs, decision tables, structure charts and so on, do not always provide sufficient detail for the needs of programming. They are therefore supported by descriptive programming specifications into which all additional points are incorporated. These specifications are particularly necessary for new programmers and for amendments to existent systems.

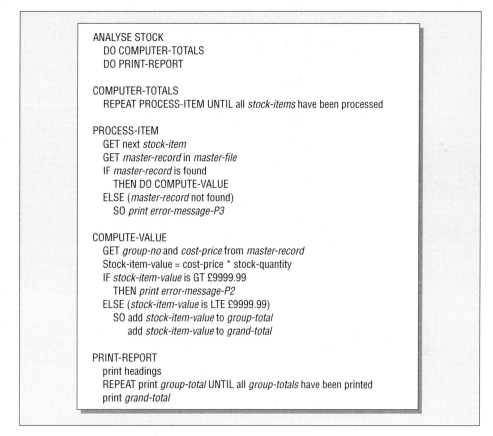

```
ANALYSE STOCK
    DO COMPUTER-TOTALS
    DO PRINT-REPORT

COMPUTER-TOTALS
    REPEAT PROCESS-ITEM UNTIL all stock-items have been processed

PROCESS-ITEM
    GET next stock-item
    GET master-record in master-file
    IF master-record is found
        THEN DO COMPUTE-VALUE
    ELSE (master-record not found)
        SO print error-message-P3

COMPUTE-VALUE
    GET group-no and cost-price from master-record
    Stock-item-value = cost-price * stock-quantity
    IF stock-item-value is GT £9999.99
        THEN print error-message-P2
    ELSE (stock-item-value is LTE £9999.99)
        SO add stock-item-value to group-total
            add stock-item-value to grand-total

PRINT-REPORT
    print headings
    REPEAT print group-total UNTIL all group-totals have been printed
    print grand-total
```

Fig. 8.15 Example of pseudo-code

Although a programming specification generally covers the one program, this is not always the case. It sometimes turns out to be convenient to create two, or possibly more, programs from the one specification. This does not in itself affect the principle of program specification.

The list below covers the main features of program specification, although further points may be added for particular circumstances:

■ The name, reference number, and brief explanation of the program's purpose.

■ The contents and layout of all necessary input media. This should be cross-referenced to the data dictionary (if it exists).

■ The contents and layout of all output, including reports, screen displays and on other media. Layouts are usually specified on report and VDU layout charts.

■ The validation and feasibility checks carried out on input whether the input is from the user or another program module.

345

■ Calculations and logical processing carried such as accumulation of totals and arithmetic calculations. This will usually constitute the pseudo-code of the program's logic, although structured English may suffice in some cases.

■ The storage and access modes of data in logical files, and DBMS sub-schema if a database is being used.

■ Notes on the meaning of special terms, codes and symbols that are being used, e.g. debit/credit symbols which entail alternative processing.

8.7 SECURITY AND AUDIT

■ Security and accuracy controls

Security and accuracy controls are necessary for the following reasons.

To detect and prevent terminal user errors

More users are involved in on-line systems than with batch-processing systems and so there is greater scope for errors to arise in the input data. On the other hand, the data is captured closer to its source, and so errors are corrected more easily.

To detect and prevent computer-room operator errors

These errors could be the loading of incorrect disk cartridges or tape reels.

To guard against hardware failure or program errors

The failure of the computer or of the transmission lines must not be allowed to cause the loss or duplication of data. A user needs to be aware of which data has been accepted by the computer when a failure occurs, and how to proceed from then on. Similarly the operating system has to be capable of reconstructing records that have been lost or damaged.

To prevent fraud and abuse

There is always the possibility of fraudulent activities involving the computer and/or data where money or important information is entailed. A similar activity is so-called hacking; this is the theft of information by other users of a computer network. The hacker breaks into an organization's database and extracts information, often merely for the sake of so doing.

Abuse includes the overloading of the computer system by trivial demands upon it, e.g. frequent and unnecessary requests for complex information. Also coming within abuse is the unauthorized dissemination of private information.

To permit the auditing of the company's finances

It should be remembered that the obligations of the auditors are not changed by the computer system. The auditors' methods are different, however, and the controls should facilitate the methods adopted.

To prevent the insertion of viruses

A computer virus is a program which has been inserted maliciously into a system for malicious intent. Procedures should be put into place to continuously monitor the software configuration.

Security and accuracy methods

Physical security

The security measures employed for cash, etc., also apply to information. These include locks, safes, security officers and burglar alarms. Also involved are protection against wire-tapping, and the identification of terminal users. Wire-tapping is frustrated by the use of cryptography; terminal users are validated by using badges, passwords and interrogation questions.

Other aspects of physical security are protection against fire and flood. The latter is not usually a problem but it must be remembered that damage by water can be a consequence of fire-fighting.

The computer hardware itself should obviously be housed in a fire-protected area, and specialist advice obtained regarding this requirement. Back-up files are stored in fireproof and waterproof accommodation well away from the computer. In special circumstances the back-up files are held in a separate building in case of damage to the building housing the computer.

Staff security

Employees who have access to computers and data need to be screened in order to ensure that their interests are compatible with those of the organization. Those members of the IS staff who have a grudge or who are under notice or resignation may need to be given special attention.

It is also important that all IS staff and users take security seriously. This applies particularly to passwords, the use of which sometimes becomes lax. Regular reissuing of passwords helps in this respect.

Batch control totals

Pre-prepared batch totals, counts and hash totals are applicable to remote job entry (RJE) as with conventional batch processing. They are unsuitable, however, for on-line work because the input messages originate from many computers and

347

in a random manner; pretotalling therefore is impossible. Post-control totals may, however, be created by the computer for each terminal and transmitted to the terminals at intervals for comparison with the locally prepared totals.

Message controls

These include check digits, the transmission of preknown values for checking against the corresponding value in the stored record, and feasibility checks. An example of preknown data is the item price; this is input and is checked against the stored price to ensure that the correct item is being looked up in the file.

Message serial numbers

Each message in an on-line system contains a serial number in order to ensure that none is lost without this being detected. After a breakdown, the terminal and the computer inform each other of the last serial number received, and a correct restart can thus be achieved.

A more sophisticated method is where the computer allocates all the serial numbers and informs the computer of each message's number. The computer user then knows whether he or she has received the serial number; an input message does not need to be retransmitted after a failure.

Message logging

A log is maintained on disk containing a copy of every input message on its receipt and at intermediate stages, together with its serial number. Other data is held along with the logged message such as the computer number, time received and data extracted from the master files. The log is available for re-input of messages after a failure has occurred.

Checkpoints

At intervals during processing, as determined by the computer's real-time clock, the status of the run is recorded on tape. These 'checkpoint tables' are reloaded after a breakdown and the supervisor program then decides which messages are capable of completion and which need to be retransmitted from the terminals. This is not a simple exercise because real-time messages are handled by a 'multi-threading' procedure. Multi-threading means that messages are processed a stage at a time in parallel, and the partially processed messages are held in queues between stages.

Bypass procedures

Whatever the nature of an on-line system, it must be capable of continuing to function when a hardware failure occurs. The level of service is generally degraded but this is better than a total break-off.

With a real-time banking system, in the event of a failure the clerk is allowed to accept cash and to allow withdrawals up to a certain predecided amount. To facilitate this, the computer transmits a portion of each day's closing balances to the branches during the night. A printout is thus available to provide some indication of the customer's financial status.

Another bypass procedure is the recording of transactions on magnetic medium at the terminal if it becomes cut off from the computer. On recovery, the data is transmitted rapidly to the computer.

Dumping (file reconstruction)

Each file is written to a streamer medium (dumped) at intervals, and the message logs retained from the time of each dumping. If any records are lost or corrupted due to program or operating errors, the file is reconstructed from the dump and then updated by the logged messages.

Audit trails

The company's internal and external auditors are involved in a computer system during all stages of its development and use. A number of auditing techniques are available for these stages and must be integrated into the system. This is especially relevant to on-line systems.

SAQ 8.15   What are the following examples of?

1 In an airline reservations system, the writing of duplicate bookings to a remote computer different from the one that normally processes the bookings.

2 The use of check digits in a key.

3 The use of a password on a computer terminal.

4 In a banking system the writing of all transactions (debits and credits) to a disk on the computer which normally processes entities such as cheques.

Solution

The first item is an example of a bypass procedure; when a failure occurs the second computer can be used to carry out further bookings.

The second item is an example of a message control.

The third item is an example of a staff security measure.

The fourth item is an example of an audit trail.

▪ Audit considerations

It is the duty of any organization's auditors, internal and external, to make such tests and enquiries as are necessary to form an opinion as to the reliability of the records as a basis for the organization's accounts. The statutory responsibilities of the auditor are not diminished by the employment of a computer within the organization.

Nevertheless the size of the computer and the sophistication and complexity of the system clearly have some effect on the auditor's approach to his or her tasks. The main difference between a manual system and a computer-based system is the lack of human readability of the records in the latter system. Another important aspect is that a computer operates from a program of instructions, the subtleties of which are not always obvious to the non-programmer.

Audit trail

The above factors mean that an audit trail must be established. This is the means by which the details behind the summarized totals and analyses can be obtained. The auditor, having found a discrepancy in a total, wants to investigate the detail causing the discrepancy. And so the audit trail needs to be capable of being followed right through the system from source data to final output, and to provide readable evidence at each stage if required.

It is therefore desirable for the auditor to be involved in the information system during its design stage. He or she is then in a position to make his or her requirements known and to determine how these are provided. This implies that the auditor needs a general understanding of information systems, including hardware, programming, software and control methods.

▪ Audit methodology

Control totals

The concept behind control totals is that each stage of processing generates control totals to be carried forward for checking during the next stage. The purpose of control totals are to detect the following:

- the loss of documents or data;
- the accidental or intentional insertion of records or data;
- fraudulent alterations to data; and
- manual, hardware and software errors occurring at any stage of processing.

The various types of control totals are explained in Section 8.4.

The sources of control totals are as follows:

- the originating point of a batch of documents;
- the data preparation department;
- initial computer input;
- each stage of processing; and
- computer output including batching of output documents.

Check digits

These are used in conjunction with auditing to detect errors (see Section 8.4).

File inspection

Any selected portion or all of a file has to be available in human-readable form so that the auditor can examine the records if errors or fraud are suspected. This requirement is covered by audit packages but may need to be programmed specifically in certain cases.

Operation logs

Every job carried out on the computer is logged manually and/or by the operating system. The log notes the data volumes, times taken and storage devices that are loaded. This enables unauthorized use of the computer to be detected, for instance in order to perpetrate fraud or for private purposes.

Staff control

Only authorized staff are allowed to operate the computer and only during permissible working hours. Programmers are not to be allowed free access to the computer. Terminal operators must identify themselves; the computer then verifies that the terminal operator is entitled to the data requested. Several sophisticated methods of identification are now available.

Documentation

System and program documentation should be fully adequate for the auditor's needs (see Section 8.9).

Audit test data

Sets of input data are kept by the auditor together with the corresponding output results. Their purpose is for spot checking to detect unauthorized amendments to programs. This technique sometimes proves difficult in practice owing to authentic amendments to programs and files causing the output to be different from the auditor's version. In order to prevent an audit test from damaging file records, a copy is made of the file for audit test purposes.

Audit packages

In order to cope with the complexity of computers and systems, the auditor utilizes the computer itself, i.e. by means of an audit package.

The features of an audit package are:

- extraction of selected file data by means of variable parameters;
- extraction of data by sampling;
- totalling of data items from records meeting certain criteria, e.g. total debit amounts to all foreign customers;
- computational functions to enable checks on extension, interest, discounts, etc;
- subtotalling at certain points in a file;
- matching of files in the same sequence to detect omissions and replications (see Section 5.1); and
- open-endedness to allow the auditor to extend the package to meet his or her further needs.

8.8 COMPUTER JOB SCHEDULING

An important aspect of systems design is the scheduling of the computer job within the operational hours available. The use of an operating system in multiprogramming mode results in the actual jobs being loaded onto the computer for subsequent execution according to the dictates of the operating system. Nevertheless the systems analyst must schedule the jobs during the design stage mainly to ensure they can be done within the times dictated by external factors. Allowance is also made for data preparation, ancillary operations, and reruns.

The following factors are taken into account in scheduling a routine:

- the times at which source data is available;
- the time taken for data preparation or entry;
- the time for each processing run;
- the time for handling and distributing output documents; and
- the times by which output must be ready (deadlines).

Figure 8.16 is an elementary example of the scheduling of a number of jobs on the computer. Several points are worth noting in this example:

- The data entry is staggered so that only one type of source data is dealt with at a time although in some IS departments it is usual to mix the data preparation work. Much depends on the hardware used for data preparation or entry.
- A certain amount of computer processing is scheduled simultaneously, e.g. stock updating and invoice preparation. This is because the former demands input peripherals and the latter output peripherals.
- The handling of computer output and data entry are scheduled in only normal working hours (9 a.m.–1 p.m. and 2 p.m.–5p.m.).

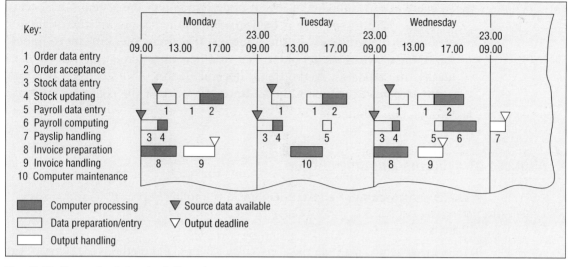

Fig. 8.16 Computer job-scheduling chart

- The computer and its operators work two shifts, i.e. from 9 a.m. to 11 p.m. in all.
- Scheduled computer maintenance is allowed for.

The activities in a routine may be scheduled either forward from a start point, i.e. when the source data becomes available, or backward from a deadline, i.e. when the final output is needed. The latter tends to apply to jobs that must be completed by the deadline; the former to jobs that are less urgent.

8.9 SYSTEMS DOCUMENTATION

The setting and maintenance of standards and documentation applies throughout the work of the systems analyst and programmer. These aims are particularly relevant to the designing of a system for a number of reasons:

- To facilitate designing the system by working to prescribed standards and having available clear descriptions of all the work covered so far. This implies that the documentation must be ongoing throughout the systems design and programming stages, and not left until the work has been fully completed.
- To maintain a good level of communication between everyone concerned with the system, i.e. designers, programmers, operators and users. Some persons are obviously more deeply involved with certain aspects than others but nevertheless they should all have access to whatever information is applicable to their involvement. This then allows for any misunderstandings or disagreements to surface before they become deeply entrenched in the system.

∎ To guard against subsequent loss of understanding of the system. This occurs only too easily when there is inadequate documentation and the originator of some of the work has left the company. It is sometimes difficult to understand one's own work, let alone that of others after some time has elapsed. Ideally the standards, methods and documentation should be such that any qualified person can smoothly continue the work of the colleague who has left.

∎ Aspects of systems documentation

The documentation of a system embraces the following:

Reports

To management explaining the aims, costs, savings, and methods of the proposed system. These, especially methods, should be couched in terms that managers understand: it is counter-productive to 'blind them with science' by including unnecessary jargon. Where jargon cannot be avoided, an explanation in clear English should be provided.

Instruction to users

∎ The contents and layout of each document on which source data is recorded.

∎ The times and dates at which the various source data are to be available, either as documents or keyed into the system by the users.

∎ The procedures for notifying source data errors detected by the system and their correction by users.

∎ Details of batching and control totals associated with source data.

∎ Explanations of all outputs from the system to the users; these are mainly printed documents but also include VDU displays.

∎ The procedure for obtaining on-request output (Section 1.3); this includes the actual information required and precisely when it is required, i.e. time and date or at a certain stage of the processing.

∎ The procedures for making on-line enquiries together with an explanation of the delays that might occur at unscheduled times.

Specifications

For programmers covering the runs to be programmed.

Instructions

For computer operating and control staff in relation to the running of operational jobs, acceptance of source data, usage of stored files, and the distribution of output.

354

■ Contents of a documented system

- A description of the purpose of each routine.
- A DFD or the equivalent for each routine.
- Specifications of all processing runs.
- Source data requirements and input procedures.
- Contents of logical files and database requirements.
- Layout and contents of all output prints and displays.
- Operating instructions for the processing runs.
- Systems amendment procedures.
- All aspects of user participation.

8.10 SYSTEMS DEVELOPMENT METHODOLOGIES

Several structured system development methods have been created since structured systems analysis and design started the movement towards a more systematic approach to systems development in the late 1970s. The essence of structured methods is to offer procedures to make software design modular and well constructed. The promise of structured software is improved maintenance and software reliability.

Most structured methods have the following components:

- A set of models usually expressed as diagrams which contain the specification and designs.
- Techniques for conducting analysis and specification.
- Guidelines and procedures for conducting the analysis and design.
- Procedures for managing the process of system development.

Three perspectives are present in many methods and these represent different approaches to systems analysis. These perspectives are:

1 *Data*: models which describe the data aspects of systems as found in Entity Relationship diagrams. Methods with a data perspective tend to emphasize data analysis and be more suitable for database implementations.

2 *Process*: these models describe the functions in the systems and the data passed between processes. Data Flow Diagrams are a typical means of representing process models, and methods with this perspective are more suitable for development of programs.

3 *Event*: these models describe the dynamic aspects of how things change within the system. Event models may be represented as entity life histories or state transition diagrams.

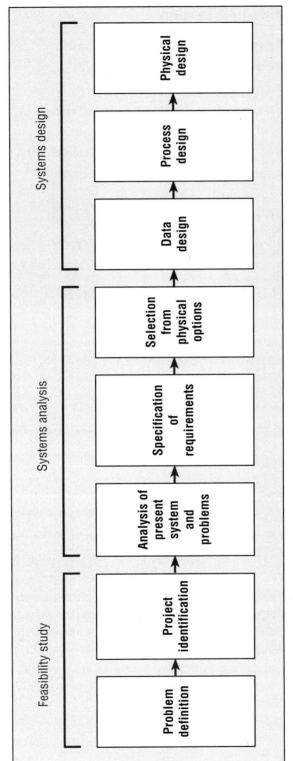

Fig. 8.17 Framework of SSADM

Many methods now incorporate all three perspectives so all aspects of a system – data, process communication, and event/control – are covered. More recently, object-oriented methods have attracted considerable attention. Object orientation essentially merges the three perspectives in one modelling component, the object.

There are a number of structured methods; two of the more important methods are described in the following sections.

SAQ 8.16 Give some examples of the data, processes and events which occur in the part of a banking system which processes cheques.

Solution

Typical data are the number of the cheque being processed, the identity of the bank and the identity of the customer. Typical processes are checking the validity of an account number and updating the account associated with a cheque. Typical events would be the transition from being a cheque which has been received to one which has been processed and archived.

Structured Systems Analysis and Design Method (SSADM)

SSADM is probably the most important method in the United Kingdom because it has been adopted by the UK government as a standard for systems development in the civil service. It was originally developed by LBMS (Learmonth and Burchett Management Systems) for the government's Central Computer and Telecommunications Agency (CCTA).

SSADM is a data-driven methodology which places emphasis on data modelling but also advises analysis and specification of process views with dataflow diagrams and behaviour using entity life histories. More recent versions have placed more emphasis on event behaviour modelling by entity life histories.

SSADM is structured in three phases: feasibility study, systems analysis, and systems design, as depicted in Figure 8.17. The first phase feasibility is optional to some extent, and is likely to be omitted for small projects which form part of a large development plan. Each of the three phases is sub-divided into a number of stages which in turn are sub-divided into a number of steps. The hierarchical structure of SSADM results in a limited area of work at the lower levels. SSADM comprises eight stages, about fifty steps, and about 230 tasks in all.

There follows a brief description of each of the eight stages.

First stage – problem definition

The aim of this stage is to obtain a precise definition of the overall problem for resolution by the system to be developed. Overviews are created of the present systems and data structure, and current problems are identified.

Second stage – project identification

This stage aims to create a number of options for dealing with the problems identified in the first stage. The options are then evaluated and formalized for inclusion in the feasibility report.

Third stage – analysis of present system and problems

This stage entails analysing the existing system and documenting it in the form of dataflow diagrams (DFDs) and logical data structures (LDSs).

Another goal is the creation of lists of the initial problems and of the requirements. This involves making a more refined identification of the problems found in the second stage.

Fourth stage – specification of requirements

In this stage the user requirements determined in the previous stage are defined more closely. A data structure is developed based on the LDSs created in the third stage. Audit, security and control aspects are defined and included to form the systems specification.

Fifth stage – selection from physical options

This stage involves the users and systems staff in selecting a suitable physical system. In most cases it is possible by this time to decide on the configuration of the hardware required and on the characteristics of the appropriate software. These requirements are related to the performance objectives, which are also defined in this stage.

Sixth stage – data design

Data structures for the proposed system are designed by combining the top-down view of the organization derived from the third stage with the bottom-up view of data groupings, i.e. composite logical data design.

Seventh stage – process design

This stage is carried out in conjunction with the data design stage, in particular it utilizes the composite logical data design. The logical processing associated with enquiries and updating is defined (all processes are seen as these). Following this the logical design is validated by means of a quality assurance review before proceeding with the physical design.

Eighth stage – physical design

The logical design is translated into programs and the database contents. The data dictionary is updated and the design tuned to meet performance objectives. Programs and system are tested. Operating instructions are created. Implementation plan is drawn up and the manual procedures are defined.

Other examples of structured methods are Jackson systems development (JSD), structured analysis and design technique (SADT), information systems analysis for change (ISAC), and structured systems analysis and structured design (SSA/SD).

Object-oriented methods

Object-oriented (OO) programming has been known for some time in languages such as Smalltalk which was developed in the 1970s. More recently object-oriented systems analysis and design methods have appeared and object orientation is being taken seriously in commercial systems development. As yet few business information systems are developed with OO methods; however, these methods may become increasingly important and major vendors of structured system development methods have now produced object-oriented versions.

The fundamental motivation of OO methods is to facilitate development by reuse of software. The idea is to create objects in class hierarchies. Objects in the upper levels of the hierarchy are very general and abstract. Lower-level objects are added by specialization, that is, the progressive addition of detail to higher-level objects. Upper-level classes can therefore be reused by specialization. An example of a class hierarchy is shown in Figure 8.18. All ships belong to the top-level class and have attributes of name and owner but little else. The subclasses of ship are passenger ship and cargo ship; these in turn are specialized into sub-subclasses, and add further details at each sub-class level. The question of 'what is an object?' revisits the 'what is an entity?' question. Fundamentally, objects model things in the outside world that have to become part of the computer system; these may be physical or conceptual objects. Furthermore, objects may vary from more data-like things such as 'order' and 'bank-account' to active agents such as 'clerk', 'salesperson', etc.

The following are the main concepts in object-oriented development:

Objects

These are a composite of data and process. It is essentially a combination of entities with processes, procedures or functions which would be specified separately in methods such as structured analysis. Objects are composed of attributes, as

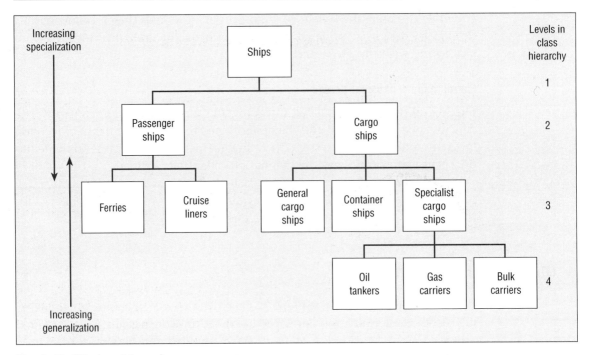

Fig. 8.18 OO class hierarchy

in entity attributes, and operations which are actions that operate on the data attributes. In OO analysis procedures are usually called 'methods' and sometimes referred to as 'services'.

SAQ 8.17 Can you think of any objects which might be contained in a purchasing system which processes telephone orders from customers for clothing and then dispatches the orders?

Solution

The following would be typical objects: customer, order, clothing item, dispatch note and credit card.

Classes

Objects are organized in classes. Lower-level classes are specializations of higher-level classes, e.g. pro forma order is a specialization on a general order. Higher-level classes are generalizations of lower-level ones, e.g. ship is a generalization of cargo ship, bulk carrier, and oil tanker.

Attributes

Data items which describe the object. Similar to entity attributes.

Operations

Actions which create, change or delete the object's attributes.

Methods (also services)

Procedures, algorithms or calculations which carry out the activity inside objects. Methods are the processing part of objects and may be specified in structured English or pseudo-code.

Messages

Objects communicate via messages. These may contain data to be processed by the object or requests for services, i.e. messages which trigger an object to carry out some processing.

Inheritance

Lower-level classes inherit properties (attributes, operations, methods) from higher-level classes; hence bulk carriers inherit all the properties of their super-class ship (e.g. name, country of registration, tonnage). By organizing objects in class hierarchies, lower-level objects can receive properties from higher-level objects. This facilitates reuse of more general higher-level objects by specialization which creates a new class by addition of further detail. Two forms of inheritance may be supported: *hierarchical*, in which a child object can inherit only from its parent object; or *multiple inheritance* when an object can inherit properties from several objects. Multiple inheritance may result in 'polymorphism' with one component having different properties in several new locations, as it is specialized in child objects.

Encapsulation

Encapsulation is the concept that objects should hide their internal contents from other system components to improve maintainability. This is similar to the idea of local variables in procedures as found in many (non-object-oriented) programming languages. By making part of the design local, the overall complexity of the system is reduced and objects limit the volatility of change in the system. The encapsulated parts of objects are hidden to insulate them from the effects of modifications to higher-level objects.

One virtue of OO methods is that the specification becomes the implementation. Analysis and design proceed by gradual addition of detail to the objects. The system is modelled as a collection of objects connected by message-passing channels. Objects pass messages to each other to request a service, e.g. to update an object's attributes, query its status, request a report, etc. More detail is added to objects until the specification becomes sufficiently detailed to be programmed in an object-oriented language such as Smalltalk, or C++.

As objects are a mixture of programs and datastructures, their storage is a problem for relational databases. This has given rise to object-oriented databases that store data and programs as one component, i.e. object-oriented storage and retrieval.

Object-oriented methods

Since 1996 there has been an increase in the use of object-oriented technology such as the programming language Java. However, one of the major problems that has faced companies who wish to use such methods has been the variety of development methods which have been available. This, allied to the fact that the proponents of the methods often devoted their time to religious wars involving the methods rather than proselytizing object-orientation, has meant that the spread of object-oriented methods has been a little slower than one would expect.

Happily this period is over. A new development notation known as UML has virtually taken over from all the previous methods. UML was developed by a team of people who were intimately involved with three of the most popular object-oriented methods prior to UML: OOSE, OMT and Booch. It is worth stressing that UML is a notation, or rather a set of notations; the way that the notations are put together to form a development method is up to the developer, although a number of well-defined methods are already in existence.

The components of UML

The first component of UML and one which is usually dealt with early in a project is the use case. A use case is a description of system behaviour which involves entities known as actors. These are usually human agents who carry out some action to which a computer system has to respond. An example of a use case is shown as Figure 8.19.

This shows a single agent in a banking system (a customer) and three use cases in which the customer participates. They are use cases which:

■ order a bank statement for the customer;

■ query the customer's balance;

■ request a statement for the customer's account.

Each of these use cases uses another use case called *Validate user*. This checks that the user's supplied personal details match those provided by the user.

A use case represents a functional view of a system. There are a number of other views of a system which can be expressed in UML. In the previous section we described how the notion of class is an important one and it will come as no surprise that it has a notation for describing such entities. Figure 8.20 shows a typical class in UML; it is represented as a compartmentalized rectangle with the top compartment containing the name of the class, the middle compartment

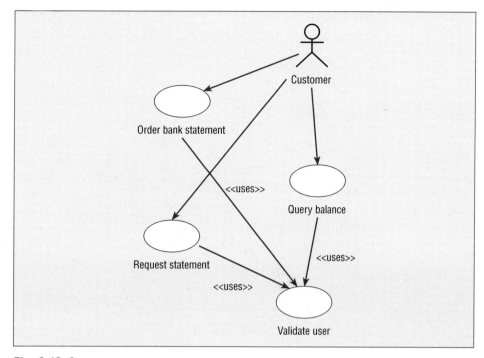

Fig. 8.19 A use case

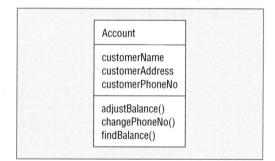

Fig. 8.20 A class box

contains the attributes of the class and the bottom compartment the methods of the class. Figure 8.20 represents a highly simplified view of a class which is used to represent bank customers (there would normally be more attributes and operations). It shows that there are three attributes which represent the name, address and phone number of the customer and operations which adjust the customer's bank balance, change his or her phone number and find the balance of his or her account. The name of the class is *Account*.

Another notation in UML describes the interaction between objects in a system which are defined by classes as in Figure 8.20.

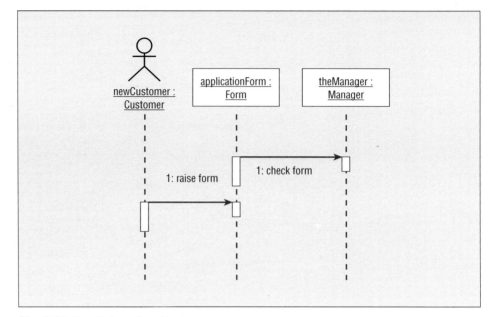

Fig. 8.21 A collaboration diagram

This notation is known as a sequence diagram; it shows the interaction between objects defined by classes. A short extract from one such diagram is shown as Figure 8.21.

It shows a new customer of a bank raising a form which contains his or her application. This creates an object *applicationForm* defined by the class *Form*, this form is then checked by a manager and is passed on to the *remainder* of the collaboration diagram which is not shown here in Figure 8.21.

An important component of UML is the diagram which represents the relationship between the classes in a system. These so-called class diagrams show how each class in the system is connected to other classes. Figure 8.22 shows an example of such a diagram.

The arrow represents the fact that one class will be associated with what is known as an association relationship with another class. For example, the class *Customer* is associated with the class *Account* by virtue of the fact that a customer can own a number of accounts. The number 1 at the base of the arrowed line indicates that an account is associated with a single customer, the string 0..* at the other end represents the fact that a customer will be associated with zero or more accounts.

The class *Account* is also related to the class *Transaction* which represents financial transactions against the account. Accounts will be associated with zero or more transactions and transactions are associated with just one account.

The developer will build up a specification of the system using these notations. There are a number of stages; one possible sequence is:

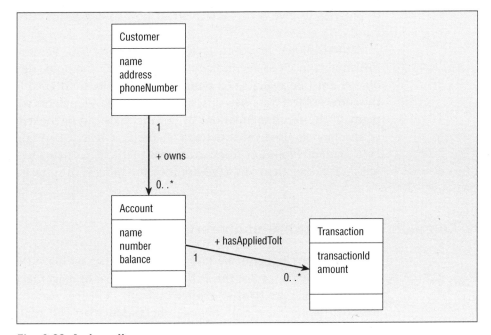

Fig. 8.22 A class diagram

- Read the specification of the system produced by a customer. This specification is usually written in natural language.

- From this specification build up a series of use cases.

- From these use cases identify the classes that are involved and specify them using a set of class diagrams.

- Finish the specification by developing the collaboration diagrams. These diagrams could then act as a check that enough information has been collected in the other diagrams.

Conclusions

Object-oriented methods are still immature. Whether they develop further and replace structured methods remains to be seen but they have some formidable obstacles to overcome. One is conservatism in system development practice. Structured methods were developed in the early 1980s yet they are only now becoming commonplace in information systems development (estimates vary: between 20–35 per cent of UK companies use some sort of structured method). Acceptance of OO methods is therefore liable to take some time. Furthermore managers will be reluctant to invest in a new method unless they can see the payoff. Object orientation only really offers a significant advantage over structured methods if reuse becomes a reality. For this to happen OO methods have

365

to progress to give better advice about how to design for reuse and develop by reuse.

Methods alone, however, rarely solve the problem. Reuse entails significant investment in CASE technology to support OO methods. Reusable libraries of objects must be available on demand. CASE tools are needed to find and match the correct object for a new application. These developments will take some time, many of the recommendations of OO methods can be looked upon as revision of the models used by structured methods. Object orientation may therefore be a gradual evolution from structured methods to new OO variants of those methods rather than any revolution in the process of software development.

▪ Computer-aided software engineering (CASE)

Structured development methods are hard to use without a computer support tool. A simple reason for this is the large number of diagrams produced by such methods which are hard to change by hand. A deeper reason is that structured methods alone may not increase the productivity of systems analysts and designers. However, computer tools which generated code automatically to form specifications would promise a large increase in productivity by eliminating programming. These needs have given rise to CASE tools.

CASE tools are a set of diagram editors, list handlers and possibly code generators that communicate with a common database which stores the specification. One classification is for tools which support different parts of the life cycle.

Upper CASE

These tools help system analysis and specification. Typically they are diagram editors for dataflow diagrams, entity relationships, entity life history, etc. List processing tools may be included for recording data structures and data analysis. Some semi-automatic support may be provided for analysis such as normalization, and some tools cross-check different parts of the specification for consistency, e.g. entities are cross-referenced to data stores on DFDs.

Lower CASE

These tools support system design and programming. They are diagram editors for structure charts, syntax-directed editors for pseudo-code, or programming languages and automatic code generators. More advanced lower CASE tools automate some of the designer's work. Transformation tools automatically, or more frequently semi-automatically, change a specification into a design. Code generators are rarely completely automatic but most create the program structure from a specification (e.g. procedure calls, modules hierarchy), while the detail of algorithms and calculations has to be entered as a pseudo-code or in the chosen target language (e.g. COBOL, C).

CASE tools usually come as a collection which either supports a particular structured method directly, for instance Information Engineering Factory (IEF TM), or the information engineering methodology. Alternatively tools may be open-ended and support a variety of methods but the trade-off is that conformance to any one method is sacrificed. CASE tools can be classified into different types according to their approach to method support and whether they are open tool sets or not.

CASE shells

These are templates or the basic building blocks for CASE which may come with ready-made tools or the purchaser may develop their own tools. The shell provides the basic facilities such as a database, a means of providing inter-tool communication, and possibly a set of user interface services. CASE shells are open environments which may be tailored by a user for their purposes. Shells do not usually support a specific method.

CASE environments

These are a more complete product comprising a database, a set of tools, the user interface and support for inter-tool communication. CASE environments, however, may be method specific, in which case they have a set of rules and procedures embedded within them to guide and sometimes enforce use of a particular method. Alternatively, they can be open environments which support a generic set of diagrammers and other tools which can then be customized.

If a CASE tool does support a method then it should do so flexibly. Furthermore the tool should support the analyst-designer's way of working rather than rigid conformance to a particular method. There are often conflicts between the management view of CASE as an effective way of policing conformance with a structured method and the designer's way of working, which is less rigid. Many CASE tools have not been successful and consequently either not been used at all or at best their use has been as documentation aids after the system has been developed.

The following is a list of desirable features of CASE tools:

- The tools should cover all phases of the life cycle, both upper and lower CASE.
- Code generation facilities are essential.
- Tools should automate development steps where possible, so that the transformation from a specification to a design is formal, i.e. if all the steps are predictable, then it can be automated.
- Tools should be usable in a flexible manner so that different approaches to system development (e.g. prototyping) can be followed.
- Diagram tools should be configurable so that the tool can be customized to the user's favourite diagramming convention.

■ Tools should be flexible to support the analyst's way of working, e.g. if the analyst wants to work bottom up rather than top down, then the tool should allow this.

■ Tools should prevent errors where possible by pre-validation, i.e. allowing the user to enter only correct choices, or if this is not possible, by consistency checking and *post hoc* validation.

■ The user interface needs to support high-resolution graphics and direct manipulation.

■ Method advice and guidance may be necessary.

■ A central database is necessary to store specifications and designs.

The need for a database is vital as CASE tools have to communicate with other data stores in the organization such as data dictionaries, documentation databases, and program source code files. This need has given rise to the concept of a super-development database for all these needs, the repository.

Repositories

These not only fulfil the database need as a common file store for specifications, designs, program code documentation, data dictionary, etc., but also standardize the storage of specifications. CASE tools produced by separate vendors are rarely compatible with each other or with data dictionaries already installed in many organizations. The CASE tools could lock development to a particular method, or prevent interchange of data between methods. Repositories attempt to overcome these problems by providing a database and a standard means of storing and accessing specification, designs and all the other related information produced during software development.

Repositories should have the ability to store specifications from a wide variety of methods and provide support services expected of CASE shells. To date, repositories that have been produced – IBM A/D Cycle and DEC's CDD – have not met with widespread acceptance.

Intelligent CASE

The current generation of CASE tools are largely passive and do not support the designer. Some tools give advice and help organize work by suggesting procedures to follow, but the advice is limited. Future tools may be more active because they have embedded mini-expert systems that contain design for good design, and method guidance. Some tools may also have knowledge of certain types of applications and hence may be able to guide fact acquisition by suggesting what type of information needs to be acquired. Intelligence may be used to reason about the specification already captured, so that the tool could automatically detect imperfections in the design and suggest improvements. Indeed, one way in which formal methods may be brought into more widespread practice is to

368

hide the formalism inside CASE tools that reason about a specification and then guide the developer to fix inconsistencies and inaccuracies by a dialogue of questions and examples. Intelligent CASE tools have the potential to improve much software development practice and realize the productivity gains which have so far eluded current systems. Currently these tools are still at the research stage.

CASE and reuse

A further advance in CASE tools could significantly improve the prospects for software reuse. Reuse is practised today in a limited fashion. For instance, many programmers re-use data definitions in COBOL programs by loading the file definitions from an existing library. However, reuse promises much more in productivity gains if libraries of programs, modules and other software components could be made generally available. Software reuse is a subject which cannot be treated in more depth here; however, the CASE-related problems are about providing a reusable library of software components and some means of retrieving them. Repository technology is necessary, although to date few reuse repositories have been created. One barrier is the critical mass problem, i.e. a library is only useful once it contains a sufficient number of useful items. Other barriers are the lack of management incentive for development by rules, and technical problems in designing software components that are easy to re-use (see Reference 8.3 for further information on CASE).

EXERCISES

Exercise 8.1   Job costing

On the basis of the situation and activities described below prepare an output analysis chart along the lines of Figure 8.5 (page 303).

At the end of each week job progress forms are received from the factory. Each of these holds the following data:

- Job number, week number, material code and the corresponding quantity used for the job, employee number and the hours worked in the week.
- Each material has a standard cost up to £99.99, and is coded with picture AA999.
- The employees each have a normal rate and an overtime rate depending on their trade and grade.
- The current jobs are numbered from 1 to 999, and the weeks from 1 to 53.
- The quantity of a material used on a job in a week is not more than 800 units, and is measured to the nearest tenth.
- The employees are numbered from 1,000 upwards, the highest number being 4,500. The hours worked are recorded to the nearest quarter below. The normal and overtime rates are expressed in pence per hour; the maximum is 350 pence per hour.
- There are forty trades (picture AA) with up to six grades each (coded 1–6).

It is required to prepare a job cost summary each week. This is to show for each job (in job number sequence):

(a) the material cost this week and to date;

(b) the labour cost this week and to date;

(c) the total (material labour) cost this week and to date.

No total exceeds £5,000 expressed to the nearest penny.

Exercise 8.2   Information system development

The financial director of your organization has asked for a report on the way in which an information system development project may be managed.

You are required to draft a report to your financial director briefly describing the phases which make up an information system development project.

You may base your answer upon an actual information system development methodology (e.g. SSADM) or upon the generalized 'system life cycle'.

(CIMA, stage 2, May 1992)

Exercise 8.3   Sales orders

Harvey Agricultural Ltd is a cattle-food merchant with a manual accounting system in which sales orders and invoicing are carried out as follows:

■ Orders are received by the sales order/invoicing department, mainly by telephone.

■ On receipt, an official order form is prepared and passed to the sales ledger clerk to be vetted for creditworthiness of the customer.

■ From there it is passed to the dispatch department to prepare prenumbered dispatch documents and assemble the goods for dispatch.

■ The dispatch department then prepares a delivery schedule for delivering the various orders by its own transport. After delivery, a copy of each delivery note, signed by the customer, is returned to the dispatch department for matching with the original.

■ The dispatch department accounts for all dispatch note numbers and passes executed notes to the sales order/invoicing department for invoice typing.

■ The invoicing department dispatches top copy invoices direct to the customers and the second copies go to the ledger clerk for entry in the sales journal and posting to the sales ledger.

Requirements:

(a) Prepare an overview manual procedures flowchart showing, in outline, the activities of the various departments described above and the movement of documents.

(b) Compare and contrast the way the operations depicted in the flowchart are carried out manually with the way they would be carried out in a computerized order entry and invoicing system with remote terminals in each of the departments involved.

(c) Comment on the ways in which you consider that a computerized system would provide greater control over the transaction cycle.

(ICAEW PE1, Aud, sys. & DP, Nov. 1988)

Exercise 8.4   Customer's orders

A company supplies frames to the building trade. The complete frames have five components, each of which is available separately. The frames are available in ten different heights and twelve different widths. Orders are received for separate components and complete frames, the latter orders only being met if all five components for that frame are in stock. If a given size frame is not available the customer is usually prepared to accept the frame having the next larger or smaller height or width. Prices vary according to height and width.

Describe, using appropriate diagrams, the following aspects of a computer program to determine how a customer's order may be satisfied:

(a) the method of allocating complete frames and components to customers' orders;

(b) the data structures used; and

(c) the invoicing.

How would you modify the data structures to include stock reorder levels and quantities?

(BCS part I, Gen. paper II, April 1986)

Exercise 8.5   Seat reservation system

A travel organization specializes in charter flights. The seats aboard their aircraft have the following attributes:

- row number;
- seat number;
- window, aisle or centre; and
- non-smoking or smoking.

Customers are allowed to reserve seats by specifying:

- the number of adjacent seats required and non-smoking or smoking; *or*
- a combination of the attributes of window/aisle/centre and non-smoking/smoking.

Reservations are processed on a 'first come, first served' basis.

Requirements:

(a) By means of an appropriate diagram, or otherwise, describe a computer program that will input a customer's choice, select the best fit to that choice, and output boarding cards detailing the seat allocation.

(b) Describe in detail all data structures used.

(BCS part I, Gen. paper I, April 1987)

Exercise 8.6   Entity relationships

Construct an entity relationship diagram showing the relationships between users, data items, programs, logical files and records. Prepare a list of reports that might stem from these relationships, giving briefly their user and purpose.

Exercise 8.7   Sales order entry

Your company is a wholesaler supplying a wide range of DIY products to supermarkets, shops and DIY centres. Orders are taken by sales representatives, received by post and over the phone.

A minicomputer with on-line access and of adequate power is currently in use.

Describe the procedures and draw a dataflow diagram of the sales order entry system.

Exercise 8.8   Library system

A public library controls its stock of books and lending records by means of a computer system.

Specify the logical files, input data, processing and outputs that you think appropriate in such a system.

Exercise 8.9   Loan accounting

A finance company uses a manual accounting system to handle its 1,000 loan agreements. It is planned to expand these to 5,000 agreements over the next two years. A decision has therefore been taken to install a computerized loan ledger to ensure controlled growth.

The loans, sometimes secured, are repayable by either twelve or twenty-four equal monthly instalments. One borrower may have several loan agreements. Because the capital element of each repayment increases with time, the respective elements of each repayment are scheduled on each loan ledger account at the outset. All loan repayment instalments are made by standing order direct to the loan company's bank account.

Requirements:

(a) Describe the features and capabilities you would be looking for in selecting suitable software for the above system.

(b) State the data fields which you consider should go to make up a loan agreement record in the new computer loan ledger.

(ICAEW, PE1, Aud, sys. & DP, May 1988)

Exercise 8.10   Systems development methodologies

When developing a new information system most methodologies follow the traditional life-cycle approach.

Using a methodology of your choice, describe:

(a) the stages included;

(b) the deliverables for each stage;

(c) the quality assurance techniques for each stage.

(BCS, GPI, April 1992)

Exercise 8.11   Prototyping/structured design

Describe the use of the following development strategies at the design stage; include in your answer the advantages and disadvantages and the deliverables:

(a) prototyping;

(b) structured systems analysis and design.

(BCS, 1C, April 1992)

Exercise 8.12   Object-oriented design

Object-oriented design is currently being proclaimed as the solution to the complexity problems of large software design.

Discuss the arguments for and against this position.

(BCS, 2D, April 1992)

Exercise 8.13   Human–computer interface

You have been asked to brief management on the opportunities presented by new forms of human–computer interface (HCI).

Draft a short report for your manager describing four data input mechanisms that have become available recently and/or that are expected to become available in the next few years, and identify how each one might be particularly useful in bridging the human–computer interface.

(BCS, 1D, April 1992)

Exercise 8.14   CASE

The use of computer-aided software engineering (CASE) has not increased as rapidly as had been predicted.

Requirements:

(a) Identify the important features of CASE that would justify its use.

(b) Discuss the likely problems of the use of CASE for software development, particularly in smaller organizations.

(BCS, 1C, April 1992)

Outline solutions to exercises

Solution 8.1   Refer to Figure 8.23.

Solution 8.2   Refer to Chapters 7 and 8.

Solution 8.3
(a) Refer to Figure 8.24 (page 375).

(b) Orders entered directly into system via terminals and stored in file; official order forms are unnecessary since these are simulated on the screen.

■ The computer vets orders automatically as it can deduce customer's creditworthiness from data in the stored sales ledger.

■ Computer prints dispatch documents for dispatch department, and stores a record of each set.

Ref. no.	Data item	Picture	J	F1	F2	F3	F4	P1
1	Job No.	999	✔				✔	F4
2	Week No.	99	✔					J
3	Material code	AA999	✔	✔				
4	Material qty	999.9	✔					
5	Employee No.	9999	✔		✔			
6	Hours worked	99.99	✔					
7	Trade code	AA			✔	✔		
8	Grade No.	9			✔	✔		
9	Normal rate	999				✔		
10	Overtime rate	999				✔		
11	Material std. cost	99.99		✔				
12	Material cost this week	9999.99						c 4.11
13	Labour cost this week	9999.99						c 6.9.10
14	Total cost this week	9999.99						c 12.13
15	Material cost to date	9999.99					c 12.15	c 12.15
16	Labour cost to date	9999.99					c 13.16	c 13.16
17	Total cost to date	9999.99						c 15.16

Column headers (diagonal): Job progress form, Material file, Employee file, Trade/grade file, Job costs file, Job costs summary

Fig. 8.23 Output analysis chart for Exercise 8.1

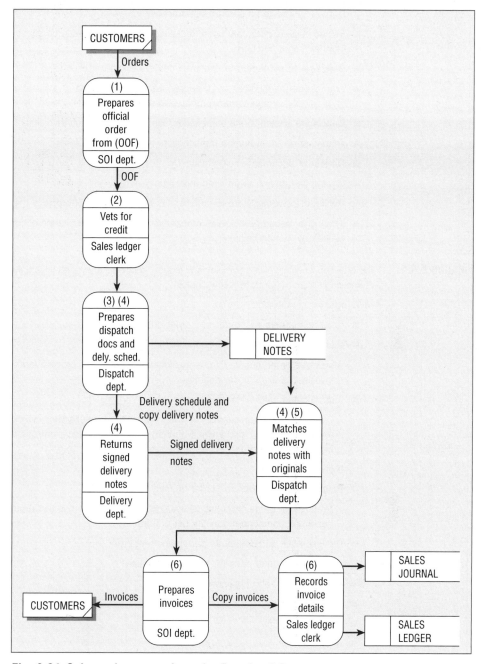

Fig. 8.24 Sales orders procedures for Exercise 8.3

- Delivery schedule prepared as at present.

- Signed delivery notes are used for ticking-off the stored dispatch documents simply by entering their serial numbers.

- Computer has a record of executed deliveries and corresponding orders.

- From these the invoices are prepared automatically using (i) a file of cattle food prices, and (ii) a file of customers' names and addresses.

- Details are posted to the sales ledger file concurrently with invoicing.

(c) Avoidance of errors in human transcription, typing and calculating.

- No fear of documents being mislaid.

- Control figures are determined more easily.

- Vetting of customer's creditworthiness is absolute, i.e. not prone to misunderstandings.

- The higher speed of the sales orders cycle minimizes the chance of confusion, especially as caused by human frailties.

Solution 8.4

This question would involve a complex and lengthy solution if answered taking alternatives into account; a simplified solution therefore follows below.

(a) If order for component:

- Enter component no., order no., order quantity, customer no.

- If unallocated stock $\geq$ order quantity:

  subtract order quantity from unallocated stock, add order quantity to allocated stock, invoice quantity = order quantity.

- If unallocated stock < order quantity:

  Outstanding order quantity = order quantity − unallocated stock, invoice quantity = unallocated stock, add invoice quantity to allocated stock, put outstanding order quantity into outstanding order record, make unallocated stock zero.

  If order for frame:

- Enter frame no., order no., order quantity, customer no.

  Break down frame order into its five components, i.e. component requirement = frame order quantity × component quantity per unit frame.

- Process each component as per component order.

(b) Component stock and price file (frame file similar):

- component no. and description;

- allocated stock;

- unallocated stock;

- reorder level;

- reorder quantity; and

- price per unit.

Frame constituents file:

- frame no.;
- component no.; ⎫ for each of five components.
- quantity per frame; ⎭

Invoicing file (outstanding orders file similar):

- customer no.;
- order no.;
- date of order;
- frame/component no.; and
- invoice quantity.

(c) Read invoice record:

- Look up customer name and address, print headings on invoice.
- For each item:

  Look up price and extend by invoice quantity = invoice item value.

  Print item line and add item value to invoice value.
- After last item:

  Print invoice value. Calculate discount, carriage, VAT and print along with net amount.

Solution 8.5   (a) Refer to Figures 8.25–8.28; these relate to the chosen flight. Figures 8.25 and 8.26 are conventional flowcharts for the two alternative types of reservations. Figure 8.26 is segregated into subroutines as indicated by the coloured boxes. Figure 8.27 depicts the interrelationships of the subroutines and lists the constituent modules. Figure 8.28 shows the processing within each of the modules.

(b) Seat availability file:

- flight number;
- row number;
- seat number (in row);
- seat type (window, aisle or centre); and
- status (reserved, earmarked or unreserved).

Row area file:

- flight number;
- row number; and
- area (smoking or non-smoking).

Individual requirement:

- seat type;
- area; and
- flight number.

Block requirement:

- flight number;
- area;
- number of seats.

Boarding cards:

- flight number;
- date of flight;
- boarding time;
- airport;
- row number;
- seat number(s).

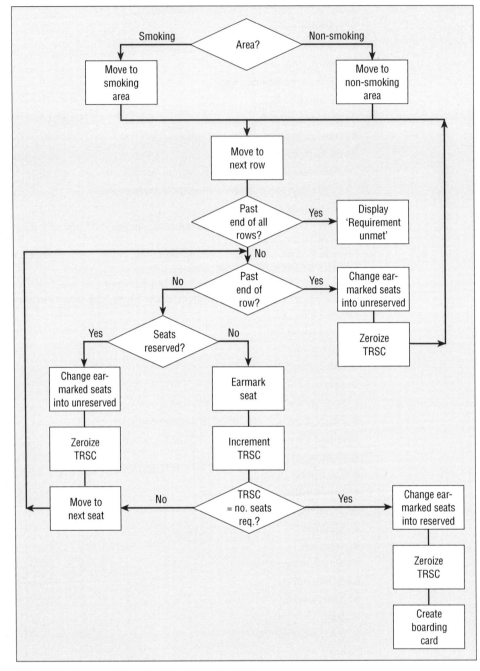

Fig. 8.25 Process flowchart for Exercise 8.5

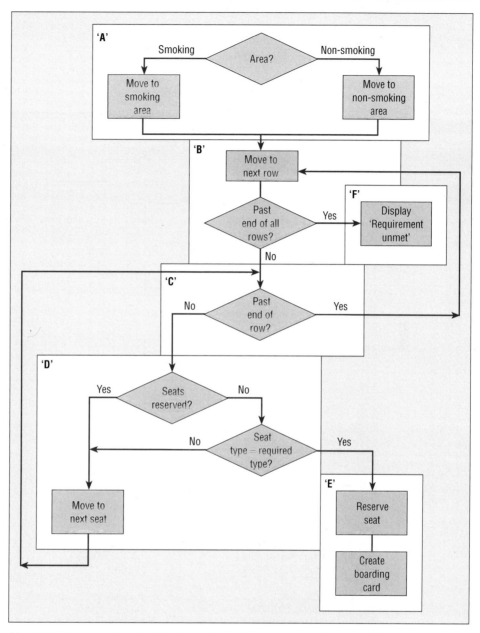

Fig. 8.26 Process flowchart for Exercise 8.5 segregated into subroutines (capital letters are subroutine names)

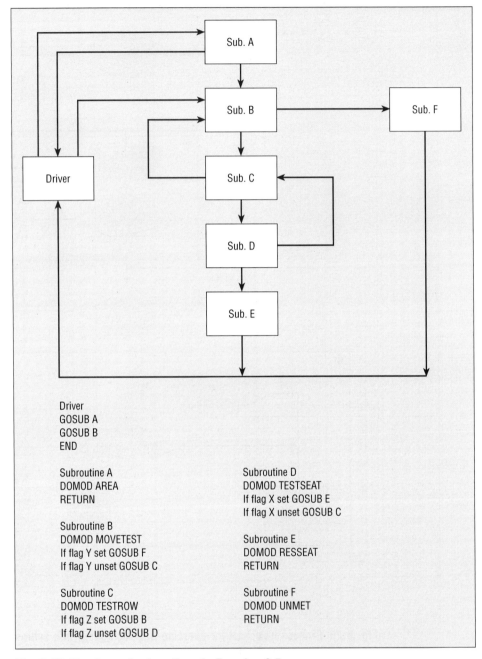

Driver
GOSUB A
GOSUB B
END

Subroutine A
DOMOD AREA
RETURN

Subroutine B
DOMOD MOVETEST
If flag Y set GOSUB F
If flag Y unset GOSUB C

Subroutine C
DOMOD TESTROW
If flag Z set GOSUB B
If flag Z unset GOSUB D

Subroutine D
DOMOD TESTSEAT
If flag X set GOSUB E
If flag X unset GOSUB C

Subroutine E
DOMOD RESSEAT
RETURN

Subroutine F
DOMOD UNMET
RETURN

Fig. 8.27 Structure of subroutines in Exercise 8.5

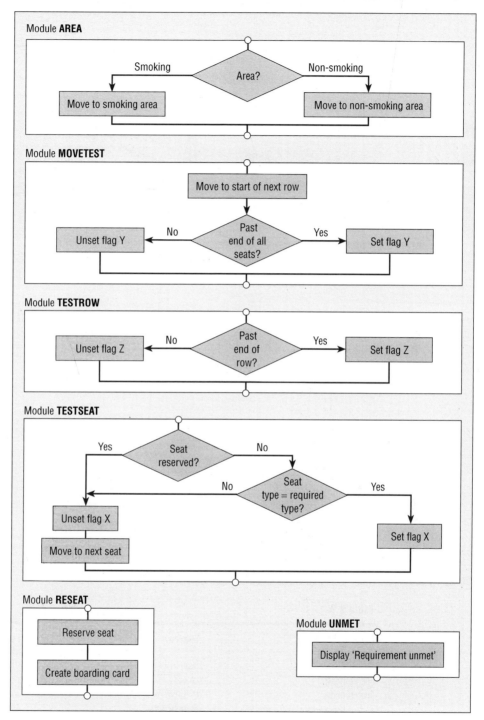

Fig. 8.28 Modules in Exercise 8.5

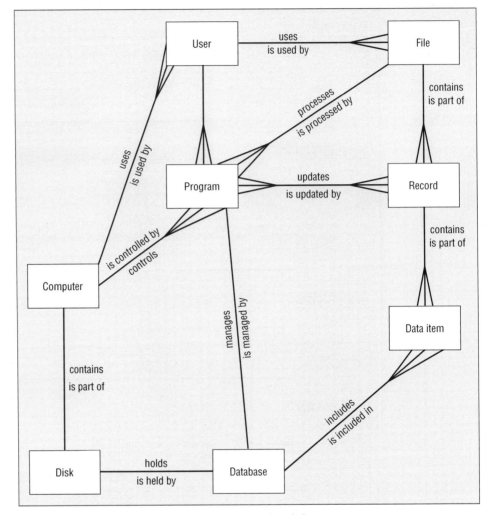

Fig. 8.29 Entity relationship diagram for Exercise 8.6

Solution 8.6   Refer to Figure 8.29, and Table 8.3 below.

Table 8.3

Report	Used by	Purpose
List of files	User	Checking existence
File contents	User	Applications development
File contents	Programmer	Program writing
List of programs	Programmer	Control of software
Program statements	Programmer	Program updating
Data items in database	Database administrator	Control of database
List of computer users	DP manager	Computer job costing

Solution 8.7   *General points*

Pharmaceutical/medical products are extensive in number and hence an early problem is identifying the customer's precise requirement. The products are obviously code-numbered but nevertheless measures must be taken to minimize errors. Possibilities are:

■ OMR order forms (negates in-line entry, however);

■ product lists on screen for selection by mouse and pointer, touch-screen or action codes; and

■ validation checks such as echo checks, range checks and probability checks;

■ check digits.

Procedure (see Figure 8.30)

■ Each order item validated against product master file.

■ Order quantity compared with unallocated stock (similarly to Solution 8.4), and recorded in invoice and to-follow files.

■ Shortages are held in to-follow file until next cycle when they receive priority (by which time more stock may have arrived).

■ Invoice set is prepared prior to dispatch as quantities are known to be available, i.e. packing list and delivery note.

■ During the invoicing run, the sales ledger is updated for subsequent use for statements and accounting.

Solution 8.8   *General points*

■ The library would be computerized and have machine-readable, i.e. bar-coded, books and members' tickets (plastic cards). Each book would have a book identification number (BIN), each publication, edition of a book, would have an international standard book number (ISBN) and a classification code, i.e. subject of the contents.

■ The BIN of a book is printed in a label stuck inside the front cover in the form of a bar code. This gives complete identification of the individual copy itself not just of the edition of the book, since this is the purpose of the ISBN.

■ Each member of the library would have a ticket on which is printed his or her member number in bar-code form. These are used merely for borrowing books and are not handed over at this time but retained by the member. Control over the number of books in the hands of a member at a time could be by the computer checking at each take-out time but this would be unusual and so is not included here. Members can reserve a book for future borrowing.

The four types of procedures considered here are acquisitions, loans, returns, disposals and reservations.

Logical files

Book master file (BMF):
    BIN
    ISBN
    Author name
    Title
    Classification code
    Date of acquisition
Member master file (MMF):
    Member number
    Name and address
    Date joined
    Ticket number (for reference only)

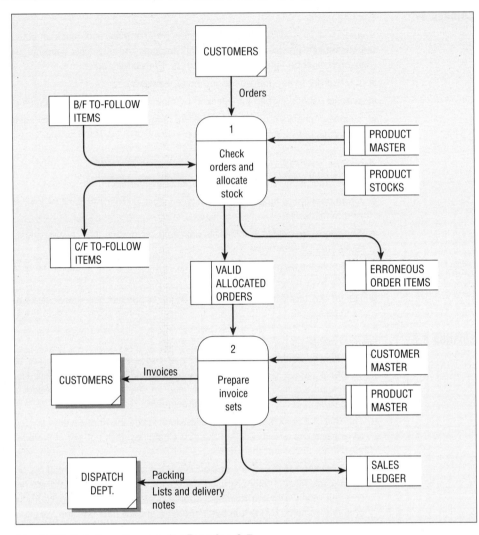

Fig. 8.30 Dataflow diagram for Exercise 8.7

Loans file (LF):
  BIN
  Member number
  Date due for return
Reservations file (RF):
  Member number
  BIN
  Date reserved
Loans analysis file (LAF):
  BIN
  Number of loans this year to date
  Number of loans last year
  Number of loans pre-last year

Inputs:
  Acquisition – all data items in BMF
  Loan – member no., BIN, date due (automatic)
  Return – BIN
  Disposal – BIN
  Reservation – member no., BIN, date reserved (automatic)
Processing:
  Acquisition – create new record in BMF
  Loan – create record in LF, update LAF, delete record from RF if present
  Return – delete record from LF, check whether in RF
  Disposal – delete record from BMF
  Reservation – create record in RF
Outputs:
  Overdue loan list (for reminders)
  Reservations available list (for notifications)
  Stocktaking list (in classification code or any other order)
  Loans analyses (according to ISBN/BIN, classification code/BIN)

Solution 8.9

(a) The software must be capable of the following:

- accepting, and checking loan repayments transmitted from the borrower's banks via EFT;
- updating the loan records in the loan ledger with repayments and new loan agreements;
- providing control totals and audit figures;
- searching for a specific record, e.g. by using the loan account number, bank account number, borrower's name, etc.;
- searching for records of a certain type, e.g. overdue accounts, large loans, unsecured loans, etc.;
- producing reports showing various lists, tabulations and analyses based on the current loan situation;
- analysing the cash input flow over the next twenty-four months so as to assist with decisions regarding the future pattern of loans.

(b) The following data items would form the records in the loan ledger. It is probable that even if a DBMS is not used, there would be several files, e.g. a loan file, a repayment file, and a name and address file.

- Loan account number
- Borrower's name and address.
- Secured/unsecured designation.
- Number of instalments.
- Date of agreement, i.e. state of loan.
- Borrower's bank account number.
- Initial capital element of loan.
- Initial interest element of loan.
- Current capital element outstanding.
- Current interest element outstanding.

Also for each repayment:

- Date of repayment.
- Amount repaid.
- Transaction number.

385

Solution 8.10   Refer to Section 8.10.

Solution 8.11   (a) Refer to Section 7.9.

                      (b) Refer to Sections 6.1 and 8.6.

Solution 8.12   Refer to Section 8.10.

Solution 8.13   Refer to Sections 4.1 and 8.5.

Solution 8.14   Refer to Section 8.10.

References and further reading

8.1 Avison, D.A. *Information Systems Development* (McGraw-Hill, 1992).

8.2 Bell, D. *The Essence of Program Design* (Prentice Hall, 1997).

8.3 Bergin, T.J. *Computer-aided Software Engineering* (Idea group, 1993).

8.4 Bernstein, T. *Internet Security for Business* (John Wiley, 1996).

8.5 Booch, G. *Object-oriented Analysis and Design with Applications* (Benjamin-Cummings, 1994).

8.6 Brown, D. *An Introduction to Object-oriented Analysis* (John Wiley, 1997).

8.7 Carroll, J.M. *Computer Security* (Butterworth Heinemann, 1996).

8.8 Castano, S. *Database Security* (Addison Wesley, 1995).

8.9 Gollman, D. *Computer Security* (John Wiley, 1998).

8.10 Haramundanis, K. *The Art of Technical Documentation* (Digital Press, 1998).

8.11 Horton, W.K. *Designing and Writing Online Documentation* (John Wiley, 1994).

8.12 Martin, J. *Object-oriented Methods* (Prentice Hall, 1996).

8.13 Penfold, R.R.C. *Computer Security: Businesses at Risk* (Robert Hale, 1998).

8.14 Pooley, R. and Stevens, P. *Using UML* (Addison Wesley, 1999).

8.15 Pressman, R.S. *Manager's Guide to Software Engineering* (McGraw-Hill, 1996).

8.16 Quatrani, T. *Visual Modelling with Visual Rose and UML* (Addison Wesley, 1998).

8.17 Russell, D. *Computer Security Basics* (O'Reilly, 1991).

8.18 Schultz, S.I. *Digital Technical Documentation Handbook* (Digital Press, 1992).

8.19 Senn, J.A. *Analysis and Design of Information Systems* (McGraw-Hill, 1990).

8.20 Sinclair, I. *Essentials of Computer Security* (Bernard Babani, 1997).

8.21 Skidmore, S. *Introducing Systems Design* (MacMillan, 1996).

8.22 Winograd, T. *Bringing Design to Software* (Addison Wesley, 1996).

9 Systems implementation

AIMS

After reading this chapter you should:

■ be familiar with the main categories of IS staff;

■ understand the function of systems testing;

■ be familiar with the main processes that comprise system changeover;

■ be familiar with the main processes that comprise system appraisal and maintenance;

■ understand how cost savings are calculated for a new business information system;

■ be familiar with the various tasks which comprise project planning;

■ understand the role of project planning;

■ understand the role of quality assurance within a modern software project.

9.1 IS STAFF

As with other departments, IS departments' structures and staff complements vary from company to company. The size of this department has a general relationship to the size of the company but also reflects the complexity of the company's operations. Whereas a mail-order company with a huge turnover may have a quite small IS department owing to the relative simplicity of its operations, a manufacturing company may need a larger IS department because of its complex production control activities.

Most IS departments are organized on some sort of hierarchical basis but perhaps not in quite the rigidly structured fashion shown in Figure 9.1. In broad terms, a largish IS department is sub-divided into systems, programming and operations. The dividing line between systems and programming is not so clear as between the latter pair, and, as mentioned in Section 7.1, the two functions are sometimes merged into one.

It is also possible that systems are seen as a function outside the orbit of the IS department, perhaps as a department in its own right or part of a management services function. Another possibility is where the systems analysts are on the staff of user departments, either on a temporary basis or permanently. This

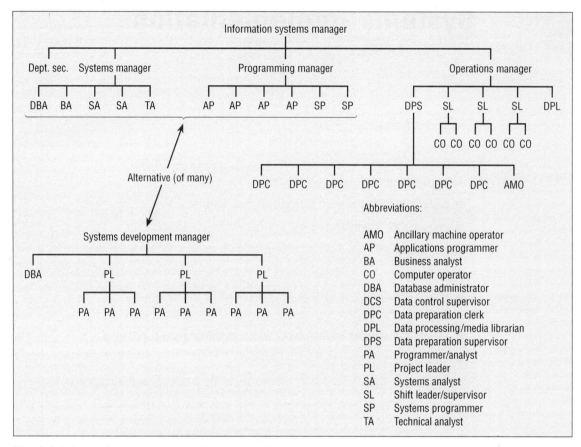

Fig. 9.1 Information systems department organization

arrangement has the advantage of keeping the system analysts close to actual user department activities.

It is likely that the programmers work as a closely knit group; this is useful in that they can assist one another, especially in the early days of new software. The danger is that they may become a closed society, impervious to outside influence, and then no longer see themselves as true servants of the company. Another problem that sometimes arises is the career structure of programmers. As more packaged software becomes available, there is a lessening demand for application programmers. This means that some programmers should be retrained into systems work for their continued careers. As explained in Section 7.1, this is not always a straightforward matter.

■ Decentralization and downsizing

With the increasing use of computer networks, end-user computing and workstations, the concept of large monolithic IS departments is being replaced by more decentralized structures. Furthermore computer services are being seen as part of a corporate resource that encompasses information, data services, and communications as well as the more traditional functions of developing computer-based information systems. As a result, IS departments are being redesigned. The top managers tend to be corporate information officers (CIOs), or corporate system managers (CSMs). They define the overall policy for information and systems management but rarely have large development teams under their control. Development opportunities are planned and analysed by business analysts and specialist IS planning staff. Opportunities arising from user requirements are also co-ordinated and negotiated by central staff and some standards may be enforced. But development itself is the responsibility of local teams of end-users who employ 4GL tools, and these may be augmented by local experts and IS experts who use application generators to construct more complicated applications.

The central IS resource still maintains responsibility for the corporate database, basic transactions processing applications, and the increasing burden of communications. These facilities require expert staff in database maintenance and communication programming. Besides fulfilling this central role, IS staff act as advice centres in the tradition of information centre analysts. The idea is to see computing as a service function for users rather than as a control. Many existing systems are transferred from mainframe applications into network systems. This process of downsizing often requires some re-engineering to split up large suites of programs into sub-systems, some of which will run on the central computer resource (the server) while others will run in a distributed system on many client workstations. Downsizing requires attention to the following factors as the application is transferred from the central IS department and batch processing to client-server applications within end-user departments and on-line operation:

- The central database remains on a large machine as a server. This may be the mainframe or a large powerful workstation which hosts the database.
- Only totally automated non-interactive programs remain on the server machine.
- Applications programs for data entry, display and data retrieval are downsized and reimplemented on client workstations.
- Communications protocols have to be worked out to control client-server data interchange.
- The user interface for client applications has to be upgraded or substantially improved.
- Applications written for batch processing on a mainframe have to be changed for workstation operation systems environment.

The impact of client-server applications, downsizing and end-user computers are still being felt on the reorganization of IS departments. While large-system steering committees are becoming part of IS history some large-scale developments still require centralized IS resources. Computer systems development is becoming partitioned into large strategic systems that may follow the old IS model while the majority of information systems are being decentralized under end-user control.

■ Staff positions

Operations covers two main aspects: mainframe operators and data preparation operators. The former also have limited career prospects with the rapid decline in the numbers of mainframes. A similar problem arises here as with programmers, i.e. how can they be retrained for other work? It does not necessarily follow that a high level of skill at mainframe operating makes a person suitable for retraining as a programmer or systems analyst.

The continuing fluidity of the IS scene has not yet allowed the careers of IS staff to settle into the formalized and qualification-based structure of other professions such as accountants. There follows below a brief description of the duties of the various kinds of IS staff.

Management services manager

An organization with a management services department is usually quite large, consequently the management services manager is a high-level position. Management services are normally taken to include IS in all its aspects, O & M, OR, and special services such as investment, insurance and legal. Certain management services departments do not incorporate IS, and others include systems but not programming and operations.

IS manager/computer manager

These titles are somewhat ambiguous. In one organization this person is in charge of systems work, programming and the day-to-day running of the computer. In another, he or she may merely be in charge of the operational aspects. Administrative ability is needed for either arrangement.

Business/information analyst

This post entails working with the end-users of an IS system to define their information requirements. A good knowledge of business and manufacturing procedures is desirable as the business analyst is often the prime instigator of new systems.

Systems analyst/designer (technical analyst)

This person requires more technical knowledge than the business analyst, and is sometimes a specialist in areas such as real-time systems, data capture or data transmissions (see Section 7.1).

Applications programmer

This person writes computer programs for use in business (or scientific) applications, and also tests, documents and maintains application programs. An increasing proportion of applications programmers' time is spent on familiarizing themselves and others with purchased applications software.

Programmer/analyst

Does both of the previous two jobs.

Systems programmer

This job differs from the above in that a systems programmer designs and writes computer software programs such as operating systems. This person is normally employed only by larger user organizations, computer manufacturers or software houses.

Network support analyst

This role involves configuring, trouble shooting, and systems programming for computer communications in local and wide area networks. As more computer systems are distributed, communications or network support is becoming an increasingly important role.

Database administrator

This person controls the database, and as such is the sole adjudicator regarding modifications to the database's structure. He or she need not, however, necessarily be concerned with the actual contents of the database as these arise naturally from the activities of the user departments.

Computer operator

He or she operates the mainframe. This work mostly consists of loading and unloading tape reels, disk cartridges, input media and output stationery; also simple maintenance, e.g. cleaning tape heads. Operators are also concerned with maintaining contact with on-line and batch users. This is necessary to deal with contingencies such as erroneous source data and late results due to operational problems, e.g. hardware breakdown or overloading of terminal facilities.

Chief operator/operations manager

This is the most senior operator or administrator in charge of operational staff, and possibly also the data entry clerks and ancillary machine operators.

IS librarian/data controller

This is an administrative job in a large IS department. The IS librarian logs work on and off the computer, and is in charge of the tape reels, disk cartridges, input media, and documents before and after these have been through the computer room. This function may be split into two posts – IS librarian and data controller – in the larger IS departments.

Data entry/preparation clerk

This is a keystation operator who transcribes source data into computer-sensible form.

Information centre manager

Manager in charge of the information centre which supports end-user computing on PCs with fourth-generation languages and applications packages. Responsible for education and training of end-users, provision and control of end-user access to databases, and standards for end-user computing.

User-support analyst

Systems analyst or analyst programmer who has experience of fourth-generation languages and end-user computing. Support analysts help end-users develop their own system by providing training in techniques and languages, trouble-shooting support, and help in more difficult analysis and design tasks. This role is usually located in the information centre within the computer department and may be called 'information centre analyst'.

9.2 SYSTEMS TESTING

Before an IS routine is brought into operation it must obviously be tested thoroughly so that all errors are eliminated. There are two stages to this: the testing of individual programs by their programmers; and the testing of the overall routine or groups of routines.

Program testing is the responsibility of the programmers, and is normally a straightforward procedure provided the systems analyst's specification is correct and has been adhered to.

With a structured program the modules are checked individually before being progressively combined to form the whole program. Groups of modules are often checked in combination if they follow a logical sequence of processing. The main aspect of testing structured programs is the coupling of interfacing modules.

As an added check that the program specification has been fully understood, the systems analyst also prepares a set of input data together with the corresponding output.

Program testing

The precise steps in program testing depend to some extent on the software support for the program, i.e. the level of sophistication of the DBMS, the compiler and the other associated software. The main aspects are, however:

- feasibility and validity checks on the input/data;
- correct interpretation of symbols, e.g. debit/credit;
- branching and looping brought about by program decisions and modifications;
- confusion due to the occurrence of meaningless values, e.g. negative amounts of stock-in-hand;
- logical file addressing and searching, involving the DBMS if used;
- capacity of storage areas and buffers;
- contents of carried-forward records;
- contents and layout of printed and displayed output;
- batch control totals;
- interfacing with other programs, software, database and operating systems; and
- documentation and handover after final testing.

Routine testing

The testing of routines entails the interfacing with users and the external situation in general. The main aspects of testing a routine (system) are:

- interfacing of runs within a routine;
- compilation and continuity of control totals;
- error-correction procedures including user involvement;
- user requests for amendments and output;
- timing of runs and routines for the data volumes to be actually handled;
- output preparation and distribution;
- audit requirements;
- logical and physical file housekeeping and control.

Test data

The usual procedure in testing is to create artificial data for the initial tests, and to use live data for later testing. The main points are:

393

■ both the artificial and the live data should be representative of reality;

■ the results from live data can sometimes be compared with the previous system's results, and any discrepancies must be completely reconciled;

■ logical files are usually needed to test fully the programs and routines;

■ data generating techniques are useful for simulating large volumes for input data and file records;

■ in the final trial run of the complete routine, a set of input data is passed through to the resultant output and/or file updating stage; and

■ test data should include known incorrect data in order to test the validation and control procedures.

SAQ 9.1   If you were to test the part of a banking system which checks the validity of a request for a standing order to be started, what sort of data would you use to carry out the test?

Solution

Test data should be generated which checks that an invalid customer is recognized, that an invalid account is recognized, that all the data required for the transaction has been input and that the amount specified is reasonable.

9.3 DATABASE CREATION

The designing and creation of a database is beyond the scope of this text. What we are concerned with here is the creation of the logical files needed for the complete system. These files almost certainly become part of a database but this does not affect the work involved in the initial stages.

The initial sources of the logical files are source documents, semi-automated data, e.g. embossed plates, and magnetic media if converting from another computer system. Possibly all of these are involved, and consequently, it is sometimes the case that the creation of the logical files is more difficult than the subsequent routines.

The data items needed to create logical records may have to be acquired from different departments or from data held in dispersed locations. An example of the former problem is a product master file in which the constituents of the products are derived from material specifications held in the production control department, whereas the current stock comes from the stock control department, and the manufacturing costs from the costing department. An example of data acquired from dispersed locations is the gathering together of stock levels from a large number of a company's branches in order to create a central stock file.

It is quite usual for various data sets to be amalgamated to form a logical file by the process of merging (see Section 5.1). The important questions in this regard are:

- Do the data items from the different sources correspond one to one when being merged into one record? This is important as otherwise certain resultant records will be incomplete.

- Are the code numbers of the corresponding entities on different data sets exactly the same? If there are even marginal differences, this can cause difficulty during file creation.

- Does any data conflict between data sets? For instance, individual costs might not add up to the total cost derived from another source.

- Is there any duplication of supposedly unique entity data records, or is the same code number used for two or more different entities?

- If data is to be transferred from another computer's database, have all the interface problems been found and dealt with?

Further points in file creation are:

- During the period in which a file is being created, transactions and amendments applicable to the file need to be frozen. They are then applied speedily before bringing the file into operation. It is a good idea to create files during non-working periods if possible, e.g. at night or during a weekend or holiday.

- Errors, omissions and conflict should be expected. Planned procedures for dealing with these contingencies must therefore be made beforehand.

- Complete or partial proof listing of file data derived from manual procedures is generally necessary to ensure accuracy and completeness.

- Control totals are necessary if any manual procedures are involved in the creation processes (see Section 8.4).

9.4 CHANGEOVER PROCEDURES

It is likely that changing over to a new system entails either moving from one computer system to another, or transferring from a manual system to a computer already in operation on other work.

The former case has the advantage that a substantial part of the system is already planned and only the amendments/additions are truly new. And even though changes have been made, it is usually possible to make a quantified comparison of the two sets of results. The situation in the latter case, however, is fundamentally the same as when introducing a computer for the first time, except that the IS staff are familiar with the existing hardware and operational software. This is obviously safer than when changing over to a completely new computer.

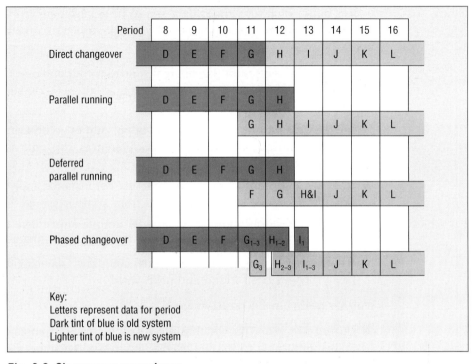

Fig. 9.2 Changeover procedures

When the new system has been thoroughly tested, the files or database created, and the hardware installed, it is possible to change over from the old to the new system. The method adopted for changing over depends on the particular applications and the company's modus operandi. There is no universal best method, only that which most suits the particular situation and circumstances at the time of the changeover.

Preparations for changeover are made over a period of time and include the activities described in the previous pages.

Fundamentally, there are four changeover methods, as shown in Figure 9.2 and described below.

■ Direct changeover

With direct changeover, the old system ceases abruptly and is replaced immediately by the new system. The changeover is best done at a weekend or during a holiday period so that it is fitted into a natural break in the stream of work. The main hazard with direct changeover is the possibility that the new system is not entirely correct or complete. This problem is aggravated by the lack of current results from the old system to compare with those from the new system once changeover has occurred. Also if the new results are incorrect, it is difficult to make corrections to the system and, at the same time, keep it operational.

Certain applications do not lend themselves to any viable method except direct changeover. This is true of on-line and real-time systems because these cannot normally have the old system running simultaneously with the new system.

Direct changeover is an efficient method in so far as it minimizes the duplication of work but demands careful planning, testing and attention to operational detail if it is to be completely successful.

Parallel running

With this method the old and new systems are run concurrently for a few periods. This means that the two sets of results can be compared for similarity, assuming that this is intended to be the case. If it turns out that there are discrepancies not easily reconciled, the old system continues until things are put right.

If the changeover is from one system to another, as is likely to be the situation nowadays, parallel running should present few problems. The input data is held on a compatible medium, e.g. magnetic tape, and fed into both systems concurrently. This assumes that the old system can remain in operation for the period of parallel running.

When the results are not intended to be identical, it is nevertheless usually possible to compare the new results with the old, making reconciliations manually if necessary.

The major drawback of parallel running is the duplication of effort imposed upon the user department staff as it is likely that the same persons have to run both the old and the new systems during the periods of parallel running. This problem is much less severe with computer-to-computer changeover.

Deferred parallel running

A variant of parallel running (deferred parallel running) involves rerunning the data from the previous period while the old system continues with the current data. Also known as pilot running, this method has the advantage of allowing more time for arranging the source data and checking the results. When the new system is proved to be fully correct, a double cycle of the work, i.e. two periods' worth, is done in order to catch up, and the old system then ceases.

The disadvantage of deferred parallel running is the necessity to carry out the double cycle of work.

■ Phased changeover

Phased changeover is similar to parallel running except that at the start only a portion of the data is run in parallel, e.g. only certain customer accounts from a sales accounting system. The portion taken into the new system increases each period and, after the final portion has been run in parallel, the old system is abandoned.

This method has the weakness of delaying the full implementation of the new system if several periods are taken for the changeover.

A situation for which phased changeover is often suitable is where a number of branches of the company are undergoing changeover. Starting with one or just a few branches, an increasing number are changed over in each successive period.

Prototype changeover

Referring to Section 7.9, it is apparent that three of the categories of prototyping – pilot, staged and evolutionary – entail parallel running and/or phased changeover. Pilot prototyping, for instance, has strong similarities to phased changeover. Staged prototyping might necessitate parallel running if the new system is an enhancement of the old system.

Changeover planning

It is evident from the foregoing explanations that a considerable number of activities are involved in the changeover procedure. In a large system these may run into hundreds, and thus the systems analyst needs a logical method for scheduling and controlling the project so that the precise progress achieved at any time is clearly known.

The two main methods of project control are bar charts and network analysis. These techniques are applicable to any project involving numerous activities, e.g. systems analysis, as well as to changeover planning.

Bar charts

Otherwise known as a Gantt chart, a bar chart consists of a list of activities on a chart marked off in time periods. Each activity is entered initially as a horizontal line denoting its duration and scheduled periods. As and when progress is made, these lines (bars) are annotated to show the achievement to date.

The major weakness of bar charts is that they fail to demonstrate the dependence of one activity upon another, nor do they facilitate the planning of resources and costs. Nevertheless they are convenient for small projects and can be managed entirely by hand. An improvement is the GASP chart, which allows activity dependencies to be shown.

Network analysis

Also known as 'critical path method', this technique can cope with a large number of activities, especially if a computer package is used. It is a powerful management tool, and in its more advanced form (known as PERT) incorporates resource allocation and cost planning.

In utilizing network analysis for changeover planning, a list is drawn up of all the activities in the changeover procedure together with their estimated times, resources needed and the activities that immediately precede them. From this information a network diagram is constructed based upon the logical relationships of the activities. By utilizing the network diagram it is straightforward to compute the span of dates during which each activity must be performed if the changeover is to be completed in the minimum time. We can also find the 'critical' activities, i.e. those with no spare time, which determine the overall time taken for the changeover.

It is also possible to obtain a picture of the amounts of resources (programmers, systems analysts, etc.) needed during each period of the changeover, and thereby to allocate the staff accordingly. The resource requirements can be converted into costs so as to give a pattern of accumulating costs throughout the changeover.

Network analysis also caters for progressing the changeover so that when activities are completed, partially completed, amended, inserted or deleted, the remaining activities can be quickly rescheduled. Regular updating of the changeover in this way provides a means of continuous monitoring so that the completion date and the resource requirements are predicted well beforehand.

9.5 SYSTEMS PRESENTATION

Part of a systems analyst's work is to present and explain a proposed or actual new system. This may need to be repeated several times through the life cycle of system development. In the early stages of the system development the proposals are tentative and somewhat intangible, and so engender many questions from management and user staff. Even when there is already a computer-based IS system in operation, proposals regarding a new system cause some degree of apprehension, if not suspicion.

The systems analyst needs to propose his or her ideas and system enthusiastically, confidently and clearly. In answering questions and dealing with objections, he or she must not talk down to nor allow a confrontation to build up with the questioners. Neither should he or she be drawn into situations that are really management's responsibility, e.g. redundancy policy.

One of the main points is that the systems analyst makes it clear that he or she fully understands the problem and that his or her proposals are made with the intention of helping the user rather than simply for the sake of change or using new technology. The prospective users should be left with the feeling that they are fortunate to be able to work with the new system. It seems to be a fact of life that whereas engineers, technicians and scientists are pleased to use new equipment and methods, business staff are in general not so. The more they can be persuaded that they are also entrepreneurs, the better.

The type of audience must be taken into account in the presentation. Top managers usually have little technical knowledge but are interested in information for decision-making. It is advisable to keep presentations to management quite short but polished and with effectual visual aids. Prospective users of the system are interested in the aspects relating to their work. The presentation to users is therefore more technical and may include some operational details.

Communication with users

Steering committee

This body (see Section 1.6), formed in the early stages, acts as the focus for maintaining communication between the systems analysts, the user departments and, later, the IS department. It is a convenient link for organizing the activities described below.

Discussion meetings

Formal and informal meetings between the systems analysts and the user staff, including trade union representatives, are a good means of maintaining communication.

The formal meetings should have a prepublished agenda so that all the participants are well acquainted with the discussion points. The informal meetings are a good opportunity for the user staff to ask questions and make suggestions. They also tie in with prototyping in that prototypes are intended as a basis for discussion. By the time the final prototype has been agreed, it is inevitable that users are well acquainted with the proposed system.

The use of project teams also facilitates communication as the project team members are in close touch with the staff of their own departments.

Training courses

Courses for user staff may be held either internally, at a training school, or at a local college. They fall into two categories: general informative and specific training. The former can be attended by a wide range of user staff, perhaps on a voluntary basis in some cases. The specific training courses are aimed mainly towards staff who will be performing operational tasks such as using terminals and end-user systems development with fourth-generation languages.

Magazine articles

Articles in the staff magazine are best if of a general informative nature rather than technical. They are made more interesting by the inclusion of photographs of the proposed hardware.

Visits

In the early stages certain staff are taken on visits to other companies using similar methods to those proposed. Later this can be extended to allow user groups to visit the company's own IS department at prearranged times. Although these visits may not convey much information, they often arouse interest and make visitors aware that people as well as machines are involved in IS.

Methods of presentation

Before giving a formal lecture, or even an informal talk, it is important to prepare the material so that it is presented clearly and logically. Very few people are capable of giving off-the-cuff talks of a good standard.

The lecturer may make use of the following audio and visual aids.

Slides

Slides are useful for showing pictures of hardware and diagrams depicting systems generally. For particular purposes slides involve a lot of preparatory work and are not usually worth the trouble of making unless they are used repeatedly.

Overhead projectors

These are effective for showing fairly complex diagrams through the use of pre-prepared transparencies. The transparencies may be laid over one another to emphasize comparisons or a build-up of ideas. Another approach is to use a continuous transparency that is 'scrolled' across the projector.

An advantage of overhead projectors is that they can be used in normal lighting, although bright sunlight is best avoided. A skilled user can write or draw on the transparency as he or she talks, but beware of attempting this without practice.

PC presentation packages

The presentation of information by means of a PC's visual display is effective provided the information is clear and understandable to the audience. The display must be capable of being seen by everyone present and so either a large screen display or multiple linked microcomputers are used.

The demonstration should relate to the audience's responsibilities and is improved by encouraging hands-on interaction by certain members of the audience. These are best chosen beforehand as complete novices can make an awful mess of things.

On the whole, graphical displays are more effective than sets of figures to a general audience. Graphical techniques are briefly described below. Almost certainly the Powerpoint package marketed by Microsoft is the most popular package available.

Films

Films are good for presenting a general background and they are usually professionally produced. They are available on hire from film libraries such as the National Film Library. Films are likely to be too mundane to be of interest to organizations that have employed computers previously.

Flipcharts

Good if well prepared, portable and flexible in use. They need some practice if the material is to be created (using a felt pen) at the time.

Whiteboard

A whiteboard is best used in conjunction with a dry marker as this is most easily erased.

▪ Diagrammatic techniques

A number of techniques make a good impact upon readers or on an audience; for example:

- *Histograms* – blocks representing amounts.
- *Line charts* – vertical or horizontal lines representing quantities.
- *Bar charts* – similar to line charts, but often used to represent time periods on a graph.
- *Graphs* – different scales and intercepts are used to emphasize the details but a graph should not be misleading.
- *Pie charts* – circles divided into sectors to illustrate proportions of the whole.
- *Pictograms (ideographs)* – different sizes or numbers of pictures of objects represent a quantity.

The above presentations can be displayed in colour on a screen by software such as spreadsheets (Section 6.4). This method gives life to the presentation and relates to the use of a PC as described above.

▪ Report writing

A report is a written and, possibly, a diagrammatic presentation of a policy, situation or plan. The following points are worth noting:

- A report is not, *per se*, the end aim of the report writer. The report merely serves as an aid in achieving an objective.

- The person requesting a report should give an indication of its intended purpose. This helps the report writer in presenting the facts and suggestions in the most suitable manner.

- Follow a logical approach, e.g. introduction, sectionalized main body, conclusions, summary and appendices. An index is useful if the report is extensive.

- The writer should present the facts and suggestions in an unbiased way, and must not omit any relevant information.

- Present quantitative data rather than qualitative explanations but beware of swamping the reader with masses of figures. Summarized figures in the main text with detailed figures in the appendices are beneficial.

- Keep to correct but not stilted grammar, avoid verbosity, jargon, and tautologies. Refer to a dictionary if you are at all uncertain of a word's precise meaning, and a thesaurus for word ideas.

9.6 SYSTEMS APPRAISAL AND MAINTENANCE

The implementation of an IS system, even though it is fully correct and complete, is not the end of the matter. Systems are ongoing so need to be reappraised continually and regularly maintained in order to keep them efficient and up to date.

Shortly after the changeover to the new system, appraisal is required to ensure that actual performance is close to that predicted during the design stage. There is always the possibility that certain operational factors were overlooked during the design stage. Similarly, circumstances could have changed between the dates of the systems investigation and its implementation since this might be a year or two.

The particular factors entering into the immediate appraisal of a system are:

- Throughput speed from the capture of source data to the delivery of output. How does this actually compare with the estimated times for various data volumes?

- Storage space occupied by the logical files. Has this turned out to be greater than allowed for and if so, what are the reasons and consequences?

- Errors, exceptions and queries arising in the system. These can be expected to be higher at first than later, but are they disproportionately high for the amount of data? If so, why is this?

- Data capture, preparation and ancillary operations. Are the costs of these in line with estimates? Alternatively, are the throughput times as low as estimated?

- Response time. Is this as short as was envisaged for each type of input message and for a given traffic level? Has the peak level of traffic caused the response time to increase above an acceptance level?

The subsequent reappraisal and system maintenance also includes the following factors:

- *Cost/benefit*: Is this still at an acceptable level in general terms? That is to say, are the end-users satisfied with the information they received or has it become irrelevant, confusing or no longer in the most suitable form?

- *Source data*: Has the pattern and/or amount changed to the extent that it is worthwhile to reconsider the methodology?

- *Technology*: Have hardware developments made the existing computer and associated equipment obsolete to the extent of seriously affecting the efficiency of the system as compared with that possible with more modern hardware?

- *Legislation*: What are the effects of impending legislation such as in regard to tax, social security, privacy, databanks and information disclosure?

- *Patching*: Has the existing software, especially application programs, been patched, i.e. amended, to the extent of jeopardizing its efficiency?

SAQ 9.2 One of the most important tasks after delivery of a system is maintenance: this involves making changes to the delivered system. Can you think of three reasons for carrying out maintenance?

Solution

They are:

- Systems development staff could have made errors in the development of the system which were discovered only after implementation; they would need to be rectified.

- The requirements of the customer may have changed; for example, a finance company's system may be affected by a change in financial reporting laws.

- The customer may have bought some new hardware and the system needs to be transferred to this hardware.

9.7 COSTS AND SAVINGS OF NEW SYSTEMS

In estimating the costs of an information system, it is advisable to look forward over a period of five years or so. A lesser period does not allow the system to become fully effective, and beyond five years the situation is increasingly unpredictable.

The factors entering into the estimation of future costs include some of those listed below. The extent to which these are significant depends largely on how radical the change is to be. Changing from a completely manual system to a large mainframe necessitates more installation costs and personnel costs than moving to a microprocessor. Changing from one model of computer to another may attract only minimal costs.

Equipment costs – purchase, lease or rental

- Processors, peripherals, terminals and communication equipment.
- Data preparation, capture and entry equipment.
- Ancillary machines, e.g. stationery handling.
- Air-conditioning equipment in the case of mainframes.
- Electric supply equipment – stabilizers and standby generators.
- Initial stocks of disk cartridges, diskettes and magnetic tapes.
- Racks, trolleys, trays, furniture, internal telephones, etc.
- Security equipment, e.g. alarm systems, safes, etc.

Installation costs

- Structural alterations to the existing building for large computers.
- Construction of a new building (unlikely but possible).
- Demolition of an old building (unlikely but possible).
- Removal and resiting of existing departments.
- Special floors, ceilings, walls and lighting, double glazing.
- Disposal of obsolete equipment (could be a negative cost).

Development costs

- Software from computer manufacture and/or software houses.
- Consultant's fees.
- Changeover activities including file creation and system testing on time-hired computer.

Personnel costs

- Salaries and allowances of all IS staff.
- Pension fund and charges of IS staff.
- Staff recruitment and relocation.
- Training, including course fees, accommodation and materials.
- Staff expenses, e.g. travelling, accommodation and meals.
- Redundancy payments.

Operating costs

- Consumable materials, i.e. stationery, print ribbons.
- Floppy disks, disk cartridges and tapes.

405

■ Maintenance of all equipment, especially computer hardware.

■ Rent, rates, depreciation, maintenance and cleaning of IS department accommodation.

■ Heating, electricity, data transmission and telephoning.

■ Standby services.

■ Insurance premiums including loss of profit through damage to the computer installation.

Each of the above costs is estimated for each of the five years from the start of the project and then totalled each year. This is repeated for each system under consideration so that a table of comparable cost estimates is derived.

■ Estimating savings

Much of the savings of a new system is associated with eliminating or reducing the costs of the existing system, as discussed in Section 7.6. It is more difficult to estimate the savings from a system than its costs. Benefits from computer systems are often said to be 'intangible'; nevertheless every effort should be made to put a monetary figure on an intangible benefit as otherwise it has dubious meaning.

Tangible savings

Equipment savings – replacement or rental

■ Obsolete computers, accounting machines, typewriters and other machines.

■ Cabinets, racks, furniture and filing equipment.

Personnel savings

■ Salaries and allowances of redundant or transferred clerical and administrative staff.

■ Pension fund, expenses and charges of above staff.

Operating savings

■ Consumable materials, i.e. stationery, ledger and stock cards.

■ Maintenance of existing machines.

■ Rent, rates, depreciation, maintenance and cleaning of redundant departments.

■ Heating, electricity, telephone and insurance charges of above.

Capital allowances and taxes

These vary from time to time depending upon government policy, but may provide significant savings.

Intangible savings

Planning information

The information needed for planning the company's activities should be available sooner and in a more accurate and comprehensive form (see Example 9.1 for examples of planning information). Much depends upon the nature and complexity of the activities.

SAQ 9.3 Which of the following are intangible savings?

1 Staff morale being increased by them not having to worry about errors which were generated by the previous manual system.

2 The saving of transport costs by virtue of the fact that all stock will be stored in one warehouse.

3 The saving of stationery costs by a system which sends electronic documents to customers rather than paper documents.

Solution

1 is the only intangible saving.

Example 9.1

Planning information

■ *Sales* Statistical analyses and forecasts of sales, e.g. sales analysed according to customer, type of customer, salesperson, area, country, product, product group, period, and any combinations of these heads.

■ *Purchasing* Suppliers' delivery promises versus actual deliveries, discount analysis, forward ordering and supplies availabilities.

■ *Stock* Forward stock requirements and valuation, stockholding costs.

■ *Materials requirements* More accurate and speedier determination of forward requirements of raw materials, bought-out components and tools.

■ *Production* Forward machine loads, labour, plant and equipment requirements; detailed job scheduling and routeing, machine allocation, and rapid replanning of production.

■ *Cash flow* Projected in and out movement of cash, and the net cash position in future periods.

■ *Investment appraisal* Comparison of various projects from their long-term profitability prospects.

Control information

Example 9.2

Control information

- *Debtors* The more rapid collection of debts through more efficient invoicing, and the detection of incipient bad debts.

- *Operating statements* Quicker and more accurate preparation of all types of operational information, e.g. running profit/loss accounts.

- *Stock control* Optimum stock levels based on demand and stockholding costs.

Personnel

- Less dependence on unreliable staff.
- Avoidance of complex training procedures for clerical staff.
- Minimal staff administration costs.

The estimated savings – tangible and intangible – are summated for each year and set against the corresponding costs. The net differences provide a basis for comparison with alternative systems by using investment appraisal techniques such as net present value in Section 7.6.

9.8 IS RESOURCES

Software houses

A software house is a company supplying products such as applications and systems software, the former either as packages or in bespoke form. To a lesser extent, services such as customized programming and systems analysis are provided, especially in an advisory capacity. There are many variations in the services offered, and many software houses specialize in certain areas of work and applications. By so doing they are able to build up a depth of experience beyond that of the ordinary computer user.

Large organizations may be able to afford the services of a software house without concern, but a small company needs to be particularly clear as to what it will be charged and precisely what it is getting.

Consultancy firms

A business consultant usually works at quite a high level, giving advice to top management and on broad systems. Some companies of professional accountants also provide a consultancy service or will recommend suitable consultants. The main point is to ensure that the consultancy firm has practical experience of solving problems similar to those of its client. This is more evident if the consultancy firm is a subsidiary of a company that has actually implemented the system being purveyed.

As with software houses, the terms of contract should be agreed at the outset.

Finance houses

These are merchant banks or finance companies, and provide the capital to assist prospective computer users.

There are four main methods of acquiring computers:

1 Outright purchase by means of a loan from a finance house.

2 Renting from the computer manufacturer.

3 Leasing from a finance house after this has purchased the client's choice of computers from the manufacturer. At the end of the leasing period, the leasee chooses: to renew the lease, probably on different terms; or to enter into a fresh contract for new equipment; or to end the agreement.

4 Industrial hire purchase by which the user hires the computers for an agreed period, after which it becomes his or her property.

The relative advantages of the above methods depend partly upon the taxation legislation in the country of contract.

Training establishments

Training establishments operate under a variety of titles but are essentially in business to provide concentrated training in IS. These may also be method vendors and often consultancy houses also offer training in a variety of methods. The courses last from one or two days to up to a few weeks, and are oriented towards topics such as systems development methods, user staff appreciation, management appreciation, programming, computer operating, real-time applications and database understanding.

Computer manufacturers also run courses; these tend to be appreciation courses, and have a natural bias towards their own hardware.

9.9 PROJECT MANAGEMENT

The previous sections of the chapter have described how systems are implemented. In this discussion we have omitted a description of one important set of tasks which occur throughout the implementation process: those tasks which come under the broad title of 'project management'.

It is worth pointing out that we shall use the terms *project manager* and *customer* in the remainder of the chapter. You may get the impression that this means that these entities are always quite separate: for example, the project manager might be employed by an external company which is hired by the customer in order to develop a system. We have not made this assumption: the terms could quite easily be used to describe staff in separate departments within the same company. Increasingly, companies who have an IT function organize themselves in such a way that, for example, formal contracts are written and budgeted for and hence our description of project management quite easily describes this form of working.

Before looking at software project management in particular it would be useful to describe the sort of tasks that any project manager has to carry out. There are five categories of task:

■ *Planning*. This is the process of deciding how a project should be organized: what technical tasks, such as programming, need to be carried out during a project; when they are to be carried out; and the resources required – what the cost of a project will be; and what risks will there be in carrying out a project and how to minimize the effect of these risks. Most of these tasks generate information which is placed in a document known as the **project plan**. Planning is a task which occurs at the beginning of a project, although, even in the most successful project, some re-planning is required when, for example, an unplanned event happens such as a key member of staff becoming ill for a long period of time.

SAQ 9.4 Can you think of some typical unplanned events which might occur on an IT project?

Solution

There are quite a number. Typical events include staff illness, the customer being affected by circumstances beyond their control such as being subject to a takeover, a new set of financial laws coming into force which invalidates some of the assumptions of the developer and a sub-contractor delivering some item late.

■ *Monitoring*. Once a project plan has been developed, the project manager then has to check that it is being followed. For this he or she receives regular reports on progress from the staff on the completion of tasks. The project manager will also receive regular reports from accounting staff charged with monitoring the financial spending on the project. The project manager also has to ensure that the project adheres to the quality standards that are adopted by the project. For example, the project may use a certain set of standards to document the testing process, and it is the project manager's task to ensure that such documentation is produced and filled in properly.

- *Controlling.* This term describes the process of carrying out actions which ensure that a project keeps to its plans. Even the best-planned projects suffer from problems which need active intervention from the project manager in order to keep a project on schedule. For example, in software projects a common occurrence is for the customer to change his or her mind about requirements for a system; often this change is outside the control of the customer: for example, a new set of tax laws brought in by a government might mean that an accounting package which is under development may need more work done on it during a project. The role of the project manager is to evaluate the seriousness of any event which threatens to throw a project off course, evaluate what needs to be done to bring the project back on course and, sometimes, calculate the cost to the customer of these actions.

- *Representing.* The project manager is the public face of a project. It will be the project manager who regularly meets the customer in order to discuss progress and to discuss any requirements that the customer has over and above those determined at the beginning of the project. The project manager will also need to represent his or her project with their senior manager. Such managers have a strategic role and are more interested in receiving broad-brush information such as how much on schedule a project is and the staffing requirements for that project.

- *Innovating.* The project manager also has a role in carrying out innovations and evaluating them. Software development is a rapidly developing engineering discipline and many projects often use some new technology or carry out tasks in an innovative way. There are a number of tasks that project managers carry out which are connected with innovation. The first is to encourage and embrace it in the planning for a project: not so much that a project becomes very risky but enough that there is a good chance that results will be generated which might encourage the company to embrace the new technology or new processes on more projects. The second is to monitor and evaluate the innovations that are used and produce reports which are then used by the company in making decisions about whether it should be adopting a novel technology over a larger number of projects. We should stress now that the remainder of this chapter will not deal with this topic but concentrate on the previous four functions of management. This does not mean that we regard innovation as unimportant – far from it – however, the topic is out of the scope of this book.

SAQ 9.5 Some tasks that a project manager carries out are shown below. Which function do they correspond to?

■ Checking the bids from a sub-contractor for some item of work.

■ Attending a meeting with the customer.

■ Replacing a designer who has left with another designer.

■ Calculating the manpower to be expended on a project.

■ Checking the time-sheets returned by staff on a project.

Solution

They are costing, representing, controlling, costing and monitoring.

9.10 PLANNING

At the beginning of the software project a manager will receive a document from the customer which describes what the customer requires from a computer system – a document often known as *statement of requirements*. There is no standard format for this document: some customers who are totally conversant with modern information technology will produce a very detailed description of their needs which describe the functions that the system should carry out, constraints such as the minimum response time to certain queries, the hardware required and services such as training which will be required. However, there will also be customers who may have little knowledge of computing and who will give the software developer a single sheet of paper with only the vaguest descriptions of the system that should be developed. The role of the project manager during planning is to carry out the management tasks which lead from the receipt of the statement of requirements to the development of a project plan.

Project planning consists of a number of tasks and the following sub-sections describe them. Normally project planning is carried out using a planning tool such as Microsoft Project. Such tools:

■ Allow the manager to specify the tasks which make up a project and specify the various items of data associated with each task, for example the start date, finish date and duration.

■ Contain reporting facilities which enable the manager to see how well or badly the project is going in terms of its adherence to its timetable.

■ Contain facilities whereby a project manager can see whether the late completion of a task might threaten the finish date of the project.

■ Often contain financial facilities whereby the manager can examine reports which give the month-by-month spend on the project and compare the figures to what has been estimated.

Outline requirements analysis

Sometimes the statement of requirements given by the customer is so poor that a further amount of work needs to be done on it in order to put it into a form whereby it can be used as a document for planning the project. Normally the analysis that is required is not carried out by the project manager, but by a systems analyst who will meet with the customer, ask a number of questions and then develop a version of the statement of requirements which can be used for the remainder of the planning activities. The role of the project manager here is to monitor the analysis activity and step in as a referee when any problems occur. For example, poor statements of requirements are often produced by customers who have limited knowledge of IT development and who might feel intimidated by an analyst asking piercing questions about what they want from a system. The role of the project manager then would be that of a diplomat: soothing the customer and explaining what further analysis is required.

Feasibility analysis

Feasibility analysis is the process of determining whether a project can be carried out successfully. Often a customer will ask for something which just cannot be implemented. For example, he or she may have read a book about artificial intelligence and assumed that the developer could easily write some program code for a system which takes natural language input. More likely, though, the mismatch between customer expectations and what can be delivered will be smaller. Some examples of this mismatch are shown below.

■ The customer may require a very large system to be built which is capable of responding to any request within two or three seconds. The developer may feel that since the software system is so large there is little chance that this can be done with the customer's existing hardware.

■ The customer may urgently require a system within a short timescale; so short a timescale that the developer may not have the manpower necessary to deliver it.

■ The customer may require the use of a new technology such as a recently released computer within a certain time which the developer feels is unrealistic given the teething problems that have been encountered with similar models of computer.

413

- The project may be a relatively easy project to carry out, but the customer is willing to pay only a small amount of money for the development – an amount which would result in a financial loss for the developer.

SAQ 9.6 Can you think of a possible reason why many of the problems above might occur?

Solution

The usual reason is that the customer is not knowledgeable about the detailed technical aspects of computer implementation.

These are examples of typical mismatches which often turn up at the beginning of a software project. The role of the project manager is to evaluate these mismatches and negotiate with the customer to see whether the customer would be happy to scale down their requirements in order for there to be a good chance of a profit being made. In the end, though, if this negotiation fails then the project manager would have to advise his company to turn down the project.

The role of the project manager during feasibility analysis is to co-ordinate the work of the technical experts being used to check feasibility and to negotiate with the customer over the dropping of requirements.

Risk analysis

Every software project suffers from risk, even those projects which use well-established technology. Some of the events which a software project could suffer from are shown below.

- A key member of staff leaves the development team to join a competitor.
- The customer is taken over by another company which requires major changes to the software being developed during the project.
- The software customer is so inexperienced that large amounts of time need to be spent in meeting the customer to clarify both technical issues and their requirements; this results in the software developer delivering the system late.
- A project uses innovative hardware which is only partially documented. Problems with hardware lead to a late delivery time and the profit being made by the customer being reduced.
- A sub-contractor does not deliver software or hardware in time. This results in the postponing of the delivery date of a project.
- The contractor appoints an inexperienced design team to the project that develops a design which generates a lot of programming errors. This results in a late hand-over of the software.

414

- The customer requires a software system with a very tight response time: that although the response time was met it took an inordinate amount of time to design the system to achieve it.

- The IT developer is only developing one part of the project, the customer is developing one part while a developer in France is developing the other part. Communication problems add time to the analysis phase resulting in the system being delivered late.

Many of these events could not be predicted in detail. However, the computer industry has had enough experience over the past thirty years of projects which have failed for managers to be able to put some sort of probability on types of risk: for example, the use of sub-contractors makes a project very risky.

The role of the project manager in risk analysis is to list all the possible risks that could happen to a project and assign some sort of probability. Risk assessment is not an exact science so most projects tend to classify a risk into three categories of likely, medium and unlikely.

Many project standards insist that once this categorization has taken place the project manager looks at the most likely risks and then includes in the project plan what actions they feel should be taken in organizing the project in order to minimize the risk. Some examples of this are shown below.

- If the manager has a feeling that a critical member of staff, such as a designer, is likely to leave the company during a project, then the manpower plan for the project should deploy competent staff to help that member of staff. If he or she then leaves, there will be at least one person who has sufficient knowledge of the work to carry on.

- If the customer has asked for a tight response time from a system and the project manager feels that this would be met only after extensive work has been carried out, then he or she might commission a consultant who is an expert in the area to confirm that this was so. If it was, then the project manager might decide to either persuade the customer to re-specify the response time or increase the cost of the project to the customer.

SAQ 9.7 If you were the manager of an IT project which suffered from the risk that a sub-contractor would deliver some item late, how would you plan your project to take account of this risk?

Solution

Two things could be done. First, the sub-contractor would have to sign a contract with some hefty penalty clauses. Second, the sub-contractor would need to be monitored very closely in order to have early warning of any non-delivery.

These are some of the decisions that a project manager needs to make. The final bullet point above illustrates a typical response to a high-risk project: that of increasing the quoted price of a project. Normally after a project manager has carried out a risk analysis they will have a good feel about how risky the project is and, based on this, they will add a further element to the estimated cost of the project to cater for any of the likely risks that might occur.

■ Project costing

The closing paragraphs of the previous section have hinted at the fact that the project manager is involved in costing a software project. This process occurs at the beginning of the project and can be carried out in a number of ways. The conventional way is for the project manager to examine then read the outline statement of requirements produced by the customer and identify the major tasks that are needed to implement the system that is required. At this stage the tasks that are identified will be quite broad in their scope and will be described by phrases such as *design sub-system A*, *test sub-system B* and *program sub-system C*. An experienced project manager will be able to predict the rough size of each of the components of the system that is to be developed and based on past projects will then be able to predict a cost. The total of the individual costs for the whole system will be a large proportion of the total cost of the system; however, there will be some costs, for example the cost of hiring consultants, any hardware costs, the cost of travel, the cost of stationery, etc. These must all be added to the technical cost. To this must then be added any overheads which the developer usually applies. These overheads usually cover aspects such as heating, lighting, clerical services, rent of the developer's buildings and support services such as the use of the quality assurance department.

Once this total has been achieved (technical cost + non-technical cost + overhead), the project manager needs to add a contingency amount based on his or her perception of the riskiness of the project. This total will then be the full cost of the project. This will not be presented to the customer directly, but will then have added to it the profit which the developer wishes to make from the project. Normally a company will have some policy about profit: for example, it may say the profit element of a project must be 40 per cent of the final estimated cost. However, this amount might be allowed to vary in some circumstances; for example, work may be scarce and the developer might be looking for projects which just keep staff working and feels that a project with a very modest profit would stand a chance of being chosen by the customer.

This is the conventional way of costing a computing project. It mirrors the way any engineering project is costed: by accumulating the costs of individual tasks and adding a number of other elements such as overheads. There is, however, another more scientific way of costing projects which has come to the fore over the last ten years. It requires the use of historical data from past projects.

SAQ 9.8 How might you check the cost of a project?

Solution

There are a number of ways. First, the costing could be carried out independently by two members of staff and their results compared. Another technique is to use two different methods of costing. The most common technique, however, is to check the cost prediction of a project against the final cost of a similar completed project.

The development of the project plan

The main document that is produced by the project manager connected with project planning is the overall project plan. Many of the outcomes from the activities detailed above will be embedded in this plan; for example, the results of the risk analysis. However, the main component of the plan will be a list of tasks which are to be carried out by staff on the project: these range from early tasks such as producing a design for the system to later tasks such as applying tests.

There are a number of technical tasks which need to be placed in the project plan:

- *Requirements analysis*. The analysis of the customer's requirements for a system.
- *Requirements specification*. The documentation of the customer's requirements in a document known as a requirements specification or system specification.
- *System design*. The specification of the system architecture in terms of its software and hardware components. The software components would be subroutines.
- *Detailed design*. The process of defining the processing that occurs in the software components.
- *Programming*. The process of transforming the detailed design into program code in some programming language.
- *Testing*. The checking that a system meets the customer's requirements.

It is worth stressing that at the beginning of the project most of the tasks will be expressed at a very high level, for example 'Design sub-system A' or 'Test sub-system B'; however, as a project progresses these tasks will be split into smaller and smaller components such as 'Test module X from sub-system C'. The implication here is that the project plan is a flexible document: although a version of it is produced at the beginning of the project, it will be subject to continuous change throughout the project.

417

The part of the project plan concerned with the tasks is known as the resource plan. It will contain a list of tasks, when they are to be started and completed, the resources required and who is to carry out each task. Many resource plans also include details of the cost of a particular task. It is this part of the plan which will be consulted most during a project when the manager monitors its progress.

9.11 MONITORING AND CONTROLLING

During a project the project manager switches to tasks which form part of his or her monitoring and controlling functions. During the project the manager will receive reports from his staff on which tasks have been completed, which tasks have been started and which tasks have not been started. The manager will then compare these reports to the project plan which will specify the required timetable for the project. Usually some tasks will be completed early while others will be late; if they balance out, the project manager will write a weekly progress report to his or her manager which would state that the project is roughly on track. However, there will be times when the project manager will need to carry out some action which is needed to bring a project back to schedule. Some examples of this are detailed below.

■ A designer on a project becomes ill and is likely to be away from work for three weeks when a critical part of a system is to be designed. The project manager assesses this as something which could adversely affect the completion date of the project, a junior designer is assigned to the work and the project manager asks his manager for some temporary resources to cover the work that the junior designer was to carry out.

■ A sub-contractor reports that the software that he or she was going to deliver will be four weeks late. The project manager reassigns the staff who are to carry out the testing of this software to other testing activities in order that they do not become an idle resource.

■ The customer for an accounting system asks for some changes to the system because of recent government legislation. These changes are quite large. The project manager examines the changes, costs them and checks with the customer that he or she is prepared to pay for them. The customer agrees and the project manager modifies the project plan and requests extra resource from his manager in order to carry out the new software tasks which are the result of the changes.

SAQ 9.9 How should a project manager react to a change in the customer personnel that he or she has to closely liaise with over the detailed requirements for a system?

Solution

This is potentially a serious problem as the new members of staff could have a different perception of what a system is to do. The project manager should arrange for these staff to be thoroughly briefed on the functions of the system. Documentation such as signed letters or memos should also be generated which commit these members of staff to the current requirements.

These are the main activities that the project manager carries out during a project. If a good project plan has been constructed and the manager has carried out an adequate risk analysis, then the vast majority of problems encountered during a project can be handled smoothly.

The other function that the project manager carries out during the project and after the main effort on project planning has been completed is that of representing the project. He or she will represent the project to both the development company and the customer. This will involve attendance at meetings which are mainly to do with progress reporting. Typically a modern computer project will have major progress meetings with the customer every three months when customer representatives meet with the senior staff of the project. There will also be internal progress meetings which take place with a greater frequency, for example once per month. Normally, if a project is proceeding to plan these meetings are very small and may just involve the project manager, his or her manager and one other member of staff; for example, this might be the senior analyst attached to the project if the project is currently in the middle of the analysis phase or the senior designer if the project is in the middle of design.

There will also be a number of other meetings that the project manager will attend. For example, he or she may need to meet with any sub-contractors to check on their progress; in this respect the project manager behaves as the customer for whatever is being delivered by the sub-contractor. There will also be meetings with other departments that provide services to the project, for example translation services, clerical services and software quality assurance.

9.12 QUALITY ASSURANCE AND THE SOFTWARE PROJECT

Introduction

One important function which occurs during a computing project is that of quality assurance. Quality assurance tasks pervade the whole of a modern computing project. Before looking at these activities it is worth examining the whole concept of quality assurance.

A software product embodies a number of what are called *quality factors*. Some examples of these quality factors are shown below.

■ *Portability*. The ability of a piece of software to be moved from one operating system to another.

■ *Reusability*. The ability of a system or, more likely, parts of the system to be used in another system.

■ *Maintainability*. The ability for a system to be easily modified once it has been delivered to a customer.

■ *Usability*. The degree to which the user can easily employ the system.

■ *Correctness*. The degree to which a system reflects the requirements of the customer.

These are just a few of the quality factors associated with a modern computer system; there are many more that quality assurance activities address but the list above is exhaustive enough for the purposes of this book.

SAQ 9.10 Maintainability is regarded by many system developers as an important factor. Why do you think this is?

Solution

It is because many systems in operation will be subject to large amounts of change: for example, to rectify development errors, to speed the system up or in order to respond to changes in requirements from the customer.

One of the key software assurance activities that occurs in a project occurs at the beginning of the project. The project manager will consider which quality factors are important for a system. Much of the information needed for this will come from the customer. For example, if the customer thinks that the system is going to have a long life and will require a large amount of change during its lifetime in order to cope with a changing world, then the project manager will mark maintainability as an important factor.

The system may be one which is used by a number of non-technical staff, where a large number of transactions need to be carried out in a fairly short time with a minimum number of errors being committed; for example, a system for air-traffic control. Such a system has to have a high usability quality factor embedded in it.

Before looking at how quality assurance helps with the process of including quality factors in a system and also validating their presence, it is worth saying something about correctness. This is an important quality factor which you might imagine is either present or not present: after all, a system is either correct or not correct. In an ideal world this would be true. However, in the real world correctness is something which lies on a spectrum between the endpoints of broadly correct and totally correct. Systems should normally be broadly correct in that all their main functions will work but that there will be minor functions which may not work or only partially work. Typical of systems which need to be broadly correct are word processors and other clerical applications.

On the other side of the spectrum are systems which need to be totally correct and where any failure would be disastrous. Typical systems which have to be totally correct are air-traffic control systems, hospital systems which monitor the state of a patient and systems which control nuclear reactors. Given that a project manager has identified certain quality factors as important in a system, how does a quality system ensure that they are embedded in the system? In order to answer this we need to look at quality tasks.

Safety critical systems normally have a large amount of effort spent on quality assurance, much more than would, for example, be expended on a clerical system. In some projects this could be as high as 35 per cent of the project budget.

■ Quality tasks

A quality task is one which demonstrates that a particular quality factor is present in a software system and its documentation. Some concrete examples of such activities are listed below.

■ *System testing*. This is the process whereby the developer checks that the functions of the system described in the requirements specification work correctly. For example, in a requirements specification for a banking system there may be a function which describes the fact that when a customer requests a balance from an automatic teller machine then the correct balance is displayed on the machine within a specified time. This will give rise to a number of system tests which check that different customers with different accounts have their correct balance displayed within the specified time. System testing addresses the correctness quality factor.

■ *Technical reviews*. These are meetings of technical staff which examine a document for correctness. For example, one of the most important reviews that is carried out on a software project is the requirements review. This review, which will normally also involve the customer, checks the requirements specification for completeness and correctness. There are a wide variety of reviews which are carried out on a software project; most of them address the correctness quality factor.

■ *Software tools*. There is a class of software tools which certify that a particular quality factor is present in a system. For example, there are tools known as metrics processors which take a chunk of program code and produce a number which describes how badly structured the code was. Such tools produce information on the testability quality factor. Another example of a tool used in quality assurance is a test monitor. Such a tool monitors the execution of a chunk of software and provides information on what parts of the software have been executed. For example, they provide information on whether all the statements have been executed by some test data. Such tools provide information about the efficiency of testing and hence contribute to the correctness quality factor.

■ *Usability testing*. This a form of testing whereby the real users of a system are allowed to access the system. Any errors they commit such as mistyping a command are monitored and analysed with a view to modifying the inte face of the system to eliminate the errors. This form of testing addresses the usability quality factor.

SAQ 9.11   Would you regard the task of attending a progress meeting as one which addresses a quality factor?

Solution

No, this is an important task but one which does not address any of the factors above, such as maintainability. If the meeting was one where the results of a usability study were examined then this would be an example of a quality task.

Quality factors lie at the heart of a quality assurance system. Once a project manager has decided what quality factors are important for a particular system, then he or she carries out two types of planning. First, there is normal project planning which will list all the developmental tasks such as system design and when they are to occur in the project. The second form of planning is quality planning. Here the project manager will specify those tasks which provide evidence that a particular quality factor is present, for example a task such as system testing. As well as specifying the quality tasks the project manager will also specify what documentation is generated which shows that a quality factor is present. For example, some projects have portability as a major concern: the ability to move a software system from one computer to another. One way to do this is to develop the system using a portable subset of a programming language such as C. There are a number of software tools which will notify the programmer when he or she deviates from this type of subset. The project manager will decide that in order to demonstrate portability the project will produce documentation of the proofs that there are no deviations from the standard subset used.

Thus, if you look at the quality plan then you will find a list of quality tasks such as system testing, when those tasks should be carried out, who should do them, the resources required for the tasks and the quality documentation that is produced.

The quality system

The part of a company's development effort which addresses quality is known as the quality system.

The quality system consists of a number of elements. At its heart there is a document known as the quality manual. The quality manual will contain standards, procedures and guidelines.

- A standard is text which describes how program code or documentation is to be laid out and displayed. For example, a standard for the requirements specification would detail how the functions are to be displayed and how they are to be numbered. A standard must always be adhered to.

- A procedure is text which describes the steps that need to be carried out when a particular software task is executed. For example, a procedure for a requirements specification review would describe who attended, who would chair the review, how long a review normally lasts, what documents are produced by the review and what preparation is needed for the review. A procedure must always be adhered to.

- A guideline is a statement of good practice. It describes what needs to be done in order to create a high-quality product. For example, a guideline for system testing will describe how to develop test data which thoroughly exercises a system. Guidelines are not normally mandatory; however, it is expected that they would be adhered to most of the time.

A major part of the quality manual will be a description of how to use its facilities on a project. The major decision that faces the project manager when developing a quality plan is how much quality assurance to add to the project. If, for example, the project was a safety-critical one then almost all the facilities of the quality manual would be used in order to ensure that there would be no catastrophic errors. However, if the project delivered a software system which was not critical but where the time to release the software to market was, then the project manager has to decide what quality tasks to include so that the software is broadly correct but doesn't delay the software's release.

One component of the quality plan is the test plan. This is that part of the quality plan which deals with those quality tasks which are related to the execution of the software. It describes system testing, acceptance testing and module testing tasks.

The role of the quality assurance function

Quality assurance tasks will be carried out by various staff depending on the size of the company. In a large company there will be a separate quality assurance department. This will be independent of the software development part of the

423

company and members of this department will report to a quality manager who will normally have a senior status in the company.

In smaller companies there will be a number of specialized quality staff who would be part of the software department. In the smallest companies the quality assurance tasks will be carried out by staff such as the project manager or senior analyst on a project.

There are a number of tasks which the quality assurance department needs to carry out. First, they are the guardians of the quality system. They are responsible for its maintenance and will be the initiators of change. A quality system will change over time. There are three reasons for this: first, the software technology used by the company will change; for example, the company may change from a third-generation language to an object-oriented programming language and will need a new programming standard. A second reason for change is that the company developing software may change. For example, a change in company policy from developing bespoke software to packages might necessitate a change to testing standards and procedures. The third reason is that an initial version of a quality system is rarely totally correct and changes need to be made to it early in its life as staff gain experience in using it.

The quality assurance function is responsible for gathering the information which suggests change, proposing the changes and then carrying them out. The quality assurance function is also responsible for checking that projects adhere to their quality planes: that is that a particular quality task such as a technical review has been carried out at a particular time and that adequate documentation is produced by that task. This is a vitally important role, but is not that popular a one since it asks staff to behave as police.

The final task for the quality assurance function is to provide advice on new quality assurance technology. For example, they might provide advice on the latest testing techniques or recent testing tools.

EXERCISES

Exercise 9.1   System testing

A sales ledger system is under development, and is intended to include batch data entry and on-line enquiry facilities. The design specification has been accepted and implementation is under way. To date, all program specifications, user documentation, screen and forms design have been completed, and the appropriate hardware installed. However, before changeover begins it is the system analyst's responsibility to test the component parts of the system, both individually and collectively.

List the various tests on inputs, outputs, files, clerical and computer procedures which would be carried out prior to changeover.

(ACCA level 2, Sys. an. & des., June 1988)

Exercise 9.2   Data Protection Act

The Data Protection Act 1984 has had an effect on almost every organization and individual using computers; indeed the CIMA has published a 'Practical Guide' to the Act. Many accountants have become alarmed at the publicity received by the Act and the implications of installing a new computer. You have been requested to give advice on a number of points relating to the interpretation of the Act.

You are required:

(a) to state *three* major requirements of the Act;

(b) to state *four* data protection principles contained in the Act relating to the processing of personal data;

(c) with reference to an organization with which you are familiar, to

   (i) state *two* exemptions which might apply;

   (ii) describe *four* files and systems which might need to be registered.

(CIMA, stage 2, Inf. tech. man., May 1988)

Exercise 9.3   Systems documentation

'However efficient the processing and well satisfied the users, a system cannot be considered adequate unless it is documented to a very high standard.'

Requirement:

(a) Why is so much emphasis placed on the quality of system documentation?

(b) For each of the following, state the purpose and briefly describe four issues which should be included in the specification or manual:

   (i) program specification;

   (ii) computer operations manual;

   (iii) user manual; and

   (iv) system changes manual.

(ACCA level 2, Sys. an. & des., Dec. 1987)

Exercise 9.4

Write down the factors which you think will affect the cost of an IT project.

Exercise 9.5

What qualities do you think would be required of a good project manager?

Exercise 9.6

You have been asked to develop a procedure for a meeting which deals with reviewing the requirements for an IT system. What would you put in this document?

Exercise 9.7

What sort of data do you think would be needed in order to keep adequate track of the progress of a project?

Exercise 9.8

You are developing a system for a company which has been taken over during the middle of the design stage of a project. What sort of problems would this raise and how would you, as a project manager, respond to them?

Outline solutions to exercises

Solution 9.1 Refer to Section 9.2. This section may be enhanced by mention of the following:

- checks on customer details, e.g. names and addresses;
- checks and controls on opening balances in sales ledger;
- checks on batching and control-totalling of input, especially customers' payments; and
- checks on the accuracy and completeness of sales statements.

Solution 9.2 Refer to Section 9.9.

Solution 9.3 Refer to Section 8.9.

(a) ■ To facilitate making changes to the system at a later date.

- To allow the users to operate the system with the minimum of trouble, particularly in its early stages.
- To enable any errors (bugs) detected subsequently to be rectified quickly.
- To facilitate auditing of the system and auditing in general.

(b) (i) Input specification, file contents, processing specification, output contents and layout.

(ii) Security arrangements – checkpoints, spooling, backup files, physical security, passwords.

Control section – release of disks and tapes, handover of output, receipt of input media.

Operations – disk and tape loading, operating system messages, error messages, printer stationery and set up.

Job scheduling – times of start and completion of computer jobs.

(iii) Refer to 'Instructions to users' in Section 8.9.

(iv) Details of changes, authorization for changes, target dates for completion, test data and test results covering changes to system.

Solution 9.4 There are a large number: internal staff costs, equipment costs, travel costs, cost of software, cost of consultants, stationery costs, and hardware costs.

Solution 9.5 A good project manager should have good interpersonal skills and be able to deal with a wide variety of staff and customer representatives. He or she should have the skills necessary to detect trouble in a project in advance. The effective project manager should also be open to new ideas and be able to try out such new ideas on his or her projects. A project manager should also be able to interpret project monitoring data quickly such as data on task completion.

Solution 9.6 The type of items that should be in such a document are: details of who should attend a requirements meeting, how to arrange a requirements meeting, what documents need to be processed, how to communicate the results of such a meeting, what to do when a requirement is not clear, when such meetings should take place and the circumstances when a meeting needs to be abandoned.

Solution 9.7 You would need the details of each task in a project: the name of the task, how much effort is required for the task, the earliest time the task can be started, the latest time the task can be started, the earliest time a task can be finished, the latest time a task can be finished and its duration. This is planning information. The project manager will also require information about a task as it is carried out, for example the actual time when the project was started and the actual amount of resources that were required to complete the task.

Solution 9.8   The major problem is that the requirements for the system would have changed and also the customer staff that the project dealt with would have changed. The major task the project manager has to carry out is to identify which requirements have changed and ensure that for these requirements an analysis is carried out and the requirements re-specified. He or she will also need to determine how this affected the design tasks that were being carried out. Those which were affected would need to be stopped while the other design tasks could continue.

References and further reading

9.1   Bennatan, E.M. *Software Project Management* (McGraw-Hill, 1995).

9.2   Craig, S. *People and Project Management for IT* (McGraw-Hill, 1995).

9.3   Hambling, B. *Managing Software Quality* (McGraw-Hill, 1996).

9.4   Humphrey, W.S. *Managing Technical People* (Addison Wesley, 1997).

9.5   Ince, D.C. *An Introduction to Software Project Management and Quality Assurance* (McGraw-Hill, 1993).

9.6   Purba, S. *How to Manage a Successful Software Project* (John Wiley, 1995).

9.7   Ricketts, I.W. *Managing your Software Project* (Springer Verlag, 1998).

9.8   Shah-Jarvis, A. *Inroads to Software Quality* (Prentice Hall, 1996).

9.9   Smight, D.J. *Achieving Quality Software* (Chapman and Hall, 1995).

9.10 Watkins, J. *Information Technology, Organisations and People* (Routledge, 1998).

9.11 Wysocki, R.K. *Effective Project Management* (John Wiley, 1995).

10 Case studies

AIMS

- To illustrate some of the technologies that were described in previous chapters;
- To describe some of the problems encountered by software developers that did not fit naturally within previous chapters;
- To dispel some of the current myths of IT development;
- To describe some emerging technologies which are so novel that they are not yet in the main stream of information systems development and hence cannot be placed in the main body of the book.

10.1 INTRODUCTION

This chapter brings together many of the topics described in previous chapters. Its prime function is to describe some examples of current software and hardware technology employed in modern information systems. It also has a number of subsidiary aims:

- To describe emerging technologies which are so novel that they cannot yet be placed inside the main body of the book; an example of this is the use of a network computer at Scottish Telecom.
- To dispel some of the myths that are sometimes associated with information systems, such as that all these systems do is to process large databases. For example, the IBM Olympics system is a heavy mix of a number of different technologies that are associated with both data processing and realtime processing.
- To show that information systems development is often a process of modification or integration, rather than one of developing from scratch.

10.2 CASE STUDY 1: INVESTMENT ADVICE AT SMITH BARNEY

Case study relevance

This case study is relevant to Chapter 2 since it describes a company using information technology in a way that is strategic in that it alters some of the ways in which the company works. It is also relevant to Chapter 5 as it describes the use of Internet technology.

The company

Smith Barney are a conventional and very successful financial services and brokerage company which provides financial products to a wide variety of individuals and companies. It competes with many other financial services institutions in an increasingly crowded American market. One of the problems that Smith Barney faced was that they felt that their services and products were not sufficiently differentiated from those offered by other American companies such as Merrill Lynch and Dean Witter. The management of Smith Barney discovered that customers could not give a specific reason why they should use the company's expertise over and above other similar companies and decided to use new technology to differentiate their services from other similar companies.

The solution

The solution adopted by the company was to offer potential customers and existing customers a service which enabled them to tap into some of the databases that the company maintain and also to provide some financial services at no cost.

Normally, existing customers are allowed access to Smith Barney's Web site; however, non-customers are also allowed access providing they fill in a registration form. This form is used by Smith Barney's consultants to determine leads for business and also carry out research into what financial products customers are interested in.

A typical example of the use of the Smith Barney Web site are those pages which deal with education provision planning. A major growth area in American financial services is that of investment products which are geared towards providing funds to enable a customer's children to complete college. Smith Barney have Web pages which allow customers to input data such as the date of a child's first year in college and the name of the college that the child might attend and then receive a report which will detail information such as the annual tuition fees, the annual boarding fee and the annual increase in these expenses over the last five years; it will then provide a financial plan for the customer which includes the customer paying a lump sum and an annual amount, and then specifies the cash earned at the time when the child enters the first year of college.

Another example of the services offered by this company is the ability to interrogate the financial history of a company before, say, investing in that company's shares. A typical set of information that is provided from a query would be data which described the history of the share prices over a period, a number of moving averages which describe the company's financial performance, pre-tax profits and selected income and balance sheets. Users can add companies to a 'Watch list' which allows them periodically to consult financial details of those in which they are interested.

Technical details

The Smith Barney system stores data on a relational database system and employs servers running AIX (a UNIX variant) on IBM servers and Solaris (another UNIX variant) on Sun workstations. Pages are created using the mark-up language HTML and the programming languages C and Perl.

Teaching points

This is a typical use of an Internet application bringing information closer to a customer; previously such information was paper-based or was provided by companies such as Reuters for a fee.

It is an example of a company using IT technology in order to gain a competitive edge over its competitors. For example, many financial service companies have Web sites, but the amount of information provided is much more diverse. The educational planning part of the Smith Barney Web site allows potential investors to target a *specific* college for their son or daughter's education, and the financial reporting part of the Web site outperforms many Web sites which just feature facilities such as the ability to interrogate a current share price.

The system is an example of a strategic system which changes the way that a company does business. In the past, much of the work of the financial services industry was carried out face to face with customers. Much of this business was fairly routine: for example, calculating the rate of return from a particular investment portfolio. With this Web site the time devoted to customer interaction by staff at Smith Barney is much more focused and involves less mundane processing. This elevates the consultant's role to that of a financial adviser.

10.3 CASE STUDY 2: DATA MINING AT FIRSTSTAR BANK

Case study relevance

This case study is relevant to Chapter 2 since it describes a company using information technology in order to achieve a productivity gain for one of its core functions: that of marketing via direct mailing. It is also an example of the intelligent use of a software-based technology that is described in Chapter 6: that of data mining. This case study is also relevant to Chapter 9 because proper project management disciplines were employed even though the project was not strictly one in which new software was developed.

The company

Firststar Bank is an American bank based in Milwaukee which has assets of over $20 billion. It serves both individual and corporate customers in the states of Illinois, Iowa, Minnesota and Wisconsin. The company uses data mining technology in order to increase the response rate of direct mailings that they make to their customers.

The solution

Firststar Bank had a huge database of customer data which had been built up over a considerable number of years. This data came in a number of forms: some of it were held in simple sequential files, other parts were held in indexed files, while some of it was held in relational databases. The main problems with the data were that it was unstructured and there was a lot of it. It was unstructured because it had been gathered by a large number of separate, unco-ordinated application programs. Because of its size and unstructuredness marketing staff at Firststar found the data very difficult to use in one of their main tasks: that of marketing financial products such as loans, credit cards or savings accounts. The main strategy which Firststar used to elicit new business from existing customers was that of direct mailing: sending brochures through the post which attempted to elicit further business from existing customers. Firststar found that, in common with other businesses, it was very difficult to increase the response rate by a few per cent.

The strategy that they adopted was to use a data mining tool called Marksman to look through their corporate database in order to determine what groups of customers were attracted to particular financial products. In order to do this they sub-divided their customers into various financial categories depending on whether they had bought financial products such as mortgages, charge cards, savings accounts or investment plans.

The data mining tool then processed the raw data about such customers and built up a profile about which customers bought which specific products and enabled the bank to target specific groups of customers to specific financial products such as investment portfolios. The results from the data mining project were startling: many direct marketing organizations feel very happy when a technical solution enables them to increase their response rate by a few per cent; at Firststar Bank the increase in responses to the direct mail increased by a factor of 4.

Technical details

The tool that was used was Marksman; this is a computer program that employs artificial intelligence technology to carry out the data mining process and which runs on a high-end PC augmented by fast parallel processing hardware. The PC acted as a client to the bank's databases; these resided on an IBM ES/9000 mainframe computer.

Teaching points

An introductory teaching book such as this one has to concentrate on principles and ideal situations. However, the world is never perfect and one manifestation of this lack of perfection is the fact that many companies have built up large and heterogeneous collections of data. Such collections have often been developed by separate projects which, for very good financial and delivery date reasons, have not been integrated with existing databases. New technologies based on artificial intelligence enable problems associated with such data to be solved.

One of the features of this case study was that the developers at Firststar organized the development processes required to produce a system for mining their data as a proper software project. An important consequence of this was the fact that they started off with a clear set of objectives: that of identifying categories of customer and the products with which they were associated. Many data mining projects have failed because of the fact that the developers using the technology have had no fixed plan or any fixed set of aims over and above the rather general one of finding interesting relationships in the data that was mined. The clear message is, that even for projects which are out of the ordinary, a discipline of project management such as that outlined in Chapter 9 still needs to be adopted.

10.4 CASE STUDY 3: THE USE OF WRAPPER SOFTWARE AT *AT AND T*

Case study relevance

The case study is relevant to Chapter 8 which describes HCI – particularly direct manipulation interfaces. It is also relevant to Chapter 6 which describes tools for wrapping existing systems in a layer of software that allows them to have an extended life.

The company

AT and T are a major provider of telephone systems in the United States. In the early 1990s many of their systems were based on simple computer terminals which only allowed rather rudimentary menu-based access to their core applications. The company wanted to move to direct manipulation interfaces in order to enhance terminal staff's productivity: for example, AT and T's network order management system handles 2 million transactions and a billion dollars of revenue each month. Clearly a more efficient interface would provide major savings.

The solution

One solution that the company could have adopted would be to redevelop all the core systems so that they could work in a client–server environment and use direct manipulation interfaces. This was clearly impractical as it would have taken considerable effort and time: the company had invested a large amount of capital in its core systems and did not yet want to write them off. The solution adopted involved a form of software wrapping, where an extra layer of software is added to existing applications in order to transform them. In Chapter 6 we describe the use of software wrappers to enable a third-generation software system to interface with object-oriented technology. The example here is similar: the main difference being the fact that the new technology (new in 1995) was that of direct manipulation, windows-based software.

The software that was chosen was a package called FlashPoint that enabled users of the mainframe system to interact with the AT and T core systems using the standard windows and icon-based facilities found in operating systems such as Windows 95.

FlashPoint also enabled the developers at AT and T to tidy up many of the functions which users of the mainframe terminals initiated. Many of these functions often required terminal users to navigate through a number of screens in order to satisfy some simple query. FlashPoint contained a scripting language which enabled staff at AT and T to automate repetitive procedures and reduce many of the tedious tasks to a process of entering data on a single window.

A third achievement was that as well as migrating their core systems to a direct manipulation interface, the developers at AT and T were able to migrate to using a different network protocol (a protocol is a set of rules which enable computers to communicate with each other). The protocol chosen for AT and T was TCP/IP. This is the protocol used for the Internet; it provided a firm base for the company to migrate in the future to newer technologies such as those based on a client–server architecture and Intranets.

Technical details

The company used FlashPoint which interacted with a variety of applications on an Amdahl mainframe computer which stored data in a variety of formats ranging from simple files to relational databases.

Teaching points

This book describes the main developmental steps required to produce an IT system. The impression you may have got from the description of these steps is that all IT development is *new* development. This is far from the truth: many companies have large software systems which have been built up over the years and are quite rightly reluctant to ditch them and develop a new system every time a requirement changes.

Recent studies carried out by the University of California at Los Angeles and by IBM have shown that as much as 85 per cent of the budgets of IT developers devoted to software development is consumed by maintenance: the process of modifying a system after it has been developed, with a very large proportion of this figure being devoted to responding to changes in customer requirements over time.

This example shows that even a major change in requirements, that of moving from one form of interface to another more productive form, is still usually implemented in an evolutionary way rather than by new development.

10.5 CASE STUDY 4: A HETEROGENEOUS SYSTEM AT TRACKER NETWORK

Case study relevance

This case study describes an example of a relatively simple data processing system which computerizes a novel application. It is an example of an information system which contains both real-time and data processing functions.

The company

TRACKER Network is a new company which fits tracking equipment to cars. This equipment can be activated when a car is stolen and the position of the car located using a large number of tracking units located around the United Kingdom. The company required a computer system which handled the basic real-time functions concerned with activating the tracking equipment and finding cars, and also data processing functions which involved processing basic customer information.

The solution

In 1993 TRACKER Network started operating its stolen vehicle service. In the period until 1996 it had recovered 1,478 vehicles worth £14.8m; the system has led to over 440 arrests. A customer of the company buys an electrical device known as a transponder which is fitted to one of 30 points in a car. This transponder is capable of being switched on remotely by means of a wireless signal being sent to it.

When a theft is reported to the police, staff at TRACKER Network activate the transponder via a network of radio transmitters. Once the transponder is activated the car can be tracked via a network of monitoring computers which are located at major roads and ports. Police in all the British police authorities also have computers which enable them to track stolen vehicles equipped with the transponder.

TRACKER Network required a system which carried out functions associated with these transponders and also to carry out basic data processing functions associated with customers who have bought or wish to buy a transponder.

The system which was built on top of a relational database system carries out the following functions:

- The activation of the transponder when a car is reported as stolen.

- The tracking of cars which contain activated transponders.

- The deactivation of a transponder when a car has been recovered.

- The processing of sales orders from new customers who want a transponder fitted to their car.

- The scheduling of the installation of a transponder to a car.

- The maintenance of basic database information such as changing the name of the owner of a car fitted with a transponder when the car is sold.

- Stock control: keeping track of the existing stock of transponders.

- Management reporting: providing information such as the monthly sales of transponder units.

- The carrying out of system management functions such as periodically backing up databases and providing authorization to staff using the system.

- The transmission of test messages which check that the system is working correctly.

Technical details

The central system runs on a Sun workstation which is connected to a wide area network that handles most of the functions associated with the transponders. It is also connected to a local area network. This is used by TRACKER Network staff to carry out functions such as sales order processing; such staff use PCs.

Teaching points

This is an example of a system which has a wide range of functions. For example, it carries out very complicated processing of real-time signals from transponders and yet also carries out relatively mundane activities such as providing financial reports. It is typical of many of the newer information systems that are being developed.

10.6 CASE STUDY 5: NETWORK COMPUTERS AT SCOTTISH TELECOM

Case study relevance

This case study describes a novel way of implementing information systems which requires the use of a computer known as a network computer. It also describes the use of the programming language Java which is outlined in Chapter 6.

The company

Scottish Telecom was granted a licence to provide telecommunication services in 1995; during its initial period of growth it concentrated on providing services to businesses via a fibre-optic network. Recently it has moved into the residential market and required systems to handle customer care and the marketing of products such as cheap call packages. It decided to adopt a solution involving a computer known as a network computer.

The solution

Most of the initial development at Scottish Telecom involved the programming of software for a network of over 400 PCs running the Windows 3.1 operating system. These systems presented Scottish Telecom with a number of problems. The first was that PCs are often configured differently; they may have different versions of application software or communication software running on them with different configurations of the operating system; this can lead to compatibility problems where a PC can freeze up and not be used again without restarting it. Another problem arises from the same source: that of different versions of software being run on a large number of computers. Keeping track of the versions and keeping copies of the software which match the version in stock can be a nightmare for staff charged with maintaining the system. For example, when a PC suffers hardware problems and has to be replaced with another, Scottish Telecom found that it was a major effort reinstalling all the software that was used on the defective PC.

The final solution that Scottish Telecom adopted was to use a network of computers known as network computers. These are cut-down PCs with very little associated hardware. The average network computer has memory, a keyboard and a monitor, but no permanent storage. For permanent storage of files network computers rely on a central computer which acts as a file server. All the applications which the network computer uses such as spreadsheets and word processors reside on that central computer. If a network computer malfunctions all that is needed is to plug in another one.

The programming language that Scottish Telecom used was Java. This is a language which, as well as being designed for portability, was also designed for deployment in a network of network computers. A major problem that the company faced was that it had made a large investment in its conventional data processing systems and it could not afford to redevelop these systems from scratch. The solution adopted was to use wrapping software which converted the direct manipulation interfaces on existing systems to look like those systems which were to be developed using Java. In this way the existing systems had the same look and feel as the newly developed network computer systems.

Technical details

The network computers that were initially used by Scottish Telecom were Javastations supplied by Sun MicroSystems. The software which allowed the look and feel of applications to match those of Java-developed systems was Insignia's Ntrigue software. The database used was relational and was supplied by the Oracle Corporation.

Teaching points

Again, this case study illustrates the fact that even new development is never really carried out in isolation: that a company which is developing systems from scratch has to consider existing systems, not only from a technical viewpoint but also in terms of whether they will match in terms of their interaction with users.

The case study also described a novel technology which could well transform the development of information systems in the first five years of the twenty-first century.

10.7 CASE STUDY 6: SYSTEMS INTEGRATION AT WAL-MART

Case study relevance

This case study is relevant as it describes a typical stock control system touched on in Chapter 1. However, it also is relevant to Chapter 2 as it describes a strategic use of an information system where information is shared with other enterprises: in this case companies who supply a retailer with goods that are sold in a number of stores.

The company

Wal-Mart is a $93.6bn consumer goods retailer based in Bemtonville, Arkansas. One of its main systems is one which controls stock. In order to increase the efficiency of this system Wal-Mart decided to integrate it with the stock control and stock replenishment systems of its suppliers.

The solution

Stock control is one of the oldest applications of information technology. The computerization of this type of application has saved retailers many millions of pounds since the 1970s. However, even with a sophisticated system there are inefficiencies. Currently a retailer will develop its forecast of the demand for a particular product from its customers and the suppliers will make forecasts of the demand for a product from the retailers. Often there is a mismatch between the forecasts: for example, the supplier might not be aware of a sudden surge of demand for a product from the retailer because there is a large amount of the product in the retailer's stock. Such mismatches can mean not enough stock being available, too much stock being available, lost business opportunities and a reduction in profits.

Wal-Mart discerned that this was a problem and decided to share its internal information with all its suppliers. It placed these suppliers in an equivalent position of internal Wal-Mart staff who carry out the function of ordering products from a supplier. For example, suppliers can access sales figures for items and also look at the sales history of individual stores. Using these figures Wal-Mart staff then discuss requirements for future orders with the suppliers and hence have two sets of minds working on the problem: one set with expertise specific to Wal-Mart and the other specific to the products which are ordered.

This has resulted in more products being available to Wal-Mart's customers and increased their ability to retain existing customers (in the retail industry figures suggest that 30 per cent of the customers that stop shopping at a store do so because they find products are no longer available).

Technical details

The Wal-Mart system is based on an NCR WorldMark computer which administers a collection of databases known as a data warehouse. Wal-Mart suppliers and its internal staff access stock and sales information via a relational database system known as Informix via database servers. The stock databases are linked via satellite to server computers at each Wal-Mart store. These servers are further connected to point-of-sales terminals in each store.

Teaching points

The Wal-Mart system at one level is a conventional stock control system: data is kept on current stocks in the Wal-Mart stores and typical functions include the ability to query past deliveries, past levels of stock and the speed at which products are sold. However, what distinguishes the Wal-Mart system from other stock control systems is the fact that they have been integrated with the systems of its suppliers. This is an example of a strategic use of an information system where information is shared with other enterprises in order to maximize the profits of all parties.

10.8 CASE STUDY 7: MEMEX AND CRIMINALS

Case study relevance

In an introductory book such as this we have to concentrate on simple descriptions. Consequently, one impression you may have got from a reading of the early chapters is that systems development is usually a process of creating information systems from scratch. This is often a long way from the truth: many information systems grow via processes of integration and modification; sometimes new systems are even created just by a process of integration. This case study describes a simple description of an integration strategy. It is relevant to Chapter 6 which partly describes software which enables existing systems to be extended.

The company

Greater Manchester Police is a police authority which has the responsibility for law enforcement and crime prevention in Manchester and its environs. One of the problems that the authority faced was the fact that when investigating a potential crime officers had to consult a wide variety of files.

The solution

In the United Kingdom there are a large number of databases which contain information that is relevant to a police authority. For Manchester these include:

- HOLMES, the Home Office Large Major Enquiry System
- PNC, the Police National Computer system
- CTO, the Central Ticket Office
- MANCRO, the Mancester Criminal Records Office system.

A typical query carried out by police staff might involve a check on criminals, their addresses and the cars they own or have owned. This often meant that a staff had to consult four or five databases, each with a different query language. This process was very time consuming and error prone.

Memex is a system which integrates a number of diverse databases and enables the same query language to be used in order to retrieve information. Moreover it has advanced facilities for searching where some form of inexact matching is required: for example, a query such as 'Find all the criminals who have been prosecuted for bank robberies in the north of Manchester who are associated with a car whose first two registration letters are RT'.

Teaching points

This short case study shows how an information system can be developed using a form of integration and employing software which allows a uniform level of access to a large number of diverse databases.

439

10.9 CASE STUDY 8: IBM AND THE OLYMPIC GAMES

Case study relevance

This case study is an example of a modern information system which features the sort of real-time system described in Chapter 3 and also a conventional data processing system. It also demonstrates a large-scale example of the sort of Internet technology described in Chapter 6.

The company

IBM is one of the largest information technology providers in the world, manufacturing a wide range of computers from microcomputers to massive mainframe computers. IBM also provides a large number of software systems ranging from operating systems to application packages. The case study describes one of its most prestigious projects: that of providing an information system for three Olympic games.

The solution

In 1993 the International Olympic Committee asked IBM to provide all the computer systems for the Olympic games in Atlanta and Sydney and the Winter Olympics in Nagano in Japan. The system that IBM developed carried out the following functions:

■ The provision of results. This involved the capture of real-time information from each sporting data before feeding it to judges, spectators, commentators and scoreboards.

■ The provision of communication facilities which, for example, enabled athletes to keep in touch with relatives and families at home.

■ The provision of images such as photographs which enabled judges to separate participants in an event.

■ The capture and dissemination of images taken from an event, for example a photograph of an athlete celebrating his or her victory. These images were loaded into the Internet so that they could be viewed throughout the world.

■ The provision of general information on Atlanta, statistics on events such as the current world record, histories of the past Olympics and results from events in Atlanta.

■ The provision of a results service to the world press.

The IBM system for the Olympic games is, without a doubt, one of the most heterogeneous systems ever built: it is a mix of real-time, data processing and information system functions and stores a wide variety of media ranging from simple records containing, for example, the names and country affiliations of the athletes who came first, second and third in an event, to data such as video images.

Technical details

The Olympics system employed 4 System/390 mainframe computers, 80 AS/400 minicomputers and 7,000 PCs. The networking was achieved via 250 local area networks. The project also employed 13,000 telephones, 6,000 pagers, 9,500 radios and 11,500 televisions.

Teaching points

This is an example of a trend in information system provision which has accelerated since the mid-1990s. The typical information system in the 1980s was dominated by database functions such as those associated with users querying the contents of a database. The IBM system is an extreme example of a massively heterogeneous system where systems development, real-time development, data processing development and information systems development were all deployed.

10.10 CASE STUDY 9: BOEING, STOCK QUERYING AND THE INTERNET

Case study relevance

This is an example of the sort of strategic system detailed in Chapter 2; it also describes the use of the Internet technology detailed in Chapter 3.

The company

Boeing Commercial Airplane Group is an arm of the Boeing company which sells over $1bn worth of aircraft spare parts to major airlines which either buy or lease Boeing aircraft, airlines such as United Airlines, American Airlines and British Airways. Boeing wanted to bring all of its customers closer to its stock control system for spare parts and do it in such a way that it did not overburden them in terms of hardware and software.

The solution

The majority of Boeing's customers ordered the spare parts for planes by means of a variety of paper-based methods using thick catalogues. The process was very cumbersome, heavily bound in paper and was error prone.

Boeing used an Internet-based system to provide a solution to this problem. It involves making the interface to their spare parts system Web-based. Customers who require spare parts can now use an Internet browser to log in to the Boeing system and carry out a number of functions:

441

- They can examine the current stock of a part.

- They can order a part from Boeing.

- They can track the progress of the part through the delivery system used. The Boeing Web site has links to carriers such as Federal Express and United Parcel Services who provide this facility as a routine part of their business.

- They can receive a report on the current status of the order they have made.

- If the closest Boeing warehouse is too far away then the Web site allows the customer to look for alternative parts from one of Boeing's suppliers.

The system has resulted in a major decrease in paperwork and problems; for example, Boeing's experience was that some of their customers could easily clog Boeing's fax machines with the amount of paperwork that was generated by the previous manual system. One of the key reasons for Boeing's choice of browser software was the fact that it enabled its customers to access spare parts databases with very simple software: all they effectively required was a PC running a World Wide Web browser.

Technical details

The parts system is built on a Sun UltraSPARC server using Internet software developed by the Netscape Corporation.

Teaching points

This is an example of Internet software being used in order to provide a uniform interface to an existing system. One of the advantages of using a browser is that because such software has become ubiquitous it enables staff with a minimum amount of computer knowledge to interact with a very complicated system.

Glossary

Certain abbreviations and acronyms included in the glossary are intended as quick reminders of terms in the text.

3GL Third-generation language.

4GL Fourth-generation language.

access A cheap single-user database management system.

ACM Association for Computing Machinery.

acoustic coupler A device for acoustically linking a telephone handset to a keyboard or printer. Now largely obsolescent.

Ada A programming language for real-time and process control applications.

AGA address generation algorithm.

ALGOL A high-level programming language for mathematical applications.

algorithm A set of computational and/or mathematical steps for converting one value into another.

ANSI American National Standards Institute.

APL A high-level programming language.

application wrapper A software tool which enables existing legacy software to interact with more modern technology.

artificial intelligence A term used to describe the discipline which aims to make computer systems behave more like human beings.

ASCII American Standard Code for Information Interchange.

ASIS American Society for Information Science.

Bahmann diagram A diagrammatic representation of the interconnections between data items in a database.

bandwidth The effective information carrying capacity of a line or radio link.

baud The rate of change of signal level, equivalent to bits per second if there are only two signal levels.

BCD *See* binary coded decimal.

BCS British Computer Society.

benchmark test Programs and data designed to measure and compare the performance of computers. *See* Whetstone rating.

binary A numbering system to the base two, thus using only the digits 0 and 1 (bits), which is the basis of computer logic.

binary coded decimal A code using 4 bits to represent each decimal digit, e.g. 93 is 1001 0011.

bisync *See* BSC.

bit The digit 0 or 1 as used in binary.

bit-mapped *See* memory-mapped.

Boolean algebra The mathematics and logic applicable to binary situations, e.g. on/off, yes/no, stop/go.

boot The initial start-up procedures of a computer.

BOS *See* business operating system.

browser A program used to display documents which reside on the World Wide Web.

BSC binary synchronous communications.

BSI British Standards Institute.

bubble chart A diagram depicting entities and their attributes in the form of circles or ellipses joined by lines.

buffer A portion of memory for holding data awaiting transfer or processing.

bug An error in a program.

bus A path (channel or circuit) along which any form of data or signals are moved.

business operating system (BOS) A 16-bit multi-user operating system for business PCs.

business process capture tools These tools are used to analyse the business processes of a company in order to streamline them.

C++ An object-oriented version of the C programming language.

CAD computer-aided design.

CAFS *See* content addressing.

CAM computer-aided manufacturing.

Cambridge ring A LAN protocol using 'slots' for holding data passing through a network.

canonical model A representation of the inherent properties and structure of data independent of any hardware or software.

CASE computer-aided software engineering.

CCD charge-coupled device memory.

CEEFAX A teletext service.

channel *See* bus.

CHAPS Clearing Houses Automated Payments System.

CHILL A high-level programming language.

chip A wafer of silicon on which is a complete electronic circuit for a particular computer function.

CIM computer-integrated manufacturing.

CIMA Chartered Institute of Management Accountants.

CISC complex instruction set computer.

clock frequency *See* clock rate.

clock rate The rate at which electrical pulses are generated in a computer in order to represent data, i.e. data transfer rate. Clock rate is a measure of a computer's processing speed.

clone Hardware or software that mimics other hardware or software so that there is no apparent difference.

closed user group A group of users who have private access and use of a viewdata system.

cluster controller A device controlling the data passing to and from a number of grouped terminals.

CMC7 A font used in magnetic ink character recognition.

CMOS *See* complementary metal oxide semiconductor.

CNC computerized numerical control.

code converter A program which converts a system written in one programming language into a system written in a (more modern) programming language.

COMAL A high-level language with characteristics similar to BASIC and Pascal.

complementary metal oxide semiconductor (CMOS) A type of chip used in the memories of computers.

concurrency A computer's ability to perform several jobs concurrently.

content addressing A concept in which a data item is stored and accessed according to its value rather than its key. Also termed CAFS (content-addressable file store).

controlling The process of applying changes to a project in order to ensure that it meets its objectives when there is a chance that these objectives are not going to be met.

conversational mode The alternate interaction of a user and a computer such that each responds to the other's input.

CORAL A high-level programming language.

CPU central processing unit, also called processor.

CSMA–CD carrier sense multiple access with collision detection.

CUG closed user group.

cybernetics The theory of communication and control in all types of systems.

cycle stealing A means of transferring data without affecting the operation of the CPU for other purposes.

data compression The reducing of storage space required by data by removing spaces, encoding commonly used items, etc.

data description language (DDL) A computer language for specifying the data structures in a DBMS.

data mining The processing of large amounts of unstructured data in order to extract important business information.

data processing system A system whose primary function is to process data and provide little information about it.

database tuning The process of restructuring a database to improve its performance.

DBMS database management system.

DDL *See* data description language.

debugging Removing errors from a program.

DFD data flow diagram.

diagnostic routine Software for detecting errors in software or malfunctions in hardware.

direct memory access The transfer of data between memory and backing storage without involving the CPU.

DMA *See* direct memory access.

DOS disk operating system.

dragging Moving an image on a screen by means of a mouse and pointer.

drum printer A rotating cylinder embossed with printing characters against which the paper is presented by print hammers. Now obsolete.

DSS decision support system.

DTP desktop publishing.

duplication Two similar units used in conjunction for backup and security.

DVI digital video interactive.

E13B A font used for magnetic ink character recognition.

EAN European Article Number.

EAROM electronically alterable read-only memory.

EBCDIC Extended Binary Coded Decimal, Interchange Code.

ECMA European Computer Manufacturers' Association.

EFT electronic funds transfer.

EIS executive information system.

e-mail Electronic mail.

emulation The imitation of one device or software by another.

EPROM erasable programmable read-only memory.

Ethernet A LAN using CSMA–CD and allowing straightforward linkage of units.

Excel A sophisticated spreadsheet system.

expansion card A slot-in card holding additional chips for enhancing the processor or memory of a PC.

facilities management A service entailing the development and operation of an IS by a service bureau for a client.

fail soft An orderly close-down of a system when a failure is detected that avoids catastrophe.

Fax Facsimile transmission.

feasibility analysis The process of checking that a project can meet its requirements.

FIFO *See* first-in first-out.

first-in first-out (FIFO) The first message, etc., arriving is the first to be dealt with, i.e. messages are formed into a queue and taken in turn from the head of the queue. *See also* LIFO.

flag A digit or character in a computer program which is given various values to indicate a situation. The flag is tested (used) at a later point(s) so that the program knows which way to proceed.

FORTRAN A set of high-level programming languages for technical work.

FTP file transfer protocol.

fuzzy theory A scientific approach to handling inaccurate data and unclear situations.

gb *See* gigabyte.

gigabyte (gb) One thousand megabytes.

graphics The display of diagrammatic or pictorial information, e.g. graphs, pie charts, etc., on a business PC.

groupware Software that allows a number of users to interact together. Typical groupware functions include diary management and meeting organization.

guideline A document in a quality system which provides strong advice on how a technical task should be carried out.

hacking Unauthorized intrusion into a computer system with the intention of stealing, altering or damaging data or software.

hash total A total made from figures which have no meaningful sum in themselves, e.g. total of account numbers. Hash totals are used for control and checking purposes.

HDLC high-level data link control.

hertz One cycle or pulse per second.

heuristic Any method or process based on trial and error, rules of thumb or intuitive decisions.

hostile software detector This is a program which detects when unauthorized software such as a virus has been inserted into a system.

housekeeping Computer operations not specifically connected with applications but maintaining the overall efficiency of the computer's use, e.g. redistributing stored files.

ICAEW Institute of Chartered Accountants in England and Wales.

icon A symbol on a screen indicating a computer's current function.

ICSA Institute of Chartered Secretaries and Administrators.

image processing (imaging) The recording of diagrams, signatures, etc., in digital form to facilitate processing and transmission.

information theory The mathematical theory of information, data and signals as applied to communication.

innovating The process of introducing and monitoring change on a project.

integrity Maintenance of error-free and accurate data through the detection and removal of errors.

interface The circuitry enabling two devices to interchange data. This may be either serial or parallel and is generally standardized.

Internet The term used to describe a huge collection of interlinked networks which are generally accessible.

Intranet A network which uses Internet technology but which is confined to one company and, in general, not available to public access.

IS information system.

ISDN international switched data network.

ISO International Standards Organisation.

Java A programming language used to develop software for Internet or Intranet applications.

job control language Control commands entered by a mainframe operator to specify a computer job to the operating system.

KADS knowledge analysis and design for expert systems.

kb *See* kilobyte.

kilobyte (kb) One thousand bytes.

LAN local area network.

large-scale integration The concept of chip technology.

last-in first-out (LIFO) The last message, etc., to arrive is the first to be dealt with, i.e. messages are formed into a stack with the last arrival being put at the top of the stack. *See also* FIFO.

LCD *See* liquid crystal diode.

legacy system Software which has been in existence for a large number of years and is showing its age.

LIFO *See* last-in first-out.

LINUX A version of the UNIX operating system which is targeted at microcomputers. It can be obtained free.

liquid crystal diode (LCD) A tube of crystalline substance that changes its appearance when an electric voltage is applied to it.

load-sharing Two or more processors sharing a load of on-line transactions.

lockout A means of preventing more than one user simultaneously changing data in a common file or record.

Lotus 1–2–3 A piece of sophisticated multiple spreadsheet software.

LSI large-scale integration.

magnetic drum A rotating cylinder holding data on its magnetic surface. Their high cost per unit of data has resulted in their supersession by magnetic disks.

mail list A collection of users who can be mailed. Usually these users have some common interest or function.

maintainability The ease with which a system can be changed.

management information system A system which provides basic management information for a company, for example the level of stocks in a warehouse.

mb *See* megabyte.

megabyte (mb) One million bytes.

megahertz (mh) One million hertz.

memory-mapped An exact replica in storage of the pixels forming a display on a screen, also known as 'bit-mapped'.

mh *See* megahertz.

MICR magnetic ink character recognition.

microchip *See* chip.

microsecond (us) One millionth of a second.

middleware A term used to describe software which sits between application software and system software.

millisecond (ms) One thousandth of a second.

mips millions of instructions per second.

MODULA A high-level programming language.

modulation A process of modifying an electric current or a radio wave so as to carry information, i.e. by varying its amplitude, frequency or phase.

monitoring A management function involving tracking progress of a project.

MOS metal oxide semiconductor.

ms *See* millisecond.

MS-DOS A 16-bit operating system.

MTBF mean time between failures.

multi-tasking The capability of an operating system to handle several independent jobs concurrently.

nanosecond (ns) One thousand-millionth of a second.

NCC National Computing Centre.

net present value The present-day value of future amounts of money taking into account the gain of interest in the meantime.

news group A set of computer users who are interested in a particular topic such as the marketing of software. These users read newsletters or bulletin boards maintained by news group software.

NLQ near letter quality.

node A point in a network where there is a junction of transmission lines.

NPV net present value.

ns *See* nanosecond.

O & M Organization and methods.

OLAP online analytical processing. This describes tools which can be used for querying databases.

OMR *See* optical mark recognition.

optical mark recognition (OMR) A computer input system entailing the making of marks on documents for subsequent reading directly into a computer.

447

ORACLE A relational database management system. Also a teletext service.

OSI open system interconnection.

palette The range of colours available on a VDU.

parameter A value that is changeable and is used to control a piece of program, e.g. a discount rate in sales accounting.

parity check A method of checking data bits on a storage medium by inserting an extra (parity) bit so as to make the number of 'ones' in a group or frame always odd or even according to the system adopted. *See* Figure 5.4.

parsing A technique for saving storage space by holding data which has the same value only once instead of repeatedly. This is a method of data compression but must be balanced against the extra memory needed for the data compression program.

password dispenser A software tool that dispenses passwords to a system, which can be easily remembered but, at the same time, cannot be guessed by a potential intruder.

PERL practical extraction and report language.

PERT programme evaluation and review technique.

picosecond (ps) One million-millionth of a second.

PIN personal identification number.

pixel A dot on a VDU screen capable of being changed in colour so as to give a graphical representation. There are thousands of pixels on a screen and they are also used for printed display.

planning The process of identifying resources and timescales in order for a project to meet its requirements.

plug compatible Capable of interchange without alteration to hardware or software.

port A point of input or output on a processor via a standard interface.

portability The extent to which a piece of software can be run on different computers.

PRESTEL A viewdata system.

procedure A document which specifies how a particular technical task is to be carried out on a software project.

PROLOG A programming language used mainly in AI and education.

PROM programmable read-only memory.

protocol The rules whereby a number of devices can communicate electronically by means of transmitted messages and data.

ps *See* picosecond.

PSNT public switched telephone network.

PSS packet switching system.

pulse rate *See* clock rate.

quality assurance The process of ensuring that the end-product of a project contains all the quality attributes determined at the beginning of the project.

quality manual The collection of standards, procedures and guidelines used by a company in their quality system.

quality system The collection of standards, procedures and guidelines used by a software company to ensure that the products it develops are of a sufficient quality.

queue *See* first-in first-out.

RAM random access memory.

real-time clock An electronic clock incorporated into a computer to provide the absolute and elapsed time for programs and persons.

re-entrant Capable of being used by several programs concurrently.

Report Program Generator (RPG) A series of declarative languages for producing reports and updating files in business applications. Now largely in disuse.

representing The process of representing a project to customers or high-level management.

resilience The ability of a system to remain in operation despite certain failures.

reusability The ability of a system or, more likely, parts of the system to be used in another system.

reverse Polish notation An arrangement whereby numbers or data items are followed by the operators applicable to them.

RISC reduced instruction set computer.

risk analysis The process of examining and evaluating all the possible events that could adversely affect the success of a project.

ROM read-only memory.

RPC *See* Report Program Generator.

SDLC synchronous data line control.

search software Software that is used to search for information on the Internet or on an Intranet.

security monitor A tool used to monitor the usage of a computer system in order to detect intruders.

self-checking number A number containing a check digit.

silicon chip *See* chip.

SNOBOL A high-level programming language.

software development environment A set of tools which enable a software developer to implement a software system more quickly than by hand.

SQL Structured Query Language.

stack *See* last-in first-out.

standard A component of a quality system which determines the format of the documents produced by a software project.

stepwise refinement The repetitive breaking down of a procedure until it can be straightforwardly programmed, probably in a structured language such as Pascal.

STRADIS A systems development methodology.

strategic information system A system which has a direct bearing on the success of a company.

SWIFT Society for Worldwide Financial Telecommunications.

system testing The process of checking that a system meets its requirements. It is carried out by executing the system with test data.

TCP/IP transmission control protocol/Internet Protocol. The protocol on which the Internet relies.

technical review A meeting during which a product of a project is validated.

teleconferencing A system enabling distant participants to take part in mutual discussion, both audibly and perhaps visually.

telesoftware The transmission of computer software over communication lines, e.g. software downloaded from a mainframe to several linked PCs.

top-down The development of a system or program so that the overall structure is completed and tested first before adding successive layers of detail.

traffic Data, video, voice, etc., carried by a line or network.

transparency The utilization of equipment in such a way that it is not evident to the user, e.g. modems are transparent to the users of communication lines.

transputer A general term for any device capable of transforming movement into electricity.

turnaround document A document output by computer, passed to users for further data entry, and then returned for direct input.

turnkey operation The design and operation of an IS by an organization external to the actual user company.

UHF *See* ultra high frequency.

ultra high frequency A radio or transmission frequency within the range 300 to 3000 megahertz. Also called microwaves.

UML unified modelling language. A language and method for developing object-oriented applications.

UNIX A PC operating system.

449

UPC Uniform Product Code.

us *See* microsecond.

usability The ease with which a user can employ a system to carry out its functions.

VANS Value added network services.

VDU Visual display unit.

VEM Viewdata electronic mail.

virtual circuit The apparent connection of two nodes in a network although there is no complete electrical connection at any time.

virtual memory A technique that facilitates programming by making it appear that a much larger amount of memory is available.

virus A hidden piece of program intended to cause errors, corruption or destruction of data or programs at a subsequent date.

volatile storage Storage that loses its contents when the power is switched off, e.g. solid state chips.

Von Neumann architecture The concept used by all computers whereby data and program instructions are regarded as the same in so far as they are alterable by program. The program instructions are executed serially and held in internal storage (memory).

walkthrough A procedure for checking the correctness of a system during development.

wand A hand-held device for reading bar codes on goods, etc.

Whetstone rating A standard benchmark of computer performance based on an appropriate mix of computer functions.

WIMP Windows, icons, menus and pointer or, alternatively, windows, icons, mouse and pull-down (or pop-up) menus.

Windows 98 A WIMP-based operating system for microcomputers marketed by Microsoft.

Windows NT A WIMP-based operating system for workstations marketed by Microsoft.

workflow software Software that is used to speed up the flow of work and documentation in a business.

workstation A point at which a user operates a PC and also has access to remote storage and other facilities, e.g. a PC as part of a LAN.

WORM *See* write-once read-many.

write-once read-many Optical disks holding information that never needs changing, e.g. signatures and photographs.

WYSIWYG what you see is what you get.

X25 A data transmission protocol.

Index